The Evolutionary Strategies
that Shape Ecosystems

Companion Website

This book has a companion website:

www.wiley.com/go/grime/evolutionarystrategies with Figures and Tables from
the book for downloading

The Evolutionary Strategies that Shape Ecosystems

J. Philip Grime FRS
Department of Animal and Plant Sciences
University of Sheffield, UK

Simon Pierce
Department of Plant Production,
University of Milan, Italy

WILEY-BLACKWELL

A John Wiley & Sons, Ltd., Publication

Registered office: John Wiley & Sons, Ltd, The Atrium, Southern Gate, Chichester, West Sussex, PO19 8SQ, UK

Editorial offices: 9600 Garsington Road, Oxford, OX4 2DQ, UK
The Atrium, Southern Gate, Chichester, West Sussex, PO19 8SQ, UK
111 River Street, Hoboken, NJ 07030-5774, USA

For details of our global editorial offices, for customer services and for information about how to apply for permission to reuse the copyright material in this book please see our website at www.wiley.com/wiley-blackwell.

Library of Congress Cataloging-in-Publication Data

Grime, J. Philip (John Philip)
  The evolutionary strategies that shape ecosystems / J. Philip Grime, FRS Simon Pierce.
      p. cm.
  Includes index.
  ISBN 978-0-470-67481-9 (cloth) – ISBN 978-0-470-67482-6 (pbk.)
1. Biotic communities.   2. Evolution (Biology)   3. Adaptation (Biology)   I. Pierce, Simon.   II. Title.
  QH541.G75 2012
  577.8'2–dc23

                                                    2011045522

A catalogue record for this book is available from the British Library.

Wiley also publishes its books in a variety of electronic formats. Some content that appears in print may not be available in electronic book.

Set in 10.5/12 pt Classical Garamond BT by Toppan Best-set Premedia Limited

1   2012

Nothing in biology makes sense except in the light of evolution (Dobzhansky, 1973).

Nothing in evolutionary biology makes sense except in the light of ecology (Grant & Grant, 2008).

Nothing in evolution or ecology makes sense except in the light of the other (Pelletier, Garant & Hendry, 2009).

Contents

Preface

One of the greatest achievements of biology in the 20th century, carried forward with vigour into the 21st, has been to translate the theory of natural selection into a practical reductionist science. This development, complemented by the 'molecular revolution' ensuing from the cracking of the genetic code, has immense potential for our understanding of the evolutionary relationships between organisms and the basis of variation within and between populations. For many ecologists these advances have provided opportunities to expand and refine investigations of the population genetics of individual species. However, this has to a large extent left unattended the urgent requirement for ecologists to bring order and predictive science to the study of ecosystems, many of which are suffering degradation and extinction in an over-exploited and rapidly changing world. In established sciences, including physics and chemistry, the revelations that eventually led to general theories were surprisingly discrete and testable. We believe that ecology, despite its reputation as an 'ambitious but ramshackle science' (Calder, 2003), is no exception to this pattern, and that it is now opportune to attempt integration of the disparate fields of animal, plant and microbial ecology within a single theoretical framework that keeps natural selection at the fore.

To address this need, this short book endeavours to understand how evolutionary adaptations govern the manner in which organisms assemble into communities and process matter and energy within ecosystems. We first chart the tentative steps by which some ecologists over the course of more than a century have attempted to establish a path from evolutionary theory to an understanding of variation in the structure, functioning and vulnerabilities of ecosystems. In marked contrast to the technical difficulties of microbial ecology and the limited insights into general principles gained by animal ecology, the accessibility of plants and their dominance of terrestrial ecosystems have revealed generalities in the constraints to evolution. These, in light of recent research, are now apparent for all organisms, from the largest whale to the smallest virus. We have distilled the essence of these constraints to evolution and ecology into a novel **universal adaptive strategy theory**, in which only a restricted set of viable approaches to survival, or adaptive strategies, are possible and in turn set limits to the performance of ecosystem functions. We explain, in terms of a **twin-filter model**, how it is possible to distinguish between mechanisms of natural selection that control the functioning of ecosystems and those that merely determine which among many candidate species actually perform these functions in particular ecosystems at individual locations. Finally, we present a detailed version of the twin-filter model that provides a conceptual framework integrating evolutionary processes such as natural selection, extinction and different modes of speciation with the assembly of communities and the functioning of ecosystems.

J. Philip Grime
Simon Pierce

Chapter Summaries

The following summaries trace the sequence of ideas and evidence that, in this book, form the path from evolutionary theory to the ecosystem.

Introduction

It has been said (Dawkins, 2004) that biology already has a grand unifying theory and there is little doubt that evolution by natural selection can explain the mechanism whereby species change over generations. However, ecologists have been slow to provide a general theory of how species traits evolve in relation to neighbouring organisms and environmental factors to exert predictable controls on the assembly and functioning of ecosystems. This highlights the need for an explicit Darwinian theory in which contrasted and widely recurring types of natural selection characteristic of particular circumstances are restricted in outcome by trade-offs in life-history and core physiology, shaping both the organisms themselves and their controlling effects on ecosystems. Subsequent chapters develop a general model in which evolutionary processes (natural selection, extinction and speciation) are reviewed in tandem with ecological phenomena (resource capture and loss, species coexistence, species-richness, competition, facilitation and the transport and storage of energy and matter within ecosystems).

Chapter 1 Evolution and ecology: a Janus perspective?

Ecological enquiry operates across a scale that extends from individual organisms to the whole earth system. Near the lower end of this scale a large volume of detailed investigation has been conducted on populations of single species. Here it is well established that progress can benefit from an evolutionary perspective that examines the past and continuing ecological consequences of natural selection on particular aspects of the organism's ecology. It is not difficult to envisage how an evolutionary focus can assist both the preservation of rare and vulnerable species and the control of undesirable organisms.

It is much more difficult to harness an evolutionary approach to the task of understanding and managing interacting bodies of living and non-living material (ecosystems) or their various multi-species living components (communities). This often requires an understanding of the essential ecology of quite large numbers of organisms and must explain how particular sets of species assemble and interact, creating predictable types of ecosystem functioning that, once understood, can guide conservation, management and sustainable exploitation.

Unfortunately, many of the genetic traits that have provided a reliable basis for establishing the evolutionary affinities of an organism do not consistently reflect the part played by that organism within the ecosystem nor its status in communities. One of the main purposes of this book is to recognize a set of essential organizmal traits that, across the world and in a wide range of contrasted situations, have a controlling effect on ecosystem functioning. Equally important we need to describe how variation in these traits explains geographical and more local patterns of variation in ecosystem functioning.

Chapter 2 Primary strategies: the ideas

Until recently, relatively few ecologists have sought to establish general theories that could identify traits of particular organisms as controlling factors in the functioning of the ecosystems they occupy.

However, at a comparatively early stage several variable attributes were recognized that had obvious potential to influence the tempo of ecosystem functioning by controlling the rate at which resources were captured, converted into biomass, circulated between organisms and eventually released. These traits occurred in both plants and animals and involved not only potential growth rates and life-spans but also the proportion of captured resources allocated to reproduction and the distribution of reproductive effort over the life of the organism.

It was also noted that variation in particular traits coincided with variation in habitat conditions and ecosystem type. Significant progress followed recognition of the influence of the frequency and severity of disturbances and associated mortalities on ecosystem development (MacArthur & Wilson, 1967; Odum, 1969). This theoretical framework was then modified by incorporating evolutionary responses to resource limitation, a refinement that brought additional traits such as the durability and defence of biomass into the foreground of debate and allowed development of the CSR theory of plant primary strategies (Grime, 1974). This theory proposes that each plant species faces an evolutionary trade-off between competing for resources, enduring resource limitation and recovery from biomass destruction. Corresponding to the various equilibria between these three selection forces each species or population of a species occupies a restricted zone within a triangular area of adaptive space.

Chapter 3 Primary adaptive strategies in plants

The period from 1975 to 2011 has seen a gradual accumulation of evidence demonstrating the existence of three main directions of evolutionary specialization in plants worldwide, consistent with CSR theory. The turn of the century has also seen the development of a practical method of classifying wild plants according to CSR theory and independent application of this method to more than a thousand plant species in a range of terrestrial habitats in Europe.

Chapter 4 Primary adaptive strategies in organisms other than plants

From the late 1970s onwards the range of organisms to which CSR theory was applied widened considerably, thanks to the efforts of biologists specializing in taxa as diverse as echinoderms, corals, ants, butterflies and fungi. Many specialists, often in apparent ignorance of biologists specializing in other organisms, recorded three-way trade-offs among their particular study organisms, be they bacteria, fish or beetles. A three-way trade-off scheme for plants has been independently proposed several times in the decades following the original publication of CSR theory. All of this disparate activity led to the creation of various three-way strategy schemes for a range of organisms, whilst empirical analysis of trait variation in mammals, birds and fish confirmed that trait evolution in these groups had been constrained to three main directions of specialization.

In this chapter we examine the idea that the three-way trade-off already demonstrated empirically for many animal groups and plants is a general feature of organisms throughout the tree of life. To this end, the life-history traits of mammals, birds, squamates, bony and cartilaginous fishes, arthropods (insects, arachnids, crustaceans), echinoderms, molluscs, annelid worms, anthozoa, fungi, archaea, bacteria (proteobacteria, firmicutes, cyanobacteria) and viruses from contrasting habitats are compared, and are found to be consistent with the following statement:

> A **universal three-way trade-off** constrains adaptive strategies throughout the tree of life, with extreme strategies facilitating the survival of genes via: (C). the survival of the individual using traits that maximise resource **acquisition** and resource control in consistently productive niches, (S). individual survival via **maintenance** of metabolic performance in variable and unproductive niches, or (R). rapid gene propagation via rapid completion of the life cycle and **regeneration** in niches where events are frequently lethal to the individual.

Evidence supporting the existence of this trade-off between the primary functions of acquisition, maintenance and regeneration, accumulated during decades of patient field and laboratory work by specialists working all over the world, forms the basis of a **universal adaptive strategy theory** (UAST) which we consider to be a progression from CSR plant strategy theory.

Since all organic life depends on utilization of carbon, and is often limited by other critical elements such as nitrogen or phosphorus, we suggest that by studying the allocation of these elements between traits involved in resource acquisition, maintenance and regenerative functions it may eventually be possible to quantify and compare the adaptive strategies of all kinds of organisms and to understand similarities and dissimilarities in the way organisms affect the functioning of ecosystems.

Chapter 5 From adaptive strategies to communities

Although our primary concern in this book is to link predictable and universally recurring patterns of natural selection on organisms to their effects on ecosystems there are compelling reasons why we must first address communities. The main reason for this is that, until recently, researches on communities and ecosystems have taken very different paths in plant, microbial and animal ecology. For more than a century, botanists have compiled standardized records of the species composition of vegetation throughout the world. This has allowed theories such as the humped-back and centrifugal models (Grime, 1973a; Keddy, 1990) to be developed concerning the way that species richness is controlled by productivity, natural disturbances, vegetation management, plant competition and the pool of species available at each location. In parallel with the continuous advancement of plant community ecology, progress in microbial and animal community ecology has been slow and contentious due in part to taxonomic uncertainties and technical difficulties in collecting and interpreting data.

We review how novel techniques, particularly for microorganisms, have recently provided opportunities for the study of the role of each species in natural settings, and we also examine evidence that both terrestrial and aquatic communities of microorganisms and animals frequently exhibit, along productivity gradients, humped relationships in diversity that resemble those documented for plants.

Consideration is also given to the concepts of speciation and adaptive radiation, the latter defined as 'the evolution of ecological and phenotypic diversity within a rapidly multiplying lineage' (Schluter, 2000). It is concluded that much of our current uncertainty about the assembling of communities arises from static interpretations of structures that in reality are experiencing continuous reconfiguration as successive waves of adaptive radiation wash across the landscape. Elucidation of this four-dimensional eco-evolutionary process is a difficult goal. Even when confined to specific taxa in restricted areas (e.g. Darwin's finches), decades of study are required to trace the impacts of adaptive radiation. Understanding the evolutionary histories that created biodiversity over continental landscapes is a challenge of gargantuan proportions that will tax our technical abilities to the limit.

Local floras and faunas often contain an abundance of species with similar traits and broadly similar effects on ecosystem functioning and it may be difficult to predict which will become persistent members of a community within that ecosystem. This calls for a second line of enquiry to test the hypothesis that there are many additional traits that do not have direct effects on ecosystems but instead operate more locally to exercise a decisive role in determining which among many candidate species are recruited into communities at specific sites and times. We propose a **twin-filter model** to explain how, despite a set of CSR-related traits of life-history and resource demands that broadly qualify it to occupy an ecosystem and influence its functioning, a species may be excluded from a community by traits with no direct impact on ecosystem functioning. According to this model the equilibrium between competition, stress and distur-bance (the **primary CSD-equilibrium filter**, or **CSD filter**) selects against traits that exert fundamental controlling effects on the acquisition, retention and investment of matter and energy utilized by primary metabolism, and which are interdependent and subject to the three-way trade-off (i.e. a decline in one func-tion is associated with gain in another). A **proximal filter** selects against traits that affect survival but which are not integral to the CSR strategy (i.e. do not co-vary with competing primary functions) and often reflect qualitative differ-ences in how and when functions are performed rather than performance *per se*. These include a wide range of single traits (or small sub-sets of interlinked traits) such as pollination syndromes, seed dispersal mechanisms, sensitivity to pathogens, behavioural adaptations governing the microsites where resources are acquired (Grubb, 1977), and capacity to respond to climatic factors, soil condi-tions and specific forms of management.

The twin-filter model describes the mechanisms controlling the entry of dif-ferent identities into the community, but it is also desirable to be able to predict the relative abundance of the admitted species. In plant communities size-related traits such as height of the leaf canopy and lateral spread (Grime, 1973a, b) and the scale of foraging responses for light and soil nutrients (Campbell *et al.*, 1991) are frequently strong predictors of species abundance. The recently developed MaxEnt method of Shipley *et al.* (2006) and Shipley (2010) is also directed toward this objective.

Chapter 6 From strategies to ecosystems

The power and general applicability of CSR theory in relation to ecosystem theory rests upon the utility of a particular restricted set of trait values to predict general characteristics of organisms and their host ecosystems. The predicted ecosystem properties include annual production of biomass, storage and loss of carbon, energy, mineral nutrients and water, susceptibility to herbivory and resistance and resilience when exposed to extreme events. The origin of the set of traits used in CSR theory relies upon speculations concerning how the mag-nitude of specific traits is likely to vary in relation to the values of other traits and the selection pressures exerted by particular combinations of ecological factors. Critically, however, proof of their existence and utility relies upon the constancy with which the expected combinations of trait values have emerged in large-scale screening experiments comparing organisms of contrasted ecology in various parts of the world.

The extent to which any particular species can affect ecosystem processes is strongly dependent upon the extent to which that species can influence the accumulation and mass movement of resources within the ecosystem. It follows from this mass-flow hypothesis (Grime, 1998) that it will be the organisms that make the largest contributions to the biomass of an ecosystem that will exert major effects on its functioning. Three propositions arise from this argument, with implications for terrestrial ecosystems. First, plants, which are the dominant contributors to the biomass of an ecosystem, are likely to be the main driver of its functioning. Second, the main drivers among the plants will be the dominant species. Third, in our attempts to classify ecosystems according to their differences in functioning and sensitivity to changes in climate and land-use, the most convenient and effective criteria are likely to relate to the dominant plants.

We also present evidence suggesting that productivity/biodiversity relationships for animals and microorganisms in terrestrial ecosystems are reliant on the ecology of primary producers, with the adaptive strategies of plants 'trickling-down' to determine viable strategies in dependent organisms.

Chapter 7 The path from evolution to ecology

A detailed version of the twin-filter model is provided in this final chapter, showing that the filters governing the entry of species into the community may act by excluding all – or just some – of the individuals from each species in the local species pool. By selecting within species, ecological filters are also agents of natural selection. As the equilibrium between competition, stress and disturbance determines the evolution of primary adaptive strategies the CSD-equilibrium is both the chief filter affecting community assembly and a principal agent of natural selection. This CSD filter is also the main determinant of ecosystem processes because it selects traits governing the movement of matter and energy; ecosystem processes are ultimately an expression of the metabolisms of component CSR-strategies. Single traits or small sets of traits differing between coexisting species with similar CSR-strategies may represent subtle evolutionary differences that increase local biodiversity. The extended twin-filter model shows the entire process as a dynamic eco-evolutionary feedback that can incorporate allopatric and sympatric speciation and thus adaptive radiation. It therefore provides a conceptual framework in which evolutionary and ecological processes can be reconciled and are viewed as part of a single natural creative phenomenon that extends above and below the level of species.

Acknowledgements

We would like to thank a number of people who gave support during the writing of this book and the events that led to it:

In 1965, as an inexperienced researcher working in Connecticut, USA, I was fortunate to publish a three-page paper on the significance of trade-offs revealed by large-scale screening of plant traits. At that time it would have seemed inconceivable that the same subject would have remained central and controversial in a book produced in 2011. It is not my purpose here to comment on this evidence of inertia in ecological research. Instead I wish to offer heart-felt thanks to the large number of colleagues, students and visiting scientists with whom I have been fortunate to collaborate over the intervening 46 years at Sheffield University, Tapton Gardens and Buxton.

Long research campaigns require, but often lack, adequate funding and sympathetic management. It is therefore a singular pleasure to acknowledge support from Ian Rorison, Roy Clapham, Arthur Willis, Malcolm Press, Bernard Tinker, Michael Usher and Tony Bradshaw. Large-scale investigations also depend upon teamwork and here the sustained, unselfish contributions of Phillip Lloyd, Rod Hunt, Ken Thompson, Nuala Ruttle, Suzanne Hubbard, Hans Cornelissen, Stuart Band, Rita Spencer, Chris Thorpe, Sue Hillier, Jo Mackey, Sarah Buckland, Barbara Moser, Wei-Ming He and Victoria Cadman have played key roles. The perspective developed in this book depends crucially on data resulting from their collaborative efforts. More recently, Rosemary Booth, Raj Whitlock, Jason Fridley, Mark Bilton and Terry Burke have participated in a unique programme in the new field of community genetics.

Both Simon and I thank Alan Crowden for his advice on publishing and for keeping an open mind during the peer review process, Ward Cooper, Kelvin Matthews, Carys Williams, Sarah Karim and Kathy Syplywczak during production, and we also thank two anonymous reviewers.

Two individual contributions have had a profound influence on the research that provides the background to this book. The first relates to the achievements of John Hodgson. For me, the experience and many benefits of the five years over which we worked together to document the plant communities in all major inland habitats of Britain are unforgettable and it is a cause for celebration that John subsequently went on to complete this inventory by recording the composition of 7,000 samples of communities harbouring rare and near-extinct vascular plants. The unrelenting rigour of John's taxonomic skills coupled with his disregard for climatic extremes and threats from enraged farm animals are legendary. As changes in land-use and climate continue to impact upon the British flora the resulting computerized database will become an invaluable record of 'how plant communities used to be'.

Hodgson's heroics are rivalled only by the technical ingenuity and fortitude of Andrew Askew. It has been a revelation to witness his application of control engineering to the task of maintaining precise manipulations of climate in large grassland plots on a steep hillside in North Derbyshire over a period of 18 years. Andrew has demonstrated that climate change research need not be restricted to modelling. We can (and must) bring 'ground truthing' to bear on current predictions of its impacts on ecosystems.

My wife, Sarah, has already appeared in this preface as ecological researcher, Sarah Buckland. This second appearance is thoroughly justified; without her encouragement, wise council and stabilizing influence half of the manuscript might never have materialized.

Finally, it has been a delight to discover a kindred spirit, accomplished scientist and inspired wordsmith in the person of my co-author Simon. His expertise and knowledge has added new dimensions to our story and it has been exciting to find a colleague who shares my vision of ecology as a compact predictive science.

J. Philip Grime
Sheffield, January 2012

A debt of gratitude is, of course, owed to Phil for inviting me to join his project, and the trust he's shown me in helping deliver his baby. It's been an interesting process of e-mail correspondence – both of us with, at times, extremely unreliable internet connections – and a completely new way of writing for both of us, but it seems to have worked! Thanks also to the guys at work who may have noticed my long absences during writing, particularly my boss Bruno Cerabolini (University of Insubria, Varese), comrade Alessandra Luzzaro and, for pointers on the relevant literature on viruses, Alberto Vianelli and Nicola Chirico. At the Native Flora Centre (*Centro Flora Autoctona*), where I conduct my practical orchid conservation work, I salute Mauro Villa, Andrea Ferrario, Daniela Turri, Arianna Bottinelli and my wife and superior officer Roberta M. Ceriani. Thanks also to Ken and Pat Thompson and John Hodgson (University of Sheffield) for enthusiasm and level-headedness.

A number of people have been particularly encouraging throughout my career, and have indirectly contributed to this book. First, my honours supervisor Adrian D. Bell (University of Wales, Bangor), who nurtured my passion for plant morphology, also Alison Bell, John Farrar and Chris Marshall. My PhD supervisor

Bob Baxter and colleague Brian Huntley (University of Durham) have provided continual encouragement. Klaus Winter, at the Smithsonian Tropical Research Institute, and Howard Griffiths and Kate Maxwell, at the University of Cambridge, were all part of an extremely enthusiastic team during my years in the beautiful chaos that is the Republic of Panama. Thank you too to Richard and Tanja Gottsberger for sharing the everyday ups and downs of that chaos and for putting up with my rusty driving skills and incessant rum-fuelled mandolin music and folk singing.

At home, my wife Roberta and children Oliver and Giulia tolerated my anti-social behaviour during writing. Giulia's birth was, for me, the defining moment of Chapter 4, and Oliver's enthusiasm whilst discovering the natural world has been a great inspiration. In fact, I would like to dedicate my half of the book to them, with all the love a proud father can give. My in-laws, Roberto Ceriani and Regina Volpi, were also particularly supportive of my efforts, as were Graham, Phillip and Joan Pierce. Finally, I would not have even started on this path were it not for my late mother, Jennifer Pierce, and her mother, Phyllis Clark – the ultimate source of my love of green things that reach for sunlight, and hidden corners of the garden filled with colour and life. You'd've been proud of my tomatoes!

Simon Pierce
Varese, January 2012

Introduction

> . . . biology, unlike human history or even physics, already has its grand unifying theory, accepted by all informed practitioners. (Dawkins, 2004)

More than 150 years after its publication the theory of natural selection (Darwin & Wallace, 1858; Darwin, 1859) provides a unique synthesis and inspiration extending across all branches of biology from taxonomy to the study of man himself. The contribution of Darwin to this achievement rested to a great extent upon his patient comparisons of hundreds of specimens of plants and animals from his own collections or supplied by colleagues around the world. On this basis alone we might expect that the main beneficiary of Darwin's legacy would be the science of ecology.

However, as many authors (e.g. Harper, 1982) have concluded, efforts to use the theory of natural selection to turn ecology into a dependable, predictive and useful science have been slow to develop and often seem to stand in marked contrast to the confidence and burgeoning promise of evolutionary and molecular approaches to biology and medicine:

> '. . . the search for generalities in ecology has been disappointing'. (Harper, 1982)

Before accepting this rather pessimistic conclusion about the state of ecological research we need to recognize the progress made in particular sub-disciplines and it is imperative also to make a realistic assessment of the theoretical and practical difficulties that limit progress in others. Significant steps have occurred with respect to understanding and managing the ecology of individual populations of domesticated animals and crops. Even in the much more challenging studies of populations of plants and animals in their natural habitats there are notable successes that, in plants at least, have provided the basis for wide-ranging synthetic reviews, some of which (e.g. Briggs & Walters, 1969; Crawford, 1989) appeared at commendably early dates.

However, these advances in study at the levels of species, populations and individual physiologies have not coincided with the development of generally accepted theories that can 'scale-up' to the multi-species assemblages of organisms that constitute communities and are the living components of ecosystems. For ecologists, this failure, occurring at a time of increasing need for informed management of resources and declining biodiversity, is an acute embarrassment. We are in urgent need of robust models that can provide a general Darwinian explanation for spatial and temporal variation in the structure and dynamics of communities and ecosystems.

It is not an exaggeration to suggest that the emergence of ecology as a reliable, useful science in the uncertain future of our planet will rest to a large extent upon our successful navigation from the theory of natural selection of organisms to the fashioning, functioning and persistence of ecosystems. In this quest we are sure that research relevant to 'each step along the path' is taking place somewhere across the academic world. However, we are not sure that the steps are adequately coordinated and sustained both geographically and conceptually.

Our purpose in this short book is to help define the research path from the organism to the ecosystem.

1

Evolution and Ecology: a Janus Perspective?

> The large scale is likely to have at least some characteristics that we cannot predict at all from a knowledge of the small scale . . . Scaling up is not part of our tradition.
>
> (Grace *et al.*, 1997)

A popular name used in universities across the world is 'Department of Ecology and Evolutionary Biology'. At many sites this title is associated with productive interactions between the two major sub-disciplines. In particular, where the shared objective is to gain a detailed understanding of population processes, there are many opportunities for fruitful collaboration. Often, however, the activities of evolutionary biologists and ecologists are so different that we may be reminded of the divergent perspectives of the bifocal Roman god, Janus (see Fig. 1.1). It has become apparent (Grime, 1993) that to address certain of their key objectives, many ecologists will not easily progress by uncritically adopting the mindsets and methods of evolutionary biology. New alignments and initiatives may be necessary if ecology is to emerge as a coherent, useful science. To see why some divergence is inevitable it is helpful to examine the recent trajectories of both sub-disciplines and to visit some of the misunderstandings between them.

Evolutionary biology

One of the most treasured of the discoveries among the Darwin papers is the notebook page upon which Darwin mused about the evolution of species by drawing a diagram resembling the branching system of a tree. Nearly two

The Evolutionary Strategies that Shape Ecosystems, First Edition. J. Philip Grime, Simon Pierce.
© 2012 John Wiley & Sons, Ltd. Published 2012 by John Wiley & Sons, Ltd.

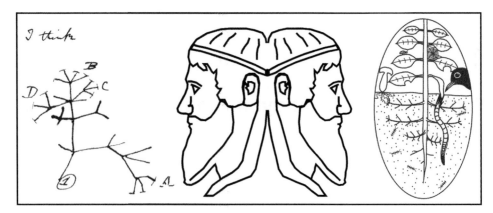

Fig. 1.1   Janus surveys Evolution and Ecology.

hundred years ago this simple sketch foretold the method that later, but with significant elaboration, would be used by taxonomists and molecular biologists to depict the origins and evolutionary affinities of the organisms that compose the world's biota. A feature of Darwin's theoretical tree, amply validated by modern investigations allowing the construction of 'real' evolutionary trees, is their highly irregular and unpredictable structure. Some branches appear to persist from early epochs to the present day whilst others have experienced bursts of recent speciation producing clusters of surviving species. The construction of evolutionary trees does not, of course, rely exclusively upon comparisons of extant organisms; palaeo-biologists through the discovery and examination of fossils continue to refine our understanding of extinct groups, some of which may provide the missing relationships between surviving taxa.

Evolutionary biology is not confined to the investigation of past connections between surviving or extinct organisms. Many practitioners are concerned with the ongoing processes of evolution within contemporary populations (e.g. Grant & Grant, 2008). Some are attempting to intervene to breed or engineer organisms with specific benefits to human society. Others are investigating evolutionary processes in plants and animals as an aid to their conservation and management; these scientists often prefer to be described as evolutionary ecologists.

Ecology

We can safely conclude that one substantial legacy from the theory of evolution by natural selection is a large college of scientists operating across the world as evolutionary biologists. Whatever the scale of their research, from processes within single populations to large-scale taxonomy, their science rests securely under the Darwinian umbrella. At present, no such unifying perspective encompasses the whole of ecology. In the opening paragraphs of this book we point to the much more diffuse nature of our science particularly where it seeks to understand the structure and functioning of communities and ecosystems. The struggles of ecologists to comprehend the community are ably summarized by Weiher & Keddy (1999) and even at one point famously provoked from Lewontin (1974) his despairing reference to 'the agony of community ecology'. The situation has often been much worse with frequent complete disjunction between evolutionary concepts and the ecosystem.

It is remarkable to record that in the highly influential textbook *Fundamentals of Ecology* by Eugene Odum (1953) the reader must wait until page 210 before Charles Darwin and the theory of evolution appear, only for them both to vanish immediately! Moreover, it is fascinating to observe that this cameo appearance of Darwin refers exclusively to palaeontology: the ongoing role of evolution in the biosphere draws no comment. It would be a mistake, however, to attribute all the travails and remaining challenges in ecology to the problems of scaling up to the complexities of communities and ecosystems. Difficulties can arise even when our research objective is confined exclusively to populations and species. In the 1990s a vivid example unfolded in the pages of the *Journal of Ecology* when evolutionary biologists objected to the comparative methods used by many ecologists in attempts to identify the factors determining the field distributions and habitat preferences of plant and animal species. Some indication of the depth of the differences aroused by this argument is apparent from the title of one of the papers published at this time: 'Why ecologists need to be phylogenetically challenged' (Harvey *et al.*, 1995). So what prompted such a critical observation by evolutionary biologists about the conduct of ecological research? Careful reading of this paper reveals that it was addressed to circumstances where ecologists had reported the occurrence of consistent differences in morphological, physiological or biochemical traits that coincided with differences in both ecology and phylogeny. Harvey and his co-authors cautioned against the assumption that such correlations were a reliable basis for ecological interpretation. In this they were likely to find support from the majority of ecologists who were equally aware that comparisons of groups of organisms differing in their ecology are often consistently distinguished by many other traits and that, at best, such comparisons merely serve as an inconclusive preliminary to experimental work in field and laboratory. Remarkably and controversially, however, Harvey *et al.* (1995) recommended the use of procedures in which attempts to establish the reliability of traits in explaining ecology should be based upon the statistical consistency with which specified trait differences were maintained in comparisons within large numbers of taxa. There can be little doubt that in some laboratories this technique, sometimes described as **'phylogenetic correction'** was regarded as a means of distinguishing between ecological and phylogenetic effects.

The protocol advocated by Harvey *et al.* (1995) drew a swift response from Westoby *et al.* (1995) who recognized that many relationships between phylogeny, traits and ecology were extremely unreliable concluding that: 'in future authors should eschew phrases such as phylogenetic effect' (Westoby *et al.*, 1995). This was a conclusion that drew strong support from many experienced ecologists:

. . . taxonomic approaches are often beset with problems . . . closely-related species often show more marked differences in response to environmental factors than taxonomically unrelated species. (Duckworth *et al.*, 1997)

Briefly this argument cast a shadow over the comparative approach to ecology and there was an episode in which some journals rejected studies in which reported differences in the traits selected for study had not been subjected to 'correction'. This was particularly unfortunate in the case of studies confined by necessity to a few species.

These arguments had exposed a fundamental difference in objectives and methods between evolutionary taxonomists and ecologists. For the evolutionary biologist and taxonomist, comparisons of DNA could meet the objective of providing a quantitative, definitive proof of relatedness between organisms. However, the same information scarcely began to address the needs of an ecologist. It was essential that the foundations of ecology should remain firmly rooted in an evolutionary perspective, but in many laboratories there was a growing conviction that to drive their subject forward ecologists would have to embark on some bold construction work on their own account. Thus, notwithstanding the interests and priorities of evolutionary biologists, many ecologists now claim the right without hindrance to recognize and explore the consequences for communities and ecosystems of universally occurring convergences in adaptive strategy even when these occur between taxonomically distant organisms.

The emergence of a science of adaptive strategies

In retrospect it can be seen that the **'phylogeny disputes'** of the 1990s originated as a well-intentioned attempt to apply some inappropriate working methods of evolutionary biology in ecology. However, such methodological differences were trivial in comparison with another substantial issue that, after persisting in the background for more than a hundred years, had begun to push slowly forward over the last quarter of the 20th century and could be recognized as a distinctively ecological initiative. The stimulus for this divergence had two main origins:

1 Recognition that classification of organisms by evolutionary affiliation did not provide all the necessary insights into the ecological role of species and populations. Functional classifications that usefully addressed communities and ecosystems did not reliably correspond to taxonomic classifications.
2 There was a need to devise a theoretical framework and database that was capable of analysing the structure and dynamics of communities and ecosystems, predicting their future states in changing conditions and eventually contributing to our understanding of biosphere functioning.

As we shall see in succeeding chapters a majority of the pioneers of the functional approach were plant ecologists, but zoologists and microbiologists have also made highly significant contributions to this rapidly expanding branch of ecology.

Summary

1 In studies of the population biology of individual species many productive interactions are taking place between ecologists and evolutionary biologists.
2 In physiological ecology and in attempts to investigate the structure and functioning of communities and ecosystems, evolutionary affinities are an unreliable predictor of the characteristics and behaviour of component organisms.
3 Because evolutionary relationships are not consistently related to ecology we require an alternative basis for prediction and elucidation of ecological phenomena. There is a need for a theoretical framework that recognizes the existence of universal constraints on evolutionary specialization that result in widely recurring adaptive strategies with predictable effects on ecosystem structure and functioning.

2

Primary Strategies: the Ideas

Ramenskii . . . was the first Soviet researcher to use long-term observations on permanent quadrats by recording the behaviour of different species: he was the first to understand the necessity to apply objective methods to study the complex relationship between vegetation and habitat.

(Markov, 1985)

Ramenskii's achievement was to begin the process of recognising fundamental and inescapable constraints . . . so integral to the core functioning of plants that they surfaced throughout the world and predetermined paths of ecological specialization.

(Grime *et al.*, 2007)

Although both plant and animal ecologists have contributed to the development of generalizing principles in ecology we will argue here that one of the most significant early insights originates from the work of the Soviet plant ecologist L.G. Ramenskii. It is only relatively recently that reviews of the early history of plant ecology researches in the Soviet Union (Marcov, 1985; Rabotnov, 1985) have brought the pioneering activities and ideas of Ramenskii to the attention of a wider audience of ecologists. The main inspiration for his contribution consisted of the observations that he conducted over a period of ten years on a set of permanent quadrats in meadow, fen and steppe vegetation. These provided a rich source of information on the development and longevity of species and on the dynamics of a wide spectrum of vegetation types. In 1938 he proposed the existence of three fundamentally different avenues of ecological specialization in flowering plants. He suggested that there are plants that behave like lions, some that resemble jackals and others that have the essential biology of camels (Ramenskii, 1938). This triangular classification of plants passed unnoticed in

The Evolutionary Strategies that Shape Ecosystems, First Edition. J. Philip Grime, Simon Pierce.
© 2012 John Wiley & Sons, Ltd. Published 2012 by John Wiley & Sons, Ltd.

the West, but in the light of more recent research it can be recognized as a truly penetrating insight. It outlines three fundamentally different ways in which plants capture and invest resources and, as we explain in later chapters of this book, Ramenskii's hypothesis leads directly to ideas about the way in which different types of vegetation exert predictable effects on aspects of ecosystem functioning such as productivity, nutrient cycling, carbon storage and responses to management and environmental change.

What was so revolutionary about Ramenskii's hypothesis? We suggest that his contribution was to make a significant addition to the theory of natural selection. Darwin's theory explained in general terms the step-by-step process by which organisms evolved but it did not analyse the extent to which paths of ecological specialization were predetermined by interaction of the physical and chemical constraints of habitats with limitations in the organisms themselves.

Throughout recorded history humans have marvelled not only at the intricacy and effectiveness of plant and animal adaptations but they have also commented on the widespread occurrence, sometimes across unrelated organisms, of similar basic solutions to the challenges of particular habitats. To some observers such convergence is evidence of an Intelligent Designer. Ramenskii provided an alternative explanation in terms of the predictable and inescapable constraints of habitat and the limited potentiality of all living organisms.

The remainder of this chapter explains the sequence of events by which Ramenskii's hypothesis eventually reappeared from an entirely independent source (Grime, 1974). This story begins with a quotation from a North American, Robert MacArthur.

MacArthur's 'blurred vision'

> Ecological patterns, about which we construct theories, are only interesting if they are repeated. They may be repeated in space or in time, and they may be repeated from species to species. A pattern that has all of these kinds of pattern is of special interest because of its generality, and yet these very general events are only seen by ecologists with rather blurred vision. The very sharp-sighted always find discrepancies and are able to say that there is no generality, only a spectrum of special cases. This diversity in outlook has proved useful in every science, but it is nowhere more pronounced than in ecology. (MacArthur, 1968)

On first consideration, MacArthur's coaxing of ecology towards repeated patterns is incompatible with what we know of the process of evolution by natural selection. To those familiar with the genetic potential of organisms to be fine-tuned by selection, evolution appears as a continuous and infinitely diversifying process (Orr & Smith, 1998) hardly consistent with a repeated typology. But this is to allow the micro-evolutionary processes, so essential to our understanding of the mechanism of evolution and the workings of populations, to dominate our perspective. Such emphasis can obscure the constrained and repetitive nature of macro-evolution and, in so doing, deprive ecology of the conceptual framework essential to its development as a predictive science. In *The Origin of Species* Darwin, rather conspicuously, does not venture far into the origin of the similar patterns of functional specialization often observed in otherwise distantly related organisms such as epiphytic orchids, ferns and bryophytes or insectivorous bats, swallows and swifts. It is inconceivable that Darwin failed to notice such widespread examples of convergent evolution, and it is tempting to conclude that these were not emphasized because they might have created confusion in a book that quite legitimately focused on the step-by-step process of evolution and drew most of its examples from micro-evolution in species such as pigeons, dogs and barnacles.

The mechanism of convergence; trade-offs

It is not difficult to envisage how selection acting on variation in members of a founding population in a new habitat might eventually cause the genetic characteristics of the new population to follow a path divergent from that of its progenitor. However, it may not be easy to predict the direction that evolution takes nor to explain the frequent observation that in the course of adaptive change some of the features of the original source population may be lost or become inactive. The answers to these questions are related to the limited potentiality of all organisms (Grime, 1965a; Wedin, 1995; Conway-Morris, 2003). First, we must recognize that the course of evolution will be conditioned by past evolution and ecological specialization; this will set limits to the changes in form and function that can be achieved in the short-term. Second, the evolutionary processes of all organisms are subject to internal constraints resulting in trade-offs:

> A trade-off is an evolutionary dilemma whereby genetic change conferring increased fitness in one circumstance inescapably involves sacrifice of fitness in another. (Grime, 2001)

Trade-offs are thus one aspect of biology about which ecologists and evolutionary biologists are in strong agreement:

> Plants have an energy economy and, as with any economy, trade-offs may favour different options under different circumstances. That's an important lesson in evolution, by the way. (Dawkins, 2009)

or, more succinctly:

> to spend on one side, nature is forced to economize on the other. (J.W. Goethe, cited by Darwin, 1859)

Examples of trade-offs of widespread occurrence in the British Flora are listed in Table 2.1. In each case an attempt is made to explain how success in one habitat is correlated with failure in another.

The trade-offs proposed in Table 2.1 apply to many species, but they are limited to individual traits and to particular ecological circumstances. This raises the question as to whether there are sets of inter-related trade-offs that engage

Table 2.1 Four early examples of trade-offs in vascular plants (modified from Grime 1965a)

Species	Trade-off
Betula populifolia	Production of numerous small winged seeds allows long-distance colonization of disturbed soil but limits seedling establishment in shaded conditions (Grime & Jeffrey, 1965).
Deschampsia flexuosa	Capacity to detoxify heavy metals on acidic soils is associated with susceptibility to iron deficiency on calcareous soils (Olsen, 1958).
Pilosella officinarum	Flattened growth-form of the shoot reduces water loss in dry habitats but prevents colonization of closed grassland (Grime & Jeffrey, 1965).
Tsuga canadensis	Slow growth and low respiratory losses permit seedling persistence in deep shade but restrict competitive ability early in vegetation succession (Lutz, 1928).

with the core functioning of all organisms (animals, plants and microorganisms) and determine much more overarching patterns of ecological specialization. A hierarchy of functional types might be expected to occur as a consequence of a trade-off between precision and generality operating across large, medium and small scales. For example, at the global scale species may be classified into small numbers of extremely general functional types, with both increasing precision and numbers of functional types moving from global, regional, local to within-population scales (Grime, 2001).

The theory of *r*- and *K*-selection

In Chapter 5 we explain how functional types recognized at regional, local or within-population scales are relevant to the investigation of communities. However, it is at the apex of this hierarchy, at the global scale, that we encounter excitement and controversy in the proposition that there are functional types of universal occurrence and fundamental significance throughout ecology. We must ask what traits, selection forces and resulting trade-offs could possibly have a functional and geographical reach sufficient to occur universally and affect all organisms? This calls for search and recognition of selection forces of an extraordinarily fundamental and inclusive nature. They must impact upon all the essential variable features of living organisms including life-history, resource capture and allocation, tissue-tolerance, reproduction and defence.

Although separated by an interval of 74 years, MacLeod (1894) and MacArthur & Wilson (1968) independently offered a solution to this quest by the daring proposal that the great variety of mortality mechanisms present in nature (natural disturbances, predation, disruptive impacts of agriculture) could, by acting at various frequencies and severity, generate sufficient habitat diversity and living conditions to select and accommodate a wealth of different organisms. This simple concept, described as the *r/K* continuum (Pianka, 1970) encapsulated in broad terms a surprisingly high proportion of the conditions necessary for plant and animal evolution and diversity. The essence of this theory was that under conditions of high mortality risk, selection favoured short-lived organisms in which reproduction occurred early (*r*-selection) whereas, at low mortality, selection delayed reproduction and promoted larger and longer-lived organisms with a capacity to monopolize resource capture (*K*-selection).

CSR Theory

The theory of *r*- and *K*-selection enjoyed considerable popularity and, in particular, had a marked capacity to predict and elucidate the sequence in which types of organisms appear during the assembly of a community or an ecosystem. However, for some ecologists (Grime, 1977; Southwood, 1977; Pugh, 1980; Greenslade, 1983) an important dimension was missing from the *r/K* model. This was manifested as uncertainty about the functional traits of *K*-selected organisms. Was the selective advantage of this proposed strategy due to pre-emptive capture of resources (competition) or to the capacity to persist in severely depleted conditions (tolerance)? Eventually, as we explain in the next chapter, this important question was resolved by empirical studies revealing that both of the two theoretical avenues of selection, confused under the title of *K*-selection, actually existed in nature as distinct, coherent phenomena and had major explanatory power in population, community and ecosystem ecology. The theory of *r*- and *K*-selection was thus transformed into a triangular continuum in which a three-way trade-off was proposed between traits permitting (1) high rates of resource capture in productive circumstances, (2) maintenance of long-lived tissues in unproductive conditions, and (3) high reproductive rates in transient disturbed habitats (see Box 2.1).

But this is to leap forward over a space of more than 20 years during which, as a potential successor to the *r/K* continuum, CSR theory (Grime, 1974) had an unusual status, perhaps best described as 'unfashionable but unfalsified'. One possible explanation for the curious status of the CSR model as a 'theory in waiting' was that many researchers during this interregnum had abandoned community and ecosystem ecology and inspired by charismatic individuals such as John Harper (1977, 1982) were investigating the population biology of individual species. The need to scale up to meet the growing challenges of global environmental change and degradation was not yet fully recognized. However, there are fragments of recorded history that remind us that beneath the surface, and occasionally right out there in the open, some ecologists were arguing quite fiercely about **how** to scale up:

> J.L. Harper: 'I can see the appeal of Professor Grime's C-S-D[1] model, but it seems difficult to make it operational. Clearly the plotting of values on a triangle in the way

---

[1]In this exchange 'C-S-D' was used to describe CSR theory in order to focus attention on the three selection forces, resource competition, stress and disturbance.

Box 2.1: Essential features of the CSR model as applied to plants

Development of the CSR model occurred in two stages (Grime, 1974, 1977). The first paper classified a large number of herbaceous plants from a range of habitats in northern England by reference to seedling growth rate and plant size. The resulting triangular distribution was interpreted in terms of a three-way trade-off between the capacities **(1)** for individuals to rapidly monopolize resource capture in potentially productive habitats; **(2)** for individuals to endure chronically unproductive conditions; and **(3)** for populations to re-establish in circumstances of frequent and severe vegetation destruction.

In retrospect it is not difficult to understand why this proposition was controversial. In particular there was resistance to use of '**competitor**' (C), '**stress-tolerator**' (S) and '**ruderal**' (R) to describe the three types of plants. Often, consideration of whether such a three-way trade-off actually recurred in nature and might be useful became obscured by debates arising from conflicts with previous uses of these terms by preceding generations of ecologists! This is perhaps a suitable place for a frank confession of the true motivations that led to adoption of C, S and R. 'Competitor' was used in defiance and despair at the confused and devalued state of the use of 'competition' in ecology (Milne, 1961) where, since the time of Darwin, it had often appeared in relation to 'fitness' or 'dominance' rather than more precisely and usefully as the process whereby neighbours seek to capture the same units of resource. 'Stress-tolerator' was used in a deliberate attempt to distinguish tolerance of resource shortage from resource competition. Finally, 'ruderal' had past association with habitat disturbance by man, but a more decisive factor was its first letter that allows a pleasing continuity with the *r*-selected organisms of MacArthur and Wilson (1967).

The second publication (Grime, 1977) ventured beyond British herbs to accommodate trees, shrubs, bryophytes and lichens and began to examine the role of the three primary strategies in vegetation succession. As depicted in Fig. 2.1, this resulted in an important clarification of relationships between resource stocks, plant strategy and plant size. Whilst ruderals were inevitably small and short-lived it was possible for competitors to attain large size by capturing extensive areas of productive habitat. A more varied and complex range of situations and sizes pertains in stress-tolerators. On impoverished, skeletal habitats such as rock outcrops stress-tolerators are pioneers that remain small. However, in mature forests stress-tolerance may be manifested as large trees with conservative traits that create tight recycling of captured resources and massive sequestration in the plant biomass.

Figure 2.1 also depicts plant succession in a range of circumstances (f–j). In (f) secondary succession in a forest clearing on a moderately fertile soil in a temperate climate is portrayed. Initially biomass development is rapid and there is a fairly swift replacement of species as rapidly growing herbs, shrubs and trees successively dominate the vegetation. Later the course of succession deflects towards the stress-tolerant corner of the triangle reflecting a gradual transition in dominance from species with high rates of resource capture and loss to those in which resources, particularly mineral nutrients, are efficiently retained. In (g) secondary succession is portrayed for a site of lower productivity. The processes are essentially similar to those of (f), but the successional parabola is shallower and the plant biomass is reduced by the earlier onset of mineral nutrient limitation. Primary succession on bare rock is examined in (h). Here the initial colonists are stress-tolerant lichens and bryophytes. Biomass development is exceedingly slow, the intrusion of herbs and shrubs gradually occurring as soil formation takes place (facilitation). The loop in (i)

(Continued)

represents the cycle of vegetation change associated with rotational burning of *Calluna vulgaris* moorland in northern Britain. A more complex sequence is depicted in (j) which attempts to explain the consequences of 'slash and burn' tropical agriculture where the declining mineral nutrient capital of the system may be expected to result in a series of arcs of progressively lower trajectory in successive cycles of vegetation destruction and recovery.

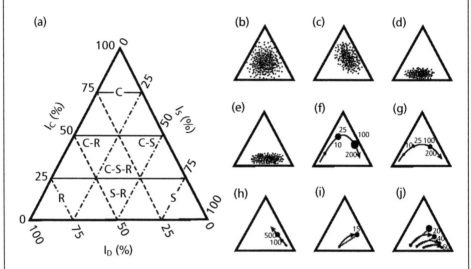

Fig. 2.1 **(a)** Model describing the various equilibria between competition, stress and disturbance in vegetation and the location of primary and secondary strategies. I_C, intensity of competition; I_S, intensity of stress; I_D, intensity of disturbance. **(b–e)** The strategic range of four life forms: **(b)** herbs, **(c)** trees and shrubs, **(d)** bryophytes, **(e)** lichens. **(f–j)** Successional diagrams: strategies of the dominant plants at particular points in time are indicated by the position of arrowed lines, with the passage of time in years during succession represented by the numbers on each line, and shoot biomass by the size of the circles (see box text for explanation of each succession). (Redrawn with permission from Grime, 1988a. Copyright © Chapman & Hall.)

proposed implies only two degrees of freedom. Any point on the C-S-D triangle determined by values of C and S immediately fixes the position with respect to D. In Professor Grime's triangle, the axes are given in percentages so that any point on the triangle adds up to 100 per cent of something. What is this quantity, and how can we measure it and allocate it between C, S and D? I worry, as Professor Grime knows, about the use of the word 'stress' in ecology. For the physicist it is 'force per unit area'. Does the word bear any corresponding precision in ecology? I am also becoming unhappy with the usage of the word 'disturbance' in ecology. Again, it is not easy to see how it can be made operational. Is there some way in which a forest fire, the track of a bulldozer, a rabbit's burrow or a falling raindrop can be compared quantitatively as environmental forces? If we have to measure 'disturbance' by the response of the organisms, there is danger of a circular argument. Ideally, I suspect we should aim to measure and compare the effects of various forces on individual fitness. Is this feasible?'

J.P. Grime: 'May I refer Professor Harper to the descriptions of the triangular model and the definitions of stress and disturbance contained in two publications (Grime, 1974, 1977). These emphasize that C-S-D is a theoretical model which proposes that the intensity with which plants compete for resources in vegetation is inversely related to the intensity of the constraints on plant production (stress) and the intensity with which plant biomass is destroyed (disturbance). It is inevitable and quite proper therefore that the axes of the model describing the theoretical relationships between intensities of C, S and D should add up to 100 per cent. The value of a theoretical model depends upon the extent to which it explains phenomena and generates testable predictions. On both these counts the C-S-D model is proving its value, not least in identifying major traits in life-history, physiology and biochemistry (now over 20 in number) which vary in relation to the axes of the model. Professor Harper asks for an operational development of the C-S-D approach. This has been taking place over the last 12 years with the result that it is now possible to begin to analyse vegetation in terms of the frequencies and functional characteristics of component strategies (Grime, 1984). Operational methods are already available where estimates of stress and disturbance are required. Stress in a stand of vegetation is inversely related to productivity, measurable as dry matter produced per unit area of ground per annum and disturbance can be estimated as dry matter destroyed per unit area of ground per annum. In recent years Professor Harper has contested the usefulness of most of the key words in my ecological vocabulary. 'Competition' was the first to fall from grace, 'stress' and 'strategy' have been castigated and now, alas, 'disturbance' has become a cause for concern. I readily concede that these terms have proved unnecessary in the increasingly specialized studies of plant demography and morphology advocated by Professor Harper. With many other ecologists I remain attached to a terminology that can assimilate information from all fields of research and can play its part in the development of a general conceptual framework for plant and animal ecology.' (Published exchange between John Harper and Philip Grime at a Royal Society Discussion Meeting, Holdgate, 1986)

We have reproduced this exchange in full because it provides a rare opportunity to consider the reservations that prevented ecologists such as Harper from accepting CSR theory. It is apparent from their own words that, in 1986, Harper and Grime were dedicated to contrasted scales of investigation, a difference that, elsewhere in the literature, is even more obvious:

. . . it is from the work of many individuals working scattered over a variety of parts of the world, but concentrating their attention over long periods on the behaviour of individual plants, that the development of ecology as a generalizing and predictive science may be possible. (Harper, 1982)

In plant ecology as in golf there is a time for precision and a time for progression. Only in fog or cases of acute myopia can the hazards of driving justify putting from the tee. (Grime, 1985)

Allied to these differing perspectives, but much more important, is the reluctance of Harper to contemplate an overarching classification for the processes of natural selection. Already we have emphasized that to achieve such a classification we must identify general constraints on responses to natural selection that are incomparably fundamental and inclusive. 'Stress' and 'disturbance' fulfil this requirement. Stress sums together the many agents that limit the quantity of living matter created per unit of space and time by constraining its production (see Box 2.2). Similarly, disturbance sums the great multiplicity of agents that limit biomass by partly or completely destroying it. Recognition of the unifying potential of stress and disturbance, as defined above, represents only the first step in describing the functional cartography of Darwin's struggle for existence. The second step is to deal with what until recently has been the highly contentious subject of competition for resources. Here for our purpose of defining competition as the third force shaping the primary adaptive strategies we will remain close to the criteria that have been already applied to stress and disturbance. For plants this can be achieved by defining competition in relation to the attempt by neighbours to capture the same particular units of resource essential for their growth and survival:

> . . . competition is defined as the tendency of neighbouring plants to utilize the same quantum of light, ion of mineral nutrient, molecule of water, or volume of space. (Grime, 2001)

Later in this book (Chapter 4) we explain how this definition of competition can be modified to accommodate organisms other than plants.

Much academic time and effort has been devoted to musing the logical implications of competition and how it can be defined, quantified and compared between situations. This, however, obscures the real reason why the role of competition in ecology is mired in controversy, and why confusion surrounding species interactions and community structure is a persistent feature of the ecological literature. The root of the problem is this: competition for resources, being the principal way in which species interact, is often touted as **the** overarching selection pressure which must necessarily be able to explain a range of evolutionary and ecological outcomes, including everything from the coexistence of species to the development of communities during succession. Interspecific competition has been invoked as the main selection pressure acting during the evolution of plant adaptive strategies (Tilman, 1988). It has even been suggested that ecological processes such as niche differentiation (the adaptation of plant phenotypes to different niches) 'require interspecific competition' (Tilman, 2001).

This popular view of resource competition as the principal driving force in ecology and evolution probably stems from the misconception of natural selection as a fierce competition between species – 'nature red in tooth and claw'

Box 2.2: Stress

Ecologists examine the bigger picture that unfolds beyond the scale of individual organisms, and view stress in terms of the constraints to productivity caused by limited resource availabilities within the habitat (Grime, 1979; Tilman, 1988). This view, whilst broadly correct, glosses over the underlying physiological processes that are essential to understanding what stress actually does and, crucially, the nature of the adaptive strategies that have evolved in low productivity niches.

Metabolism requires resources in the form of various types of matter and energy (e.g. water, mineral nutrients, O_2, CO_2, light), and a lack of resources may indeed constrain metabolic processes, causing strain (the effect of the stress). Strain due to resource limitation is a dynamic process – unproductive habitats such as deserts, tundra, montane cloud forest and alpine regions are characterized by highly variable environmental conditions during which resources may become available in substantial amounts but only during relatively brief pulses. S-selected organisms are adapted to exploit these pulses, acquiring and storing more resources during the pulse than are needed at that particular moment in order to survive the subsequent period of dearth, in a process known as 'luxury consumption' (Chapin, 1980). Luxury consumption has been recorded in a number of situations, most notably as the response of a tundra sedge community to pulsed phosphorus availability (Jonasson & Chapin, 1991; see also van Wijk *et al.*, 2003). Pierce *et al.* (2002a) recorded luxury consumption of CO_2 by the bromeliad *Aechmea dactylina*, a native of extremely wet montane cloud forests in Panama, which used crassulacean acid metabolism (CAM) intermittently throughout 24-hour periods to take up CO_2 and store carbon as organic acids whenever leaves became dry enough to allow gas exchange. (CAM is now recognized as a luxury consumption mechanism, rather than an adaptation to low availability of any particular resource; Lüttge, 2004.) The Giant Saguaro cactus (*Carnegiea gigantea*) swelling after the rain, the dromedary's hump, orchid pseudobulbs and even the human 'spare tyre' are adaptations that allow luxury consumption – consumption beyond the immediate needs of the organism to aid survival in later times of scarcity.

Luxury consumption requires complex patterns of internal resource allocation. For example, the growing points (meristems) of S-selected plant species do not depend on carbohydrates arriving directly from instantaneous photosynthesis; photosynthesis first supplies carbohydrate stores that in turn furnish sugars for growth, effectively uncoupling growth from photosynthesis, via an internal resource 'buffer' (Atkinson & Farrar, 1983). The internal stockpile is topped up during the moments when environmental conditions are conducive to photosynthetic metabolism. This mechanism has the advantage of allowing steady growth in extremely variable environments in which photosynthetic rates may fluctuate in response to changing conditions of light and temperature. The price of this dependability, however, is the relatively slow growth rates imposed by the initial accrual of internal resource pools and investment in the denser tissues that house and defend them.

Aside from resource limitation, strain may also result from damage to the cellular components of metabolism or from the impairment of translocation/circulatory systems, and thus the integration of metabolic pathways operating in different parts of the organism. Injurious stress factors may include extremes of temperature, salinity, ultraviolet radiation or, for plants, simply an excess of light – all of which induce strain by physically disrupting the form and thus function of proteins, cell membranes, genetic material and other complex structures crucial to metabolism (reviewed by Pierce *et al.*, 2005).

(Continued)

All cellular organisms respond with a 'cellular stress response', involving the production of chaperone proteins (e.g. heat shock proteins, cold shock proteins) or sugars that physically support the form and thus function of cellular components, and by investing resources in the repair, modification or replacement of damaged cellular components (Klütz, 2005). Mineral nutrient limitation is fundamental to all plant stress responses because protective mechanisms, such as chaperones or energy dissipation mechanisms, rely on protein synthesis – which requires nutrients. Even plant species with fast growth rates may be found in generally unproductive habitats at specific microsites where resources are locally abundant (e.g. where fecal matter has been deposited) probably because they have access to sufficient resources to support both growth and the production of chaperones.

Whilst resource limitation and injurious stresses represent different factors they are united in having the same effect on the organism – metabolism is strained, leading to limited productivity. Stress, be it in the form of a lack of resources or injury to metabolic components, imposes 'suboptimal metabolic performance' (Pierce *et al.*, 2005). This represents a single mechanism, despite the fact that there exist many different types of stress factor that can induce suboptimal metabolic performance. S-selected organisms are adapted to safeguard metabolic performance in the face of constraints to metabolism; both limited resource availabilities and injurious stress.

This is achieved by investing resources into sturdy tissues and internal stores that buffer and protect metabolism from the effects of environmental extremes, which is also advantageous for the protection, repair and replacement of cellular components during the cellular stress response and during aging (Pierce *et al.*, 2005). Internal stores require tissues and organs that are sufficiently capacious, durable and well defended. Thicker, denser tissues can impose greater resistances to the internal movement of resources, particularly CO_2 diffusion for plants, further limiting inherent growth rates. Thus S-adaptation is ultimately evident as robust, long-lived phenotypes with conservative functional traits that favour consistent performance at the expense of rapid growth or development. This imposes an unavoidable trade-off with regard to the ability of S-selected species to compete with neighbours or to reproduce.

These points form a major obstacle to the acceptance of Tilman's (1988) resource-ratio hypothesis, an alternative adaptive strategy theory which assumes that survival and adaptation depend primarily on the extent of foraging for different classes of resource, and the outcomes of resource competition. Pierce *et al.*'s (2005) final comment on the 'Tilman vs. Grime debate' was as follows:

> 'Cells become stressed by more than simply a lack of resources, and plants in chronically unproductive habitats cannot rely solely on instantaneous resource acquisition for survival. ... Indeed, stresses that occur over timescales as short as minutes or hours, such as those encountered during sunflecks or cold shock, cannot be countered by growing new leaves or roots. When plants experience pervasive "non-resource" stressors such as chilling, growing out of trouble is clearly not an option – the only viable strategy is resistance and readiness to act when conditions allow.'

Caccianiga *et al.* (2006) add:

> 'The contemporary ecological mindset borrows heavily from agriculture, in which periods unsuitable for growth are disregarded and thus resource acquisition and plant growth are seen as continuous processes. However, seasonal variation is superbly tangible and, for a given habitat, there may be considerable periods of the year when growth cannot occur. These periods represent an overwhelming survival risk and thus a strong selection pressure to which plants are necessarily adapted. Life forms in chronically unproductive habitats are primarily defensive adaptations against these perilous episodes, not aggressive adaptations to subtle variations in resource availability and competition during milder periods.'

(Tennyson, 1849)[2]. The basic concepts of extinction and evolution (literally, that species can change) were first inspired by Cuvier's meticulous studies of fossil mammals and reptiles, many of which had impressive teeth and claws and undoubtedly bloodied them in life. However, this bias towards conflicts between large fearsome animals had the unfortunate side-effect of tainting popular interpretations of natural selection. For example, the front cover of the Penguin Classics paperback version of *The Origin of Species* shows an ichthyosaur chomping bloodily through the neck of a plesiosaur, against a backdrop of general slaughter (see Fig. 2.2). This extract from Henry de la Beche's *Duria Antiquior*

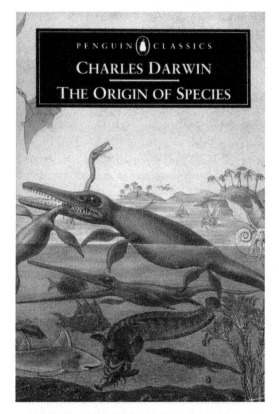

Fig. 2.2   The front cover of the Penguin Classics paperback version of *The Origin of Species*. (Copyright © Penguin Books Ltd.)

---

[2] Tennyson originally intended 'Nature, red in tooth and claw' to represent the impersonality and brutality of nature, and the philosophical dilemma that this presents if one believes in a designed, benign creation. Evolution was a popular concept at that date, although Darwin and Wallace had yet to provide a sound mechanism whereby it might occur. When they did, Tennyson's phrase was adopted to evoke the process of natural selection, and in doing so inadvertently coloured our perception of competition.

is accompanied by the explanation that the extinct animals are obeying the 'law of nature which bids all to eat and to be eaten in their turn'.

Aside from overstating the hunting prowess of ichthyosaurs, which we know from fossilized stomach contents ate squid and fish, this inappropriately sensationalist depiction has very little to do with the mechanism of natural selection laid down in the book itself. Why?

First, biotic interactions, shown in the above example as predation and competition between species for prey resources, do not constitute the only pressure that can prevent individuals from surviving. In certain situations the interactions between organisms and their abiotic environment may be more crucial to survival than biotic interactions. In these cases it may be more accurate to say 'nature white in snow and ice' or 'nature yellow in sun and sand'. Darwin (1859) put it like this: 'Two canine animals in a time of dearth, may be truly said to struggle with each other which shall get food and live. But a plant on the edge of a desert is said to struggle for life against the drought.' These alternative mechanisms of competition for resources and tolerance had already surfaced in the joint communication of the theory of natural selection by Darwin and Wallace delivered to the Linnaean Society of London in 1858. In his paper Darwin placed greater emphasis on competition between species as a driving force in the struggle for existence whereas Wallace, in his account of the process of natural selection, invoked the relative abilities of species to tolerate environmental hazards. In a sense both had seen different faces of natural selection – competition and stress are both important natural selection pressures – but whether one is more important than the other depends on the particular situation.

Indeed, there exist many situations in which organisms are physically isolated from one another and resource competition cannot sensibly be considered as an important selection pressure. In other words, competition is an important way in which species may interact, but **species do not necessarily interact**. In these situations, abiotic selection pressures are paramount. On the facing page are examples in which competition is either absent or is less important than stress or disturbance (see Fig. 2.3).

When a plant cell freezes it does not do so because of the behaviour of neighbouring plants – air or soil temperature tend to be more important. Indeed, neighbours are more likely to help rather than hinder in harsh situations: alpine plants worldwide, where they grow in close contact, have greater difficulty growing after neighbours are artificially removed (Choler *et al.*, 2001; Callaway *et al.*, 2002). This indicates that facilitation, whereby the presence of one organism favours the survival of another, is the main interaction in these environments, probably because plants growing together are less exposed to the elements. Where neighbours create niches, competition cannot be invoked to explain plant adaptations, *sensu* Tilman (1988, 2001). A further complication arises from the finding (Mitchell *et al.*, 2009) that plants in subarctic meadows readily interact but not sufficiently to influence plant survival and community structure – there is a clear difference between the *intensity* and the *importance* of competition (see also Welden & Slauson, 1986). Indeed, the importance of biotic interactions may be greatly reduced in habitats

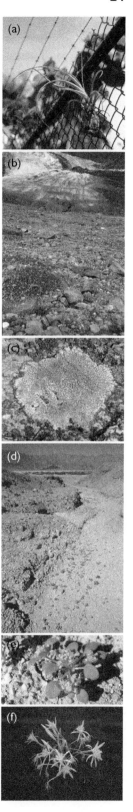

Fig. 2.3 Situations in which competition is not important. **(a) Ombrotrophic (rain feeding) epiphytes** gather water and nutrients directly from rainwater and dew absorbed over leaf surfaces, rather than from the soil. This plant of *Tillandsia flexuosa*, growing at the Smithsonian Tropical Research Institute, Balboa, Panama, has become established in a particularly unusual situation (it usually grows on tree branches where exposed to hot, sunny conditions), but it illustrates that survival depends on drought tolerance and reflection of excessive sunlight (Pierce, 2007), rather than biotic interactions. **(b) Glacier forelands** are characterized by disturbance from meltwater runoff, leading to scarce vegetation cover despite high soil nutrient contents (Luzzaro, 2005). In this case the dwarf alpine *Saxifraga aizoides* is visible among the rocks in the foreground, in front of the Cedec glacier, Valfurva, Italy: plants are widely spaced from one another. Such lack of contact between plants in pioneer stages of succession is known as a 'non-interactive species equilibrium' (Wilson, 1969). **(c) Lichens growing in isolation**, such as *Rhizocarpon geographicum* growing on a rock on Mt. Snowdon, Wales. Lichens may form communities in which space is scarce but they often grow in isolation, where there can be little doubt that survival depends more on desiccation tolerance than on competition with neighbours. **(d) & (e).** *Gilmania luteola* **shortly after a rare rain shower in Death Valley, California**, where it is endemic. *G. luteola* survives as seeds, exploiting a temporary glut of water in order to complete the entire life cycle as rapidly as possible. Its presence at this site cannot depend on competition with other species, as there are no other plants for several kilometres – just baked earth. Incidentally, this species had not been seen for 20 years before this population resurfaced in March 2003, and was thought extinct, demonstrating the exceptional resistance of the seeds. **(f)** *Pancratium maritimum* **growing in a beach of black volcanic sand** at Vulcano, Sicily, Italy. Survival depends on its ability to store reserves in a bulb approximately 25 cm under the surface, in order to avoid summer heat and drought – note that no other plant roots were visible when a bulb and root system were temporarily excavated (S. Pierce, personal observation). Abiotic selection pressures are clearly more important for the survival of this species. (a–f © 2012 Simon Pierce.)

characterized by extremes of abiotic factors (e.g. Forey *et al.*, 2010). We can be confident that competition is neither ubiquitous nor, where it occurs, consistently a significant factor, and that abiotic selection pressures and positive biotic interactions are undoubtedly more important than competition in a range of habitats. We will see this theme repeated in Chapter 5 for microbial communities that are forged by **syntrophy**.

The second reason why the struggle between species is peripheral to the mechanism of evolution is that natural selection depends chiefly on the survival of the most appropriately adapted individuals within populations of a species, which drives the direction and pace of evolution within that species:

> The individuals of a species are like the crew of a foundered ship, and none but good swimmers have a chance of reaching the land. (Huxley, 1850)

Thus natural selection is mainly concerned with intraspecific variability over generations. Although there can be little doubt that Darwin saw interspecific competition (the kind usually under scrutiny in ecology) as an extremely important selection pressure, certainly capable of favouring particular traits and influencing the direction of evolutionary trends when species compete, this is not fundamental to the mechanism of natural selection. Evolutionary change will occur even in the absence of a second species so long as the population is variable and selection pressures, such as temperature extremes, cull individuals. Thus novel adaptive traits can evolve without interspecific competition. In the words of a more recent commentator:

> For Darwin, the struggle for existence was a struggle between individuals within a species, not between species, races or other groups. . . . The misunderstanding of the Darwinian struggle for existence as a struggle between groups of individuals – the so-called "group selection fallacy" – is unfortunately not confined to Hitlerian racism. It constantly resurfaces in amateur misinterpretations of Darwinism, and even among some professional biologists who should know better. (Dawkins, 2009)

Thus the idea that interspecific competition for resources is fundamental to the evolution of the full gamut of strategies evident in plants (Tilman, 1988, 2001) is the product of some very muddled, non-Darwinian thinking.

In sum, the key concepts underpinning natural selection are survival and variability within populations: adaptive strategies are forged by natural selection, but selection does not necessarily involve competition between species. Although competition may be an important selection pressure, it is not **the** fundamental selection pressure that regulates all others. It is a common misconception that organisms persist in their habitats because they are the best adapted to compete, e.g. 'each viable morphology is a superior competitor for a particular habitat type' (Tilman, 1988), the organisms that persist are the best adapted to survive, and survival may or may not involve competition.

Thus by examining interspecific competition from an evolutionary perspective it is evident that competition cannot be invoked as an overarching, ubiquitous force in ecology. Indeed the remaining step in this review of the bare essentials of CSR theory is to consider the interaction of competition, stress and disturbance and the predictable consequences of this interaction on the process of natural selection. This can be summarized in the words used in the first account of the CSR model:

. . . there are three determinants – competition, stress and disturbance . . . each has invoked a distinct strategy . . . competition exerts its maximum impact as a determinant of vegetation in circumstances where the competition is resolved perhaps even to the extent that the habitat is occupied by one species, possibly one individual plant. Stress and disturbance together comprise those phenomena which prevent the resolution of competition. At moderate intensities this intervention has the effect of creating spatial or temporal niches; at their most severe both stress and disturbance may so suppress plant development that individual plants scarcely impinge on each other and competition is occluded. (Grime, 1974)

In Chapters 3 and 4 we examine in greater detail the sets of traits predicted to characterize C-, S- and R-selected organisms and review the large volume of empirical work relevant to testing the validity of the theory and its applicability to plants, animals and other organisms. Already we have emphasized the inevitable limitations of a general theory – one might add any theory – that sets out to explain universal patterns of adaptive specialization in organisms. Accordingly, we must not expect all organismal traits to be embroiled in the axes of the CSR model. Many are free to vary independently and influence the finer dimensions of variation in species and populations. As we explain in Chapter 5, reference to this more detailed variation is essential to our understanding of the structure and dynamics of communities.

Summary

1   It is proposed that recognition of an array of adaptive strategies encompassing all organisms, past and present, is possible by reference to two constraints that interact to determine universally recurrent paths of macroevolution. One of these constraints, described as stress, sums together the many agents that limit the quantity of living matter created per unit of space and time. The other, disturbance, sums the variety of phenomena that restrict biomass by destroying it.
2   Both stress and disturbance impede the vigour with which organisms compete with their neighbours for food. It is proposed that relaxation of their constraining effects allows the evolution and ecological success of fast-growing, monopolistic organisms.

3

Primary Adaptive Strategies in Plants

The appeal of plant traits is that there are many fewer important functional traits than there are plant species. . . . certain typical values of plant traits are systematically found in similar environmental contexts even though the taxonomic composition differs; this is part of the definition of a 'plant strategy'.

(Shipley, 2010)

The dominant tradition in plant ecology since its earliest beginnings was to describe the distribution of species and the composition of plant communities. To conduct this activity properly it was essential to be able to identify species with confidence. This was most conveniently achieved by using field or laboratory keys in which species were grouped according to similarity in structure. Experience showed which attributes were most reliable for identification and it was widely accepted that these conservative traits were not only useful for species identification but might also sometimes provide clues to evolutionary affinity. Today, taxonomies using morphological traits have been strongly buttressed by molecular techniques and these have permitted numerous revisions and refinements by establishing evolutionary relationships.

But what, we may ask, does all this interesting science do for plant ecology? First, it should be recognized that in particular plant families and in some floras taxonomy is a broad indicator of ecological specialization. Hence, in the British Flora, many sedges occupy wetlands, orchids are widely dispersed by tiny airborne seeds, the majority of legumes fix nitrogen and most *Caryophyllaceae* and *Brassicaceae* are annual weeds. However, such broad characterization does not provide the detail required to analyse the fine-scale ecology of an individual population or to predict its status in a community. Moreover, some taxonomic

The Evolutionary Strategies that Shape Ecosystems, First Edition. J. Philip Grime, Simon Pierce.

groups that have a long evolutionary history or wide geographical extent include species that differ substantially in ecology.

We should not be surprised that evolutionary trees do not map neatly over schemes that represent the ecological affinities of the same set of plant species. The conserved morphological traits (e.g. flower architecture) upon which early taxonomists relied often have little connection with the more labile attributes (e.g. leaf size) that frequently influence the ecology. Where, more recently, evolutionary relationships have been established on the basis of genomic information, connections with ecology have become even more tenuous due to uncertainty about whether the DNA sequences involved have current functional significance or are defunct relics of their ancestors.

The search for adaptive strategies

With a few additions and modifications, Table 3.1 reproduces the comprehensive account by Duckworth *et al.* (2000) of the attempts of plant ecologists to devise systems classifying plant species by function rather than taxonomic affiliation. Several themes recur. One is reliance on plant architecture and leaf form for which the required data would be available as a by-product of conventional taxonomy. Another common theme is the use of morphological traits correlated with climate: several authors draw special attention to the form in which plants survive unfavourable seasons of the year such as dry summers or cold winters. Close scrutiny of the history recounted in Table 3.1 reveals a gradual shift from dependence on morphology to the use of criteria that provide insights into phenology and dynamic properties.

Only in the last two entries in the table do we see strong dependence on life-history and reproductive traits, although the contributions of Warming (1884), Raunkiaer (1907) and Ramenskii (1938) are prescient in their attempts to embrace the dynamic properties of vegetation.

Theoretical work

All of the work included in Table 3.1 was conducted with the practical purpose of mapping and classifying vegetation, and for some plant ecologists this involved a lifetime of travel and functional taxonomy to develop then test the validity of their schemes. For some the burden of practical work must have been heavy and this resulted in an emphasis on easily measured traits that is still evident in functional ecology today (Hodgson *et al.*, 1999).

Against this background over the second half of the 20th century there were both plant and animal ecologists who took a more detached and theoretical approach to defining the functional types of organisms. For many (Cole, 1954; MacArthur & Wilson, 1967; Whittaker & Goodman, 1979) the focus was on life-histories and involved examining the consequences for fitness of alternative life-histories in specified circumstances of environment and losses to predation.

Table 3.1 A summary of the development of the plant functional type (PFT) concept (from Duckworth *et al.*, 2000, but modified with the addition of Ramenskii (1938) and Curtis (1959)).

Author	Comments
von Humboldt (1806)	First recognized relationship between plant form and function. Developed classification based on growth-form.
Grisebach (1872)	Classification of 60 vegetative forms correlated with climate
Warming (1884, 1909)	Classification based on simple life history features (e.g. lifespan and vegetative expansion power).
Schimper (1903)	Recognized convergence between plant form and function, despite taxonomic differences, between vegetation types from geographically different, but climatically similar, areas.
Raunkiaer (1907, 1934)	Life-forms system.
Kearney & Shantz (1912)	Proposed four basic strategies of plants in arid regions in response to drought.
Ramenskii (1938)	Used systematic observations on vegetation in fixed quadrats to record the growth and development of species and to propose primary functional types (see Box 2.1, page 13).
Braun-Blanquet (1928)	Added further detail to the life-forms system.
Gimingham (1951)	Growth-forms system which also considered branching of stems.
Dansereau (1951)	Classification system based on life-form, morphology, deciduousness and cover.
Curtis (1959)	Best known for surveying and classifying the vegetation of Wisconsin, but recently documents have come to light showing that just before his premature death he had NSF funding to use morphological traits in a functional approach.
Küchler (1967)	Hierarchical classification, with initial division based on whether plant is woody or herbaceous. Lower-order groups are based on life-forms, leaf characteristics and cover.
Mooney & Dunn (1970); Mooney (1974)	Investigation of form–environment relationships in the context of convergent evolution.
Hallé *et al.* (1978)	Models of tree architecture based on the underlying 'blueprint' for development rather than morphology at any given moment.
Box (1981)	Developed global classification based on structural and phenological attributes in relation to climate.
Grime (1974, 1979)	Plant strategy theory and CSR system of PFTs.
Noble & Slatyer (1980)	'Vital attributes' classification of plants on basis of life-history factors in relation to response to disturbance.

Particular attention was paid to the advantages and disadvantages of early or delayed reproduction, a line of investigation that, as explained in Chapter 2, resulted in the theory of *r/K*-selection. Other investigations had specific reference to plants and placed particular emphasis on competition for resources (Tilman, 1982, 1988) and space (Bolker & Pacala, 1999).

Opinions as to the value of this theoretical work have waxed and waned over the past 50 years. Stearns (1976) expressed his concern that ideas often achieved wide currency without ever being subjected to comparison against the realities of field and laboratory. We are inclined to a more generous view: regardless of their success or failure studies of this kind have been a talisman for the mission to imbue ecology with the priorities and predictive powers of more established sciences.

Measuring variation in plant traits: screening programmes

As we have seen from earlier parts of this chapter, the insights gained by examination of plant morphology were sufficient to stimulate some plant ecologists to begin to formulate ideas about the ways in which groups of plants achieve fitness in particular environments and consistently fail in others. However, this reliance on data accessible by field observation had the effect of leaving virtually unexplored some vital aspects of plant ecology. Little was available with which to estimate the importance of variation in physiological characteristics related to mineral nutrition, moisture relations, temperature effects on germination, frost, drought and heat tolerance and resistance to competitors, herbivores and diseases and in many other aspects of plant biology.

Despite the absence of any consistent attempt to develop a broad perspective about variation in physiological traits and plant ecology it was possible by the 1960s to detect some patterns in a very fragmented literature. Pearsall (1950), for example, had found differences in the mineral nutrient contents of leaves that were consistently correlated with the soil preferences of plants in northern England, and Parsons (1968) reported that the potential growth rates of Australian plants measured in the laboratory accurately predicted the fertility of the soils they exploited in the field. Soon afterwards Baker (1972) completed a comprehensive study of variation in seed size in the flora of California, and Bennett (1972) had begun his epic comparative screening of plant genome size that would eventually extend worldwide. In Sheffield plant ecologists using standardized laboratory conditions were screening the shade-tolerance of seedlings (Grime, 1965b; Hutchinson, 1967) and the palatability of leaves to snails (Grime *et al.*, 1970). In each of these investigations the avowed purpose was to establish the limits of variation in a single plant trait and to hunt for clues concerning the forms of natural selection to which that trait had been exposed in particular habitats.

The screening programme at Sheffield also included measurements of the growth rates of plants. The results had a particular significance in relation to development of the CSR theory of primary adaptive strategies.

Screening of plant growth rates

The screening in the Sheffield experiment involved a procedure in which potential relative growth rates were measured during the first five weeks of seedling establishment in a productive controlled environment. The experiments formed a continuous series over a period of four years and yielded data for 135 species, for most of which frequency of occurrence in each of the major habitats of the Sheffield region was available from a concurrent field survey. The results (Grime & Hunt, 1975) established a clear correlation between the potential relative growth rates of species and their habitats. Inherently slow-growing species were drawn from a wide range of unproductive vegetation types whereas inherently fast-growers were restricted to fertile conditions. It was also apparent that the fast-growers fell into two categories; one group was composed of ephemerals and short-lived perennials of disturbed habitats whilst the second was dominated by more robust, often clonal species in vegetation of low species-richness. It was this distribution of growth rates, coupled with insights from a parallel comparative study of phenology and community dynamics (Al-Mufti *et al.*, 1977) that provided the initial inspiration for the CSR theory.

Although the traits involved in the initial development of the triangular model were restricted to plant life-history, growth rate and phenology, it was possible to recognize (Grime, 1977) a number of other traits that were likely to be embroiled. These included tissue life-span, leaf nutrient concentrations, the foraging mechanisms of shoots and roots, leaf resistance to generalist herbivores and rates of decomposition. Equally, it could be predicted that some traits would not be caught up in the axes of the model but would instead relate to the finer avenues of specialization responsible for the individualities of species. The need to test these various hypotheses necessitated a screening experiment on an unprecedented scale. Funded by the UK Natural Environment Research Council this five-year experiment commenced in 1987 and became known as the **Integrated Screening Programme (ISP)**.

The Integrated Screening Programme

Following a year-long planning process that included tests of new screening procedures and collection of large stocks of seeds from widely contrasted habitats of the Sheffield region, the ISP subjected 43 common herbaceous species and small shrubs to tests measuring 63 plant traits. The choice of species ensured that vegetation from all the main landscape components of inland Britain was represented and, with valuable contributions from collaborating laboratories in York, Newcastle and Bangor, the screened traits affected many aspects of the core functioning of the vascular plant and included both juvenile and mature stages of the life-history. The main results of the ISP were as follows:

1   As we might expect, with few exceptions, plant traits are not free to vary independently of others. Despite varying widely across the 43 species many traits were shown to shift positively or negatively in relation to variation in other traits. The strongest axis of co-varying traits accounted for 22 per cent

Table 3.2 Correlations between pairs of ISP attributes with correlation coefficients (*r*) of 0.65 or above (based on data from Grime *et al.*, 1997)

Rank	r	Correlated traits	
1	0.93	Specific leaf area	Leaf area ratio
2	0.84	Leaf Mg content	Leaf Ca content
3	0.83	Leaf P content	Leaf N content
4	0.83	Shoot increment in lit quadrants (mg)	Root increment to undepleted quadrants (mg)
5	−0.81	Leaf tensile strength	Leaf Ca content
6	0.79	Yield in low nutrients	Yield in low nitrogen
7	−0.76	Leaf tensile strength	Leaf Mg content
8	−0.75	Leaf width	Leaf tensile strength
9	0.74	Life-history	Lateral spread
10	−0.72	Leaf tensile strength	Palatability index
11	0.72	Leaf K content	Leaf P content
12	0.71	Abaxial epidermal cell size	Spongy mesophyll cell size
13	0.69	Palatability index	Leaf Ca content
14	0.69	Palatability index	Decomposition percentage weight loss
15	0.67	Leaf maximum surface area	Root increment to undepleted quadrants (mg)
16	0.66	Leaf K content	Leaf N content
17	−0.66	Leaf tensile strength	Leaf N content
18	0.66	Adaxial epidermal cell size	Leaf Mg content
19	−0.65	Leaf width	Yield in low nutrients
20	0.65	Palatability index	Leaf Mg content

of the variation in the ISP dataset and revealed strong linkages between leaf nutrient concentrations, seedling growth rate, root and shoot foraging responses, the longevity, tensile strength and palatability of leaves and the rate of decomposition of leaf litter (see Table 3.2). This axis confirmed the existence of a whole plant multi-trait trade-off, predicted from CSR theory, between a set of trait values conferring an ability for high rates of resource capture and loss in productive habitats and another set leading to retention of resources in unproductive conditions. This axis of acquisitive vs. conservative 'economics' has recently been independently confirmed for leaves throughout the world flora (Wright *et al.*, 2004; Box 3.1).

2 The second axis of trait variation was largely taxonomic in character, distinguishing between monocotyledons and dicotyledons. The third axis, however, bore a strong imprint of ecology and followed in the footsteps of MacArthur & Wilson (1967) and *r/K*-selection, by reflecting variation in length of life-history. Plotting axis 3 against axis 1 produces a distribution of species within a triangular space that is highly consistent with CSR theory (Grime *et al.*, 1997). Similar results have since been confirmed for a wider flora (Box 3.2).

3 Many of the traits associated with axis 1 (leaf life-span, palatability, resource foraging, rates of decomposition) are deeply implicated in aspects of ecosystem functioning. This provides strong evidence of a key role for primary

plant strategies in controlling ecosystem processes such as productivity and the circulation and storage of resources and environmental pollutants (we will reiterate this important finding in more detail in Chapter 6, particularly in Table 6.1).

4 Variation in many other ecologically important traits, and particularly those related to regeneration from seed and juvenile characteristics were largely unrelated to the axes of the CSR model. Two important conclusions were drawn from this finding:

a The occurrence of trait variations that are not entrained in the axes of the CSR model permits plant species and populations to be assigned similar positions within the CSR array without compromising the notion that each may exhibit an individual ecology.

b Distinctive impacts of natural selection appear to operate during the juvenile phase. Fitness depends upon traits of both the juvenile and established phases.

In Chapter 5 we shall see that these conclusions are an important component of the **twin-filter model** that discriminates between the CSR traits that control ecosystem functioning and traits determining which species are admitted to particular communities at specific sites.

Box 3.1: The worldwide economics spectrum

. . . morphological plasticity and cellular acclimation reach full expression at opposite ends of a spectrum of plant functional types which in terms of resource processing range from 'the aquisitive' to 'the retentive' and correspond respectively to highly productive and chronically unproductive vegetation (Grime & Mackey, 2002)

A trade-off between acquisitive and retentive plant physiologies is predicted by CSR theory (Grime, 1979, 2001) and was recorded during the ISP as 'a trade-off between attributes conferring an ability for high rates of resource acquisition in productive habitats and those responsible for retention of resource capital in unproductive conditions' (Grime et al., 1997). Wright et al. (2004) went on to confirm that this spectrum, at least for leaves, is evident at the global scale as a 'worldwide leaf economics spectrum'. Freschet et al. (2010) confirm that, as seen in the ISP, the leaf economics spectrum forms part of the resource dynamics of the whole plant, representing a 'plant economics spectrum'.

For leaves, an economics trade-off exists because whilst thin, nitrogen-rich leaves minimize resistances to CO_2 assimilation and contain sufficient photosynthetic machinery to permit rapid growth, such leaves pay the price of not being physically durable and lacking the storage capacity to keep growth supplied when environmental variability imposes erratic photosynthesis (Pierce et al., 2005). Only a part of the leaf economics spectrum is shown in Fig. 3.1, in which the trait leaf mass per area (LMA; the inverse of specific leaf area, SLA) is intimately correlated with leaf nitrogen contents (N_{mass}) and photosynthetic capacity (A_{mass}). Other traits such as respiratory rates, leaf phosphorus contents and leaf dry matter content (LDMC), not shown here, are also important components of this trade-off. Each point on the graph represents one

(Continued)

of 2,548 species, measured at 175 sites worldwide, and the leaf economics spectrum is known to encompass 219 vascular plant families; a global effort involving 27 research groups was required to collect these data.

It is reasonable to expect that this is a truly global relationship reflecting an inescapable biophysical trade-off between acquisitive and conservative lifestyles, which is perhaps best expressed as a '**worldwide economics spectrum**'.

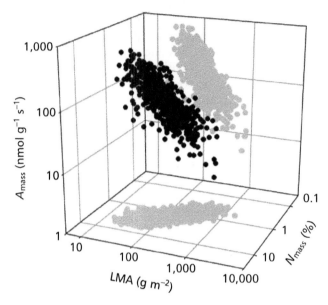

Fig. 3.1   A key part of the worldwide leaf economics spectrum – the relationship between leaf mass per area (LMA), mass-based photosynthetic capacity (A_{mass}) and leaf nitrogen content (N_{mass}). (Reproduced from Wright *et al.*, 2004. Copyright © 2004 Nature Publishing Group.)

Box 3.2: Multivariate analysis of traits

Phenotypes are compositions of traits, all of which may exhibit variation. Multiple dimensions of variability must therefore be accounted for when quantifying and comparing the phenotypes of organisms. Multivariate analysis can help simplify this apparent chaos by revealing the underlying patterns of trait variation that reflect general trends in adaptive specialization. A technique such as principal components analysis (PCA) works by determining whether traits co-vary in a consistent manner, uniting those that do and thereby reducing many dimensions into a few. These united axes of variation are known as principal components, and the two most variable are usually those plotted as *x* and *y* axes.

Díaz *et al.* (2004) measured 12 key life-history traits for 640 vascular plants species on three continents. When they used PCA to analyse the variation in these data they found a principal component that reflected leaf thickness, density and durability – an axis of investment in leaf economics. Another principal component represented a trade-off in leaf area, height and woodiness of shoots, along with seed mass. This demonstrates that plant phenotypes embody an adaptive trade-off between leaf economics, whole plant architecture and the investment of resources in reproduction, in agreement with the results of the Integrated Screening Programme and supporting CSR theory.

Figure 3.2 shows the results of another study (Cerabolini *et al.*, 2010a) confirming that the pattern of trait variability seen in the Integrated Screening Programme is evident for wild plants over a biogeographic range spanning alpine to continental Europe. Note that species are essentially scattered in a triangle, with (towards the right) tall plants with large leaves, (top) small plants with tough, carbon-rich leaves that flower late in the season, and (bottom) small plants that flower early and have leaf characteristics (high N contents and specific leaf area) indicative of extremely rapid growth rates.

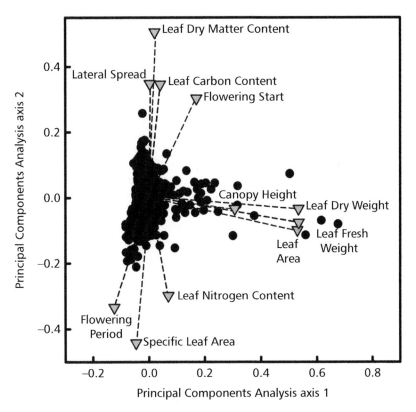

Fig. 3.2 Scatter-diagram of species positions on axes 1 and 2 of a principal components analysis (black points) conducted for 11 key traits (grey triangles, values measured from wild plants) of 506 herbaceous plant species from alpine, subalpine and lowland continental Europe. (Redrawn with permission from Cerabolini *et al.*, 2010a. Copyright © 2010 Springer.)

Further trait screening

The value of the ISP was related to the extent to which it allowed an objective review of trait variation and inter-relationships in a local flora. An unavoidable limitation was its restriction to 43 mainly herbaceous species in temperate habitats. It was extremely fortunate therefore that, in the decade following, a number of large-scale projects (Díaz *et al.*, 2004; Wright *et al.*, 2004; Cerabolini *et al.*, 2010a) have allowed aspects of the trait relationships revealed in the ISP to be re-examined and confirmed in an international context and for much larger numbers of species in the wild. These are examined in Boxes 3.1 and 3.2.

The application of CSR theory

Two enduring qualities of CSR theory as a scientific theory are, first, that it remains unfalsified – despite continually coming under attack – and secondly that it continues to remain current: it has developed, agrees with novel research findings and finds new applications. However, a perennial criticism of CSR theory is that the quantification and comparison of precise CSR strategies ideally involve the kind of labour-intensive and time-consuming experiments conducted during the ISP, and it is thus difficult to classify wild plants according to this scheme: 'an explicit quantitative protocol is lacking for positioning a species in the strategy scheme, and in consequence definite CSR positions have been attributed to few species beyond the datasets for Sheffield and the UK' (Westoby, 1998).

Just one year after these comments were made a methodology for the CSR classification of wild plants *in situ* was published (Hodgson *et al.*, 1999). As we discuss in some detail in Box 3.3, this has since been applied outside the UK to well over a thousand species in a variety of habitats including oak woodland in Turkey (Kilinç *et al.*, 2010), beech/oak woodland in Belgium (Massant *et al.*, 2009), a successional sere in Mediterranean vegetation in southern France (Navas *et al.*, 2010), 12 anthropogenic vegetation types in the Czech Republic (Simonová & Lososová, 2008) and alpine, subalpine and lowland vegetation in Italy (Caccianiga *et al.*, 2006; Pierce *et al.*, 2007a, b), most notably in a paper entitled 'Can CSR classification be generally applied outside Britain?' (Cerabolini *et al.*, 2010a). The conclusions of this latter paper can be summarized in a single word: 'yes'. Cerabolini *et al.* (2010a) did identify a small number of inaccuracies in the method and particular situations where it is not – in its current form – strictly applicable. However, these were not seen as grounds for rejection but for potential refinement, and recalibration of the method was suggested using the most extreme trait values evident in the world flora. They also proposed that woody plants could be included based on leaf traits indicative of economics and size (mass and area) and investment in reproduction. Navas *et al.* (2010) applied CSR classification to woody species by employing the revised categories for lateral spread proposed by Grime *et al.* (2007).

Box 3.3: Applied CSR classification

How can we apply CSR theory? How can an ecologist confronted with a subject species or populations of a species use CSR theory to understand these organisms? Hodgson *et al.* (1999) developed an applied CSR classification methodology that aims to be readily applicable to herbaceous species in the field, based on measures of **'soft traits'** (i.e. traits that are relatively easy and rapid to measure, allowing many plants to be investigated) that correlate with **'hard traits'** (fundamental physiological, growth or developmental characteristics that would be too time-consuming or impractical to apply to large numbers of plants *in situ*). They started by producing **'gold standard'** measures of the variability in C-, S- and R- selection evident in plant species of the flora of the Sheffield region in Britain, which was essentially used to calibrate the methodology. This was done by obtaining three hard measures for these species: **(1)** the extent of dominance by species in productive and undisturbed habitats, which provided an index of C-selection; **(2)** the extent of variability in traits known, from the ISP, to reflect a conservative lifestyle, which provided an index of S-selection; and **(3)** the frequency of species in potentially productive but disturbed habitats provided an R-selection index. During CSR classification soft trait values of a subject species are essentially correlated with the range of hard values in these gold standard indices, in order to calculate the extent of C-, S- and R-selection. Hodgson *et al.* found that the method calculated the correct CSR strategy for 96 per cent of species from a database of nearly 500, compared against a more labour-intensive classification system based on a dichotomous key of traits (Grime *et al.*, 1988).

Hodgson *et al.* provided a spreadsheet incorporating these calculations that automatically returned a CSR category for single species in the form of a simple label (e.g. S, CR, CS/CSR) when trait values were inputted. The logical next step of converting C-, S- and R- values into percentages, allowing subject species to be plotted in a ternary (triangular) graph, was taken by Caccianiga *et al.* (2006). Pierce *et al.* (2007b) then incorporated the calculations of Hodgson *et al.* and Caccianiga *et al.* into a spreadsheet structured in such a way that CSR coordinates could be calculated instantaneously for vast numbers of species, simply by pasting in columns of trait values (this spreadsheet is available to *Journal of Ecology* subscribers at www3.interscience.wiley.com/journal/118509730/suppinfo). This allowed the CSR strategies of the species forming communities or of entire floras to be visualized rapidly. Using these methods it is possible to characterize large numbers of species or conduct specific studies to compare species or communities occurring along environmental gradients in nature.

Cerabolini *et al.* (2010a) confirmed that axes of trait variation for 506 species native to southern European continental, subalpine and alpine bioclimatic zones were highly significantly correlated with the extent of C-, S- and R-selection determined by Hodgson *et al.*'s CSR classification. Thus the general applicability of CSR classification beyond Britain, at least in other European bioclimatic zones, has been clearly demonstrated. Later, we shall discuss at length (Chapters 5 and 6) examples of the use of these techniques to investigate communities and ecosystems, but here we present a simple example (Pierce *et al.*, 2007a) in which the CSR strategies of alpine and lowland grasses in northern Italy are calculated and compared using Hodgson *et al.*'s method. From the CSR triangle shown in Fig. 3.3 it is evident that whilst alpine grasses are generally S-selected some are adapted to disturbed and moderately productive niches. Furthermore, the range of strategies exhibited by lowland species is much greater,

(Continued)

implying adaptation to undisturbed productive niches, potentially productive but disturbed niches, chronically unproductive niches, and a range of niches inbetween.

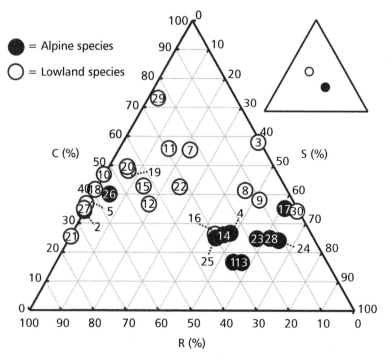

Fig. 3.3 The CSR strategies of grasses (Poaceae) in alpine (filled circles) and lowland (open circles) habitats in northern Italy. The mean strategy for each bioclimatic zone is shown in the smaller triangle above right. Numbers refer to different species, the identities of which are: **(1)** *Agrostis rupestris*; **(2)** *Agrostis schraderana*; **(3)** *Arundo donax*; **(4)** *Avenula versicolor*; **(5)** *Brachypodium sylvaticum*; **(6)** *Bromus erectus*; **(7)** *Calamagrostis arundinacea*; **(8)** *Calamagrostis epigejos*; **(9)** *Chrysopogon gryllus*; **(10)** *Digitaria sanguinalis*; **(11)** *Echinochloa crus-galli*; **(12)** *Eleusine indica*; **(13)** *Festuca halleri*; **(14)** *Festuca nigrescens*; **(15)** *Festuca pratensis*; **(16)** *Festuca tenuifolia*; **(17)** *Festuca varia*; **(18)** *Holcus lanatus*; **(19)** *Holcus mollis*; **(20)** *Lolium multiflorum*; **(21)** *Lolium perenne*; **(22)** *Melica ciliata*; **(23)** *Nardus stricta*; **(24)** *Oreochloa disticha*; **(25)** *Poa alpina*; **(26)** *Poa supina*; **(27)** *Poa trivialis*; **(28)** *Sesleria varia*; **(29)** *Sorghum halepense*; **(30)** *Stipa pennata*. (Reproduced from Pierce *et al.*, 2007a. Copyright © 2007 Taylor & Francis Informa UK Ltd, Journals.)

Virtual plant strategies

Aside from evidence from screening programmes and wild plants, another line of enquiry has also recently substantiated CSR theory – one that investigates the possible directions in which plant traits can evolve. Plant traits have, of course, evolved in response to a range of ecological situations. Unfortunately, it is extremely difficult to chart the actual evolution of specific traits over the ~450 million years that plants have occupied terrestrial environments because the fossil record is insufficiently detailed. However, the increasing processing power and speed of computers has now given evolutionary biologists an extra tool, in the form of the ability to model the evolution of life-histories in response to virtual environments. The most recent of these studies have consistently shown that when virtual plants are 'grown' in a virtual environment characterized by gradients of resource availability and disturbance intensity, three principal strategies evolve:

> What the model has shown is that differences in nitrogen availability and disturbance frequency alone can result in the evolution of three primary strategies associated with three extreme combinations of these conditions, and the intermediate environmental conditions evolving strategies that lie on a continuum between the three extremes. The fourth extreme environmental state, namely low nitrogen/high disturbance, was found to be incompatible with the long-term persistence and, hence, evolution of the plant populations. This is a significant finding because it is the first time that the 'untenable triangle' hypothesized by Grime (1979) has received theoretical support. (Mustard *et al.*, 2003)

Figure 3.4a shows the virtual space investigated by Bornhofen *et al.* (2011) thousands of generations after an initial seeding throughout (see also Bornhofen & Lattaud, 2008). Note that, as predicted by Grime (1977), sites of high disturbance intensity and low resource availability are not tenable (i.e. no adaptive strategies have evolved to cope with these conditions).

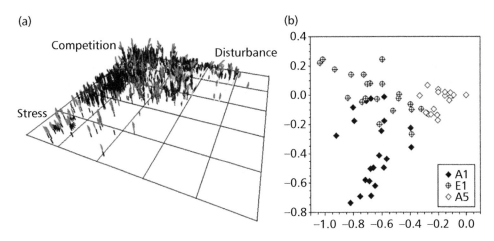

Fig. 3.4 The virtual evolution of plant life-histories: a virtual space **(a)** characterized by gradients of increasing resource availability (stress to competition) and intensity of biomass destruction (competition to disturbance) with graphical representations of the life-histories that have evolved along these gradients. In **(b)**, a principal components analysis scatter-plot showing the variability in life-histories of species from the corners of the virtual space in **(a)** is analysed (A1 = competition, E1 = stress, and A5 = disturbance). (Reproduced from Bornhofen *et al.*, 2011. Copyright © Elsevier.)

The key innovation of the Bornhofen *et al.* (2011) model was that species exhibited virtual genomes encoding a wide range of physiological and morphological trait values. Species evolving in conditions of low disturbance and high resource availability were characterized by a prolonged period of vegetative growth before reproduction, the greatest height and high seed mass. At high disturbance/high resource availability species evolved extremely high growth rates, a minimal period of vegetative growth followed by early reproduction, short life-span and low seed mass. Species of low resource availability/low disturbance exhibited the longest life-spans and slowest growth rates. When principal components analysis was used to determine the variability among these life-history traits for plants growing at the extremes of the virtual environment, a triangle was apparent (see Fig. 3.4b) that mirrored the CSR-triangle evident for wild plants in reality (e.g. Fig. 3.2; Cerabolini *et al.*, 2010a). An earlier attempt to 'evolve' strategies in response to virtual environments used much smaller numbers of traits (Colasanti *et al.*, 2001), but nonetheless found that virtual productivity gradients selected for survival strategies with essentially the same character and relative abundances as those seen in the real world, obeying the kind of humped-back biodiversity/productivity relationship that we will come across in Chapter 5. Digital tools such as these allow modelling of communities at scales that would be beyond even the largest of real-world experimental studies (Hunt & Colasanti, 2007).

To sum up, in this chapter we have explored the background, development and application of CSR theory. We have seen that, like all sound theorums (*sensu* Dawkins, 2009) it has remained unfalsified and has been found to be compatible with new discoveries such as the worldwide leaf economics spectrum (see Box 3.1) and new tools for investigating evolutionary processes, such as virtual environments. However, the focus has so far been on plants, and in the next chapter we shall discuss the evidence that an identical three-way adaptive trade-off exists in all organisms. Indeed, we shall review and collate the work of zoologists and microbiologists who have interpreted their particular study organisms in terms of CSR theory or a number of three-way trade-off theories that are surprisingly similar.

Summary

1 Plant ecology for much of its long early history was a descriptive activity and interpretation of patterns observed in the field mainly relied upon morphological traits suspected to be indicative of adaptations to climate.
2 Progress towards recognition of adaptive strategies began through the activities of individual scientists who made comparisons of particular plant attributes on specimens collected from natural habitats or grown in garden conditions.
3 Valuable insights have been obtained from investigations that involve repeated recording of the development and behaviour of plants in quadrats placed in their natural habitats. Consistent with CSR theory, these studies have confirmed the existence in productive habitats of plants capable of monopolizing

resource capture and excluding all neighbours. It has also been confirmed from both field observations and laboratory experiments that both low resource supply and disturbance reduce the expression of competition and allow less robust species to coexist with potential monopolists.

4 In a comprehensive screening of traits of 43 plant species from contrasted habitats within a local flora, patterns of adaptive specialization consistent with CSR theory were detected. Traits of mature plants were shown to vary independently of those of the juveniles. Principle components analysis of variation in mature traits revealed a first axis consisting of parallel trade-offs in a set of traits embedded in the core functioning of the plant and reflecting a transition from rapid acquisition and loss of photosynthate, mineral nutrients and water to effective protection and retention of captured resources. The analysis also detected another axis reflecting variation in the length of the life-history. Recently, this experimental support for CSR theory has been strengthened by further analyses of results from plant-screening studies collating data from widely different parts of the world (explored in Boxes 3.1 and 3.2).

5 Screening of trait data, by allowing the principal axes of trait variation to be quantified, has effectively calibrated a novel CSR classification procedure whereby an unknown subject can be assigned a position within the CSR triangle. This procedure has been found to be consistent with trait variation in a range of bioclimatic zones not included in the original calibration.

6 Virtual plants replicating in virtual environments confirm that, in agreement with CSR theory, three principal directions of adaptive specialization evolve spontaneously in response to gradients of resource availability and disturbance intensity.

4

Primary Adaptive Strategies in Organisms Other Than Plants

By definition, primary strategies, if they exist, must encompass all living matter (plants and animals) and must be fashioned by selective mechanisms operating in all habitats.

(Grime, 2001)

The purpose of this brief essay is to make two predictions: (1) that further research will expose a common pattern of ecological specialization (i.e. the same spectrum of primary strategies) in fungi, green plants and animals, and (2) that recognition of this pattern will provide a key to the assembly rules and dynamic properties of communities and ecosystems.

(Grime, 1988b)

We have discussed the evidence that plant adaptive strategies represent three-way trade-offs between C-, S- and R-selection: trade-offs between the control of resource acquisition in productive habitats (C), the persistence of individuals in unproductive habitats (involving resistance of suboptimal periods for metabolic function; S), or regeneration of the species in response to disturbance, or lethal events (R). However, as one of us has previously suggested, above, a workable theory of adaptive strategies must be broadly applicable. Do adaptive strategies of organisms throughout the tree of life truly represent varying degrees of C-, S- and R- adaptation? Is the comparison between plant and animal adaptive strategies, originally suggested by Ramenskii in 1938, really valid? Although CSR theory was developed with plants in mind, it should also be applicable to life-forms as diverse as bacteria, fungi and animals. Here we investigate the evidence that combinations of these three general directions of adaptive specialization may indeed explain the character of organisms throughout Darwin's tree.

The Evolutionary Strategies that Shape Ecosystems, First Edition. J. Philip Grime, Simon Pierce.
© 2012 John Wiley & Sons, Ltd. Published 2012 by John Wiley & Sons, Ltd.

The architecture of the tree of life

It is a source of wonder, and also of exasperation, that Darwin's tree has so many branches – mapping these branches, twigs, twiglets and myriad leaves is a daunting task that is beyond any single person. Nonetheless, some have tried. Carl von Linné (also known as Linnaeus, or simply 'L.') had to resort throughout his career to a team of almost 200 students, but together they were able to catalogue little more than 12,000 species, or less than 1 per cent of the species currently known, in the *Systema Naturæ* and *Specie Plantarum* (a Linné C., 1753, 1767).

Nowadays, internet databases such as the Tree of Life Web Project[1], the Encyclopaedia of Life[2] and the Catalogue of Life[3] represent humanity's most concerted efforts to understand the complexity of evolutionary relationships. We refer readers to these fascinating websites for the vast scientific literature in which phylogenetic relationships throughout the tree are justified and updated. The intricacy of the tree of life is immediately evident. The classification of humans, for example, has 22 taxonomic steps from 'Eukaryote', via subdivisions with names as bizarre as 'Gnathostomata' and 'Catarrhini', long before reaching '*Homo sapiens*'. Humans are also Craniata, Vertebrata, Synapsida, and many other names in between. Each name represents a branching point in the tree and a relationship with other groups, or clades, of living and extinct organisms that share phenotypic traits with us.

This taxonomic complexity presents considerable problems to those of us attempting to determine ecological differences between organisms. For example, despite their importance in contemporary ecosystems groups such as Arthropoda, Annelida, Deuterostomia and Mollusca represent just a few of the 25 twigs in the clade Bilateria. Some parts of the tree, such as Insecta, are particularly bushy, representing high biological diversity. Thus phylogenetic diversity tells us little of the ecological characteristics and relative abundance of species. Adding to this complexity is the fact that it may be difficult to precisely delimit species, especially towards the base of the tree of life where **horizontal gene transfer** mixes genomes somewhat (see the fascinating discussion by Dawkins (2004) of what constitutes a species). Indeed, despite the convenience of the term 'tree' many biologists are starting to question the accuracy of this term to describe the relationships between organisms, especially among microorganisms (here we have decided to use the term, but with this caveat in mind). Furthermore, the known tree of life is currently incomplete and almost certainly includes slight differences from the actual series of phylogenetic events that have occurred: scientific names for organisms may change with every new piece of evidence, and the apparent malleability of evolutionary relationships may sometimes defy cherished traditions.

In the face of this taxonomic complexity we have chosen to use the Tree of Life Web Project as a general guide. However, we only have sufficient space to

---

[1] http://tolweb.org/tree (Maddison *et al.*, 2007).
[2] www.eol.org
[3] www.catalogueoflife.org

investigate a few examples from prevalent branches of the tree. Nor would it be worthwhile to go into great detail. The question of whether or not C-, S- and R-selection have shaped evolution throughout the tree requires that we take a step back and look using MacArthur's blurred vision:

> . . . the objective is not to analyse each detailed pattern of specialization responsible for the "fine-tuning" of the distribution and ecology of an individual species or population. The search is directed towards common forms of natural selection and universal design constraints, recognition of which may expose recurrent avenues of adaptive specialization. (Grime, 1988b)

r, K and beyond *K*

As with plants, it is the suite of traits, including behavioural adaptations and extended phenotypes (*sensu* Dawkins, 1982), which comprise animal adaptive strategies. There is growing recognition that multiple trade-offs, not simply *r/K*-selection, are involved in the evolution of adaptive strategies for organisms as diverse as dinoflagellate algae, corals, guppies, birds and mammals (Smayda & Reynolds, 2001; Magurran, 2005; Murdoch, 2007; Sinervo & Clobert, 2008). Indeed, the terms *r* and *K*, strictly speaking, are population characteristics rather than adaptive strategies (*r* stands for rate of reproduction and *K* for maximum population size or carrying capacity). Stearns (1977) expressed the concern that: 'unlike *r, K* cannot be realistically expressed as a function of life-history traits'. In other words populations of a species may reach the carrying capacity of an ecosystem using different strategies.

Greenslade (1972) first suggested a '**beyond *K***' strategy for beetles, and shortly after Southwood (1977) incorporated this into a general three-way trade-off theory, as *r*-, *K*- and *A*-(Adversity)selection, by his own admission in ignorance of the slightly earlier CSR theory. However, he then went on to point out the similarities between plant CSR strategies and the **Southwood-Greenslade templet** for animals, whereby *K*-selection is equivalent to C-selection, and *A*-selection equivalent to S-selection (Southwood, 1988). His templet differed from CSR theory in suggesting that organisms could be simultaneously adapted to high levels of both disturbance and stress, although the fourth-trait syndrome proposed as an adaptive response to these conditions actually exhibited features intermediate between *K*- and *A*-selection, and was not explicitly nominated as a fourth strategy (Southwood, 1988). The Southwood-Greenslade *rKA* templet enjoys some popularity among entomologists, who recognize the inadequacy of *r/K*-selection theory when faced with the bewildering ecological diversity of insects (Braby, 2002), and has now become so widely accepted that it is a feature of undergraduate animal ecophysiology texts (e.g. Willmer *et al.*, 2005).

Bacteriologists call the beyond *K* strategy '*L*', which is exhibited by microorganisms that are 'well adapted to the adverse environment' (Golovlev, 2001). Ironically, plant ecologists have even come up with their own version of the *rKA* templet, in which the beyond *K* strategy is called '*I*', or 'Impoverishment' (Taylor *et al.*, 1990). Indeed, three-way trade-offs in plants are periodically rediscovered, most recently under the guise of the RVD scheme (De Miguel *et al.*, 2010). According to the RVD scheme, plants invest mainly in reproduction, vegetative growth or defence (hence 'RVD'), and are said to face two dilemmas: '**to reproduce or defend**' or '**to grow or defend**'. Although not explicitly stated by De Miguel *et al.* (2010) a third dilemma is also implied (**to reproduce or grow**) and these three main directions of adaptive specialization are essentially those of CSR theory. Similarly, the survival (L), growth (G) and fecundity (F) space of Enright *et al.* (1995) draws from CSR theory but is based on demographics (i.e. the traits of populations rather than of individuals).

CSR theory is more mature than the *rKA*, *rKI*, *rKL*, RVD and LGF schemes, not only in terms of age and robustness but also because it is more than simply a theoretical framework within which to consider evolution, having been developed into an applied methodology that is actually of practical use, at least for plants (Hodgson *et al.*, 1999; Caccianiga *et al.*, 2006; Pierce *et al.*, 2007a, b; Cerabolini *et al.*, 2010a, b; Kilinç *et al.*, 2010; Navas *et al.*, 2010). However, the similarities between CSR theory and the *rKA*, *rKI*, *rKL*, RVD and LGF three-way trade-off schemes suggest that they essentially describe the same constraints to evolution operating in a range of phylogenetically diverse organisms, and that research continues to uncover the same patterns of adaptive responses. Perhaps rather than approaching plant adaptive strategies as botanists, animal strategies as zoologists and microbial strategies as microbiologists, we should take a leaf from Darwin's book and approach nature as naturalists.

Empirical evidence for three primary strategies in animals

It is now widely acknowledged that a multidimensional analysis of functional and resource-use traits is the only realistic way of obtaining an objective view of adaptive strategies. Multivariate analysis (which we saw in Box 3.2, page 32) is a technique capable of correlating and coalescing multiple gradients of trait variation to reveal the more general adaptive themes underpinning evolution.

Multivariate analysis of bird and mammal traits reveals three main ways in which suites of traits vary: **(1)** the allometric component (variation in body size); **(2)** fast to slow rates of turnover; and **(3)** earliness and intensity of reproduction (Gaillard *et al.*, 1989). Winemiller & Rose (1992) applied multivariate analysis to 216 North American fishes and found that: 'three fairly distinctive life-history strategies are identified as the end-points of a trilateral continuum'. These strategies were characterized by the following traits: **(1)** intermediate to large size at maturity, rapid larval and first-year hatchling growth rates, delayed maturation, short reproductive seasons, and large clutches of small eggs; **(2)** small size at maturity, early maturation, rapid larval growth and a long reproductive season with multiple spawning and small eggs; and **(3)** small- or medium-size species with large eggs, small clutches, well-developed parental care, slow first-year hatchling and adult growth rates and long reproductive seasons.

The first strategy thus involves larger, fast-growing species that delay reproduction until after a period of resource acquisition, which are also characteristics of C-selected plants. The second strategy is characterized by rapid development

and completion of the life cycle involving a massive lifetime reproductive investment, typical of R-selected species. Large size and immediate reproductive effort
are not important for the third strategy, which is characterized by slow growth
and investment in the durability of individuals (adults and young), equivalent to
S-selection[4]. Winemiller & Rose (1992) concluded that: 'the suites of traits predicted by this adaptive surface appear similar to those described in the trichoto-
mous comparative frameworks proposed for plants (Grime, 1977, 1979) and
other animal groups (Greenslade, 1983)'. Winemiller (1989) found similar results
for Venezuelan fishes, and went on to describe bony fishes worldwide in terms
of this three-way trade-off (Winemiller, 1995). Other authors have also observed
a three-way trade-off in fish life-histories (Kawasaki, 1980; Baltz, 1984).

Databases of animal life-history traits are currently being compiled and these
could be subjected to multivariate analysis in order to confirm the main directions of adaptive specialization seen by Gaillard *et al.* (1989) and Winemiller &
Rose (1992), and to extend our knowledge beyond birds, mammals and fishes.
These include the AnAge database (de Magalhães & Costa, 2009)[5], which
focuses on vertebrate longevity but includes a small number of other traits, and
the PanTHERIA database (Jones *et al.*, 2009)[6], comprised of mammalian life-
history traits. However, these are works in progress: data have not yet been
recorded for many of the 5,416 extant and recently extinct mammals in the
PanTHERIA database. For example, the most widely recorded trait in PanTH-
ERIA, adult body mass, is reported for only 65 per cent of species, and only 1
per cent of species currently include data for all the life-history traits covered
by the database. The fragmentary nature of these databases currently precludes
meaningful multivariate analyses, but it is our hope that such databases will
eventually form a useful resource with which the character of animal adaptive
strategies may be confirmed.

The universal three-way trade-off

The similarities between the main directions of adaptive specialization evident
for some animal groups and those observed for plants are a matter of record.
However, is this really a general rule for life on Earth? Is the three-way trade-off
also relevant to prokaryotes and other microscopic life forms? Do ecologically
equivalent trait syndromes and patterns of resource allocation really exist for
phenotypically diverse organisms on different branches of the tree of life? Let's

---

[4]This also illustrates an important point concerning reproductive investment: R-selected fishes have a
high investment in fecundity, but reproductive investment ends with egg laying, whereas investment in
offspring by S-selected fishes mainly takes the form of supporting hatchlings and young fish. Thus there
is a key strategic difference between profligate **pre-hatching** fecundity in R-selected species and conservative **post-hatching** reproductive investment in S-selected species – a theme that we shall see repeated
throughout this chapter.
[5]These data are published online in the form of a text file at: http://genomics.senescence.info/species/
[6]Text file available at the Ecological Archives website: http://esapubs.org/archive/ecol/E090/184/
default.htm

be clear about what it is we wish to investigate, by expressing our question as a formal hypothesis. Here it is:

> We hypothesize that a **universal three-way trade-off** constrains adaptive strategies throughout the tree of life, with extreme strategies facilitating the survival of genes via: (C). the survival of the individual using traits that maximize resource acquisition and resource control in consistently productive niches, (S). individual survival via maintenance of metabolic performance in variable and unproductive niches, or (R). rapid gene propagation via rapid completion of the life cycle and regeneration in niches where events are frequently lethal to the individual.

How shall we go about investigating this idea? There is currently no study that compares life-history traits throughout the tree of life, and we have seen that trait databases have only started to be produced even for prominent groups such as mammals, and these are far from complete. Unfortunately, we are left with only one option – the rather dreary job of comparing the traits and life-histories of organisms (and how these relate to the productivity and other characteristics of the habitat) one by one. This shall be the main exercise of the current chapter and although the process promises to be somewhat long-winded, we hope that we can at least offer some fascinating, if not downright startling, examples of how organisms are adapted to survive in contrasting habitats. Take, for instance, the case of the Bone-Eating Snot Flower, which is not a plant, is not made of nasal secretions but, as we shall see later, does indeed derive sustenance from bones . . .

We will explore many such examples on our comparative tour of the tree of life, but where to start? The tree of life itself proffers no obvious point of departure. The **Last Universal Common Ancestor (LUCA)** of all organisms was an ancient organism about which we can only make educated guesses with regard to ecology and adaptive strategy (see Lane, 2009). Thus we cannot start our investigation at the base of the tree, where knowledge breaks down. The alternative is to start with one of the leaves, for which we have more information and work our way down the tree and even up and down neighbouring branches. In this case we must choose a particular leaf with which to start.

Three main branches of the tree of life are currently recognized: the *Eukarya*, *Archaea* and *Bacteria*, with the *Viruses* parasitizing organisms along all of these branches. CSR theory is already extremely well justified for flowering plants, and examples of primary adaptive strategies for these organisms were presented in the previous chapter, providing a baseline against which we can compare C-, S- and R-selection in other organisms. We have decided to start with the remainder of the *Eukarya*, and shall investigate mammals, birds, lizards, bony and cartilaginous fishes, arthropods (insects, arachnids and crustaceans), echinoderms, molluscs, annelid worms, anthozoa and fungi.

Working down this branch of the tree we arrive at the archaean branch. Unfortunately, relatively little is known about the ecology of many *Archaea*, but we discuss the strategies apparent among the organisms that have been investigated to date. The slightly more primitive *Eubacteria* (true bacteria) are more easily studied, and bacterial ecological diversity is much more apparent. We showcase the adaptive strategies evident among three of the most widespread and ecologically important bacterial groups: the proteobacteria, firmicutes and cyanobacteria. *Viruses* are not cellular life forms, but natural selection has resulted in the evolution of a range of viral phenotypes and adaptive strategies, which we shall discuss.

Here we present examples of C-, S- and R-selected adaptive strategies within prevalent clades throughout Darwin's tree.

Mammalia (mammals)

C-selected mammals are predicted to have a lifestyle based on continuous, active resource acquisition in productive habitats, with captured resources reinvested mainly in the effort to sustain or increase resource capture. Foraging activities should be relatively uninterrupted by abiotic events. We may therefore expect the C-selected syndrome to include traits that favour the control of resources (such as territorial behaviours), possibly sociality (a principal function of social behaviour is thought to be that of maximising foraging efficiency by exchanging information and optimising decision making; Danchin *et al.*, 2008), alongside senses and mouthparts that are specialized to allow efficient, active resource acquisition. In contrast, S-selected mammals are predicted to be native to chronically unproductive habitats where survival depends on accumulating a resource as soon as it becomes available, and using it to tolerate harsher conditions later on (luxury consumption). In this case abiotic constraints to growth, rather than biotic interactions, are the principal selection pressures. R-selected mammals are predicted to develop and complete the life cycle rapidly, be highly fecund and opportunistic in ephemeral niches, occupying disturbed situations.

Do mammals with these combinations of traits exist?

Here we start by contrasting the life-histories of three species from the largest mammal order, the rodents (Rodentia), and then discuss contrasting examples within other mammal orders.

The Capybara (*Hydrochoerus hydrochaeris*) from South America is the largest modern rodent, at up to 130 cm in length and typically weighing in at 50 kg (Herrera & Macdonald, 1989). Capybara are semi-aquatic and graze on rapidly growing grasses and water plants, such as Southern Cutgrass (*Leersia hexandra*), Lax Panicgrass (*Panicum laxum*) and Water Strawgrass (*Hymenachne amplexicaulis*) in highly productive grassy ponds and flooded *bajios* habitats. In the Venezuelan Llanos, a seasonally flooded region, Capybara form groups of four to 40 individuals that patrol a territory of up to 0.16 km², from which intruders are aggressively excluded (Herrera & Macdonald, 1989; Wolff & Sherman, 2007). In less productive regions capybara territories may be up to 2 km² (e.g. the Brazilian Pantanal; Schaller & Cranshaw, 1981). Territories are based around a permanent waterhole that persists even during the dry season, as the presence of water is intimately linked with the food resources that support the local population (Herrera & Macdonald, 1989). Capybara come into direct competition with other grazing species, including the White-Tailed Deer (*Odicoileus virginianus*) and livestock (Quintana, 2002).

Herding and social structure is maintained by a dominant male, reinforced using a range of vocalizations and scent markers (Herrera & Macdonald, 1993).

Herding behaviour and aggregation by herbivores such as this may help in defence against predators, but Herrera & Macdonald (1989) suggest that group grazing may also maintain a high quality and quantity of forage as grazing stimulates grass growth and prohibits the growth of plants that cannot survive disturbances, including tree seedlings: these grasslands owe their character and high productivity in large part to the fact that they are grazed *en masse* by large herbivores. A substantial 3 kg of plant food is required daily by an adult Capybara, and specialized adaptations are used to allow them to acquire resources efficiently, such as molars that continue to grow throughout the life of the animal (in most other rodents it is only the incisors that keep growing).

However, Capybara have no particular traits that allow the tolerance of stresses. Their sparse pelt is insufficient protection against strong sunlight and they may roll in mud to provide a kind of sunscreen when necessary. Capybara breed only when conditions are optimal, producing two to eight offspring and, unlike smaller rodents, they do not produce numerous litters each year. Postnatal care of the young is limited – young are able to eat grass one week after birth, and are fully independent after four months, living up to four to eight years in the wild.

The salient point here is that the lifestyle of the Capybara is geared towards resource capture and control, which must be maintained at all times. There is a lesser investment in maintenance and regenerative processes. No organism is likely to be absolutely C-, S- or R-selected, as all cellular organisms must capture resources, have systems to maintain metabolic processes and reproduce (Pierce *et al.*, 2005); thus we suggest that the Capybara is a relatively C-selected rodent[7].

In contrast to the Capybara and other large rodents of productive habitats, the Alpine Marmot (*Marmota marmota*) is native to alpine pastures in Europe at altitudes of 800–3200 m above mean sea level (asl) which have low annual productivities due to a high degree of seasonality. This species is also a large rodent, although not nearly as large as the Capybara (typically 4 kg, up to 8 kg), and is extremely long-lived, up to 18 years (Lenti Boero, 2001).

Marmot territory size, at around 0.019 km^2 (Lenti Boero, 1994), is relatively small; one or two orders of magnitude smaller than the territories of Capybaras. Groups of up to 20 individuals, usually family members, hibernate together during winter months in deep, grass-insulated burrows called *hibernacula*, some of which are estimated to be hundreds of years old (Lenti Boero, 2001; shallower burrows are used to escape heat and predators during summer months). Group hibernation ensures that the warmer adults can prevent the young from freezing, with the large size of adults allowing them to burn approximately 1.5 kg of fat

---

[7]Intriguingly, several larger but extinct rodents lived in similar circumstances to the Capybara: *Josephoartigasia monesi*, an inhabitant of tropical South American river deltas and estuaries 2–4 million years ago, probably weighed around a tonne and ate aquatic plants and fruit (Rinderknecht & Blanco, 2008). Another extinct species from South America, *Phoberomys pattersoni*, probably weighed around 700 kg (Sánchez-Villagra *et al.*, 2003). We suggest that rodents were able to achieve gigantism in South America because there were few native large herbivores of other mammalian orders in productive lowland habitats (i.e. few C-selected herbivores already monopolizing resources in these particular niches).

during hibernation (Landeryou *et al.*, 1999). Indeed, the entire social/reproductive system is adapted to facilitate the winter survival of small numbers of pups in a situation characterized by limited resources (Walter, 1990). Following mating in March or April, a litter of four to five pups arrives in July, during the middle of the alpine growth season (Lenti Boero, 2001). After a short period of lactation pups graze and fatten for winter hibernation. Offspring are nurtured for two to four years and are cared for by both dominant and subordinate adults within the family group (Walter, 1990). This imposes late sexual maturity, for a rodent, at well over two years of age.

The alpine climate imposes a relatively short period of activity each year, of between five and eight months. This annual environmental regime affects the breeding condition of Alpine Marmots because the amount of fat remaining after hibernation determines gonad development, with a lack of fat suppressing reproduction after particularly harsh winters (Landeryou *et al.*, 1999). The social structure of the family group also affects breeding conditions, as aggression between males suppresses androgen and corticosteroid hormone production, and thus testes development, in subordinate males (Arnold & Dittami, 1997). Subordinate family members must either evict the dominant pair or disperse to fresh territory in order to find sufficient resources for reproduction.

Alpine Marmot social structure and familial territorial control are thus adaptations to the unproductive nature of the environment – marmots invest in quality and durability rather than quantity in order to survive low resource availabilities in harsh and variable climatic conditions. This species exhibits a highly conservative lifestyle consistent with a high degree of S-selection.

The European Harvest Mouse (*Micromys minutus*) invests in quantity, and is of a comparatively ephemeral quality. It is a small rodent, up to 75 mm long (excluding the tail) and 6 g in weight, and is the most numerous mammal in productive but disturbed habitats in Europe, including arable fields, marshes, hedgerows and verges, with up to 93 individuals per hectare in some habitats (Haberl & Krystufek, 2003; Surmacki *et al.*, 2005).

Due to its timid nature the European Harvest Mouse is rarely seen, and even adults of the species encounter each other only during courtship. The pressure of the numerous predators that consume it can explain the shyness of this mammal; in Britain the European Harvest Mouse falls prey to almost every species of raptor and mammalian carnivore, crows and even pheasants. It avoids predation by freezing into a camouflage posture or by dropping into the vegetation (Ivaldi, 1999), and relies on good night vision and acute hearing, particularly of the low frequency sounds that betray predators.

Another major cause of mortality, particularly for the young, is cold, wet weather. The European Harvest Mouse is intolerant of climatic extremes and does not hibernate, but usually relocates to human barns in order to survive the winter. Indeed, this species has high energy requirements and must continue eating grain and insects in order to overwinter.

The adaptive strategy of the European Harvest Mouse involves maximizing reproductive output by rapidly producing a large number of young. Four to six offspring are produced per litter, with a gestation period of 17–18 days (Ivaldi, 1999), and three litters are produced over a six-month period. Offspring reach sexual maturity at just 35 days, and adult life is short, at up to 18 months. The European Harvest Mouse constructs a grass breeding nest approximately 20–30 cm above ground level among grass stalks, in which it hides its offspring from predators. A fresh nest is constructed for each litter – this represents a significant investment of energy in regeneration. The European Harvest Mouse is also adapted to avoid competition with other rodent species: a prehensile tail and feet adapted to climbing allow a three-dimensional life in the higher vegetation, which avoids direct competition with the Bank Vole, *Clethrionomys glareolus* (Ylonen, 1990). The adaptive strategy of the European Harvest Mouse is thus based on regeneration, ensuring survival of genes in the face of biotic and abiotic selection pressures that are not tolerated by individuals. This is compatible with the concept of R-selection.

Aside from these examples of extreme adaptive strategies among the rodents, summarized by Fig. 4.1, a range of mammal species in other orders exhibit

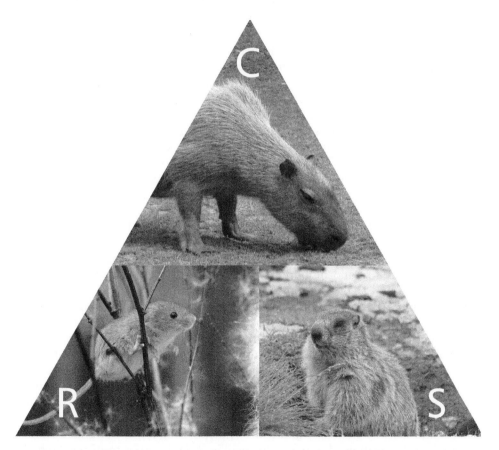

Fig. 4.1 Examples of C-, S- and R-selection at the extremes of the three-way trade-off within the largest order of mammals, the rodents: (C) the Capybara, *Hydrochoerus hydrochaeris* (© Richard Peterson/Shutterstock.com), (S) the Alpine Marmot, *Marmota marmota* (© Antonio S./Shutterstock.com), and (R) the European Harvest Mouse, *Micromys minutus* (© Mark Bridger/Shutterstock.com) (not to scale).

adaptive strategies that fit the criteria for strong C-, S- or R-selection. The even-toed ungulates (order Artiodactyla) include tiny, relatively R-selected deer, such as the 45-cm long, rat-like Lesser Mouse-deer (*Tragulus kanchil*) and small pigs such as the Pygmy Hog (*Sus salvianus*). Artiodactyls such as the Dromedary or Arabian Camel (*Camelus dromedarius*), the Oryx (*Oryx beisa*) and the Common Eland (*Taurotragus oryx*) are adapted to extremes of temperature and availabilities of food and water, and are highly S-selected. Artiodactyls also include the giant, gregarious, territorial Common Hippopotamus (*Hippopotamus amphibius*), which requires a massive 68 kg of grass per night – survival depends on the relatively consistent consumption of huge amounts of resources in productive habitats, suggesting C-selection.

Highly C-selected Cetaceans include the Blue Whale (*Balaenoptera musculus*) and other large baleen whales. Larger animals tend to have the slowest metabolisms – this negative correlation between body mass and metabolic rate is a cornerstone of animal physiology (Schmidt-Nielsen, 1997). The implication is that whales are relatively efficient consumers. However, the absolute amount of food that must be ingested to sustain an individual, which is often dismissed as 'obvious' (Willmer *et al.*, 2005), is another trait of profound relevance to animal adaptive strategies and ecology. For example, Blue Whale oxygen consumption rates, gram for gram, are three orders of magnitude less than those of the Common Shrew (*Sorex araneus*; Schmidt-Nielsen, 1997), and so the whale is indeed more efficient, but the absolute body mass that must be sustained is a disproportionate eight orders of magnitude greater (Jones *et al.*, 2009). Indeed, each individual whale consumes around 4,000 litres of oxygen per hour, with respect to the shrew's 0.037 litres per hour. Crucially, body mass is also positively correlated with food ingestion rate (Farlow, 1976) meaning that in absolute terms each individual whale consumes vast amounts of resources. This is, of course, offset to some extent by greater efficiency, but the sheer amount of resources consumed is far more evident, and thus potentially of much greater relevance to the adaptive strategy and ecology of the animal.

Indeed, the advantage of being big (at least in productive habitats) is precisely this ability to acquire and monopolize huge amounts of resources, as we have seen for plants and shall see below for the larger sharks and dinosaurs. What matters for Blue Whale survival is the ability to find and consume a massive 1.5 million calories per day (Piper, 2007) – an amount of energy equivalent to 3,600 hamburgers, 400 litres of petrol or, cumulatively over an 80-year lifetime, the energy released by the explosion of 40 tonnes of TNT. This is a colossal amount of energy to sustain an individual life, pass on genes and complete a life cycle. Whilst baleen whales are often viewed as gentle sievers of krill, the rorquals (Balaenopterinae, or baleen whales with pleated throats, including the Blue Whale and the Humpback Whale) are actually among the fastest of marine animals and actively hunt fish and krill – not just single fish but entire shoals that are literally swallowed in one gulp (see Fig. 4.2). In order to offset the high energetic costs of 'lunge feeding', during which the mouth is opened as wide as 80 degrees, causing a huge amount of drag that must be overcome, Blue Whales target high densities of prey and are obliged to feed at the highest productivities in order for this lifestyle to be sustainable (Goldbogen *et al.*, 2011).

Fig. 4.2   Resource acquisition on a grand scale: a Humpback Whale, *Megaptera novaeangliae*, demonstrates the lunge feeding used by all rorquals. Here, the throat pleats are visibly expanded whilst feeding on a bait ball (shoal of small fish) at Stellwagen Sanctuary, off Boston, USA (© Jose Gil/Shutterstock.com).

They invest huge amounts of energy in order to obtain even greater amounts of energy – this is the essence of C-adaptation.

The energy needed to sustain populations of Blue Whales is only available in the most productive, higher latitude, waters where they feed during most of the year. Blue Whales do migrate to less productive tropical waters to breed, at which point they feed less and must rely on fat reserves for three months (Willmer *et al.*, 2005). Thus rates of ingestion do vary throughout the year, and whales exhibit stress-tolerance traits such as fat accumulation. Nonetheless, the fact that the lifestyle, form and function of these giant species are based around the acquisition of vast amounts of resources indicates a relatively high degree of C-selection.

There are few strongly S-selected whales but female Orcas, or Killer Whales (*Orcinus orca*)[8], wait five years between births, have a 15-month gestation period, and the young are weaned late, at 415 days (compared to 212 days for the Blue Whale; Jones *et al.*, 2009). Relatively R-selected whales include the diminutive Finless Porpoise (*Neophocaena phocaenoides*), Haviside's Dolphin

---

[8]Genetic analysis has recently suggested that there are at least three distinct species of killer whale (Morin *et al.*, 2010), but apart from slight differences in appearance, diet and hunting techniques these have otherwise similar lifestyles and are likely to have equivalent overall adaptive strategies.

(*Cephalorhynchus heavisidii*), the La Plata Dolphin (*Pontoporia blainvillei*) and the Tucuxi (*Sotalia fluviatilis*). These have comparatively short lactation periods, small neonate body masses (several kilograms, with respect to 158 kg for Orcas and 2,738 kg for the Blue Whale; Jones *et al.*, 2009). However, with adult body masses of 33–43 kg these are still moderately large animals in absolute terms, and like many other whales only one neonate is produced every two years, meaning that reproductive investment is limited. They are unlikely to be extremely R-selected.

Carnivores (order Carnivora) include a range of adaptive strategies from S-selected ambush hunters of unproductive and abiotically challenging habitats, such as the Polar Bear (*Ursus maritimus*) to diminutive R-selected species with opportunistic diets, such as the Cape Genet (*Genetta tigrina*), the Aquatic Genet (*Genetta piscivora*), Rüppell's Fox (*Vulpes rueppellii*) or the Black-footed Cat (*Felis nigripes*). Species such as the Grey Wolf (*Canis lupus*), which actively acquires resources but has some tolerance of extreme climates and may have litters of five pups twice a year, are probably intermediate CSR-selected strategists.

Bats (Chiroptera) include small, relatively R-selected microbats that frequently produce litters. For example, the Least Pipistrelle (*Pipistrellus tenuis*) weighs 3.5 g and produces three litters of two pups each year (Jones *et al.*, 2009). Megabats, such as the Indian Flying-fox (*Pteropus giganteus*), have the greatest adult body masses, neonate body masses and metabolic requirements, produce a single pup once a year, and are thus good candidates for relatively C-selected species.

As mammals, what about our own adaptive strategy? All primates, even tiny species such as Lesser Bushbabies (*Galago* species), exhibit a combination of S-selected traits (long gestation periods, parental care, longevity) and C-selected traits (active resource acquisition, territoriality, sociality and intelligence), and most are probably intermediate SC strategists, despite extensive size variation within the group. Hill and Kaplan (1999) state that the following characteristics distinguish *Homo sapiens* from other primates: '(a) an exceptionally long life span, (b) an extended period of juvenile dependence, (c) support of reproduction by older postreproductive individuals [grandparents], and (d) male support of reproduction through the provisioning of females and their offspring'. Thus the way that humans invest in offspring is analogous to the post-hatching investment made by S-selected fish. Humans also enjoy a tendency to accumulate fat when resources are plentiful (a luxury consumption trait), and have a degree of temperature control afforded by the ability to sweat. The intelligence needed to live in large social groups and to find and exploit diverse sources of food has also allowed humans to create more effective technological barriers against natural environmental conditions. Thus humans do not need to be the most stress-tolerant of organisms in order to survive in a range of situations, including some of the most marginal of terrestrial habitats.

However, sociality and intelligence in humans has a range of advantages: not only do they facilitate investment in offspring and protection against predation (S-selected traits) but also the exploitation of a diverse range of resources and the control of territories (C-selected). Humans may vigorously defend territories and the resources they contain, sometimes leading to selection against competing social groups, or war. Many wars are either directly concerned with territorial control or resources are at least heavily implicated. For example, following Operation Barbarossa (Germany's invasion of Russia on 22 June 1941) the Nazis continued a policy of racial extermination in occupied territory, but the aim of the operation was not primarily ideological but actually intended to fulfil Adolf Hilter's stated desire for living space and to secure Russia's oil wealth to literally fuel the Nazi war machine. It was in fact simply a land-grab, albeit history's bloodiest and most shameful. In contrast, organized war over resources is not common among peoples, such as the Inuit or Australian aborigines, for whom traditions and culture reflect life in naturally harsh situations, probably because war is not sustainable in circumstances in which people are already pushed to the limits of survival. The resources gained in large-scale conflict would not balance resources spent, and the few resources available would be better employed in fighting the challenges of the environment rather than fighting other people.

Humans are competitive, territorial and social, but are also long-lived organisms with a significant investment in the maintenance of individuals. For these reasons we suggest that humans are intermediate between C- and S-selected (i.e. humans are stress-tolerant competitors).

Aves (avian therapods)

The life-history traits of birds from a range of habitats in Spain have been interpreted in terms of CSR theory by Hodgson (1991). This study concluded that large species occupying highly productive, moderately disturbed agricultural habitats or woodland typically exhibiting a single, large brood each year, moderate incubation periods and moderate fledging times, were most likely to be C-selected (e.g. the White Stork, *Ciconia ciconia*, the Common Raven, *Corvus corax*). Small species in highly disturbed, moderately productive pasture habitats exhibited short incubation periods, rapid fledging and large numbers of brood in multiple clutches, which Hodgson suggested were R-selected (e.g. the House Sparrow, *Passer domesticus*, the Barn Swallow, *Hirundo rustica*, and the European Robin, *Erithacus rubecula*, in agricultural habitats). Highly S-selected species inhabited unproductive, undisturbed rocky habitats, could be large or small, but exhibited the longest developmental times and produced a single, small brood annually (e.g. the Egyptian Vulture, *Neophron percnopterus*) (see Fig. 4.3).

Birds are extremely diverse, and a range of life-history characteristics are evident for this group that provide further examples from a range of habitats. The Common Kestrel (*Falco tinnunculus*) is an active predator that hunts using acute vision (Fox *et al.*, 1976) and like other falcons uses nictitating membranes, or third eyelids, to maintain vision during high-velocity hunting dives. The range of wavelengths visible to the Kestrel includes ultraviolet (UV) light. This is an important adaptation that aids predation, as vole urine (deposited along vole tracks as a scent marker) reflects these wavelengths. Kestrels have an innate ability to sense this reflected UV light but also learn how to develop this ability with experience (Zampiga *et al.*, 2006), suggesting a degree of mental sophistication. Kestrels thus have a range of distinctive traits that facilitate resource acquisition (C-selection).

Fig. 4.3 Examples of C-, S- and R-selection for birds, according to Hodgson (1991): (C) the White Stork, *Ciconia ciconia* (© Borislav Borisov/Shutterstock.com), (S) the Egyptian Vulture, *Neophron percnopterus* (© Vladimir Melnik/Shutterstock.com), and (R) the House Sparrow, *Passer domesticus* (© Mihai Dancaescu/Shutterstock.com) (not to scale).

Predation is a form of disturbance, or R-selection, that has resulted in the prey of the Common Kestrel being small, highly fecund, and with an adaptive strategy based on regeneration (e.g. the House Sparrow). House Sparrows are opportunists that are favoured by agricultural and urban environments, eating seeds, flowers and insect larvae. They have the shortest incubation period of all birds, at just 10–12 days, and can lay 25 eggs per season. Thus whilst Kestrels are adapted to acquire and control food resources, House Sparrows are adapted to regenerate in the face of disturbance imposed by predators.

Many similar examples are evident of predator/prey relationships between highly C-selected raptors and smaller R-selected birds and mammals. For instance, the Peregrine Falcon (*Falco peregrinus*) preys mainly on pigeons and doves, occasionally taking small mammals. Peregrines are large, agile and highly active predators, which attain the fastest speeds of any animal – a speed of $389\,\mathrm{km\,h^{-1}}$

has been recorded during a powered hunting dive (Harpole, 2005). Peregrines are not usually social, but form life-long breeding pairs that defend a well-defined territory and the food resources it contains, to which the breeding pair returns each year. These territories typically extend 19–24 km around a central nesting site and are separated by at least a kilometre from the territories of rivals, which include eagles, ravens and others of their own species.

Peregrines have a lifestyle based on active resource acquisition, defending resource supplies and aided by acute senses and a degree of intelligence and simple sociality. They do not invest heavily in regeneration and are relatively intolerant of adverse abiotic conditions. Some subspecies are found at high latitudes, but migrate to avoid the winter – *Falco peregrinus tundrius* is a summer inhabitant of tundra habitats in North America and Greenland but vacates to the tropics during the winter. Bird migrations such as this are usually triggered by a seasonal lack of food resources, rather than changes in abiotic conditions *per se* (Willmer *et al.*, 2005), and are thus more likely to occur for species that rely on consistently abundant resources. Indeed, altered territory quality and productivity has an immediate effect on the reproductive success of raptors (Newton & Marquiss, 1976; Korpimäki, 1988).

Peregrines breed at two to three years of age, and may go on to breed for a further 14 years, nesting in a simple depression scratched out on a cliff edge, known as a 'scrape'. Three or four eggs are laid, which are incubated for around 30 days (Terres, 1991), and the young are dependent on their parents for food for up to two months after hatching. Tinkle *et al.* (1970) point out that cliff nesting is particularly useful for species with longer developmental times and the greater risk of predation that this could entail. Thus Peregrines exhibit a degree of parental care, both in the choice of nest site, by feeding chicks, and because the female stays in the nest to guard the chicks whilst the male hunts. We previously suggested that parental care is an S-selected trait, but parental care is a general characteristic of birds and, as we shall see below, there exist bird species that go to much greater lengths to reduce chick mortality in harsher environments, or invest more time and energy in gathering nesting materials. Indeed, Peregrine chick mortality is actually extremely high (up to 70 per cent; Snow & Perrins, 1999), mainly because competition for food starts even in the nest as part of an intense and often lethal sibling rivalry.

Like Peregrines, herbivorous geese and ducks migrate seasonally, suggesting that these large, active birds must also be supported by relatively high seasonal productivities. The Brent Goose (*Branta bernicla*) is an Arctic species that over-winters in Northern Ireland, and must feed almost continuously, usually on the seagrass *Zostera*, to meet daily energy requirements that vary seasonally from 715 to 1656 kJ g^{-1}, attained from between 121 and 247 g of food each day (Tinkler *et al.*, 2009). This species is forced to continue eating throughout the night during the winter – activity that may provide up to 50 per cent of its daily requirements. Thus whilst the Brent Goose is tolerant of low temperatures it is first and foremost a slave to food availability and must migrate and eat continuously in order to survive, suggesting a high degree of C-selection in addition to some S-selection. Large, flight-capable herbivorous birds are perhaps not truly equivalent to large C-selected mammalian grazers, due to the weight limitations that a large 'fermentation-tank' stomach would impose, but such birds do nonetheless exist: for example the Hoatzin (*Opisthocomus hoatzin*), which grazes the tree-tops in tropical forests in South America, is infamous for the stench of its eructations.

Moving further along the gradient from C- to S-selection, penguins and the Atlantic Puffin (*Fratercula arctica*) are adapted to survive extreme seasonal variability in climatic conditions that can only be tolerated with physiological mechanisms that buffer metabolism from the environment. For the Emperor Penguin (*Aptenodytes forsteri*), this includes fat reserves (both as insulation and to maintain the availability of resources for metabolic processes), countercurrent circulatory systems that scavenge heat from blood flowing towards the extremities, warming up the colder blood that returns from the feet, and breeding behaviour that represents an investment in safeguarding small numbers of offspring during the harshest season. Indeed, male Emperor Penguins incubate a single egg for 60 days – a period twice that of the Peregrine, and six times that of the House Sparrow – holding it off the ice on top of their feet, blanketed in folds of insulating fat and feathers. After hatching, the chick also enjoys this shelter and, being an 'only child', does not experience the added pressure of competition with siblings for food. Penguins do not have to be particularly well adapted to compete for food because they are adapted to resist habitats in which competition for resources is not the main prerequisite for survival, and few species can challenge them – the main challenges are abiotic. Leopard seals (*Hydrurga leptonyx*) and Orcas predate penguins, but resistance of a harsh climate is the main force shaping the evolution of these birds, and so they are neither particularly fecund nor well adapted to monopolize food resources. They are extremely S-selected.

Penguins and Peregrines are both predators, but disparities in life-history can be attributed to differences in the extent of S- and C-selection, respectively – something that cannot be explained in terms of *r/K*-selection. Based on the above evidence, we can be confident that variation in bird life-histories is consistent with the operation of C, S, and R-selection.

Squamata (snakes and lizards) (with notes on other extant reptile clades)

The squamates are the most diverse of the extant reptile clades, but nonetheless do not exhibit the range of trait variability evident in groups such as mammals and birds, and thus 'appear rather invariant in lifestyle' (Promislow *et al.*, 1992). Indeed, examples abound of smallish, durable and spiny lizards from warm climates, but large, highly active reptiles that are prominent components of productive ecosystems have not existed since the Mesozoic Era, and these were, of course, not squamates.

Promislow *et al.* (1992) suggested that following the extinction of the giant dinosaurs the evolution of large squamates that require large amounts of resources and can potentially dominate productive habitats has been constrained by the presence of mammalian and avian taxa that have greater competitive abilities. This is demonstrated by the Komodo Dragon (*Varanus komodoensis*) which, at up to 3 m in length and 70 kg, is the world's largest extant lizard, and is thought to owe its great size to a lack of mammalian and avian competitors on the arid Indonesian islands that it inhabits (homeothermic predators of similar stature would have excessive energy requirements for survival in what is essentially an extremely low productivity environment; Burness *et al.*, 2001). Thus whilst we can be sure that mammals and birds include C-selected forms, squamates apparently have had less opportunities to evolve highly acquisitive survival strategies.

Do C-selected reptiles exist today?

Squamates do, of course, exhibit trait variation, and the trait variation in evidence suggests a predominance of S-selection in this group. Larger squamates tend to become older, exhibit delayed reproduction, and have fewer clutches per year with larger offspring – traits that are associated with lower predation pressure (Promislow *et al.*, 1992). These traits are particularly evident for the more archaic squamates, the suborder Iguania, which includes spiny desert lizards such as the Texas Horned Lizard (*Phrynosoma cornutum*) and the Thorny Devil (*Moloch horridus*) from Australia.

The latter is a modestly sized species (20 cm long) but may live for up to 20 years and has a lifestyle based on survival in an arid environment. The fluted body of the Thorny Devil includes hygroscopic grooves between its thorns, which channel dew and rainwater towards the mouth (Bentley & Blumer, 1962). Thorns also contribute to the innate defence of the animal – even down to having a false, decoy head on the back of its neck – which is an important factor for S-selected organisms that grow slowly and must protect their accumulated investment of resources and energy. Reproduction, in a nesting burrow, involves the defence of six or seven eggs, which take around four months to hatch (Pianka, 1997).

The more derived squamate group Autarchoglossa includes a range of adaptations that have been interpreted as increasing competitive ability, such as more flexible skulls (increasing the size range of potential prey), vomerolfaction (tasting the air with a forked tongue allows otherwise well-camouflaged prey to be detected, and can increase the range of carrion detection by kilometres) and active foraging (Vitt *et al.*, 2003).

Nonetheless, most autarchoglossans have a suite of S-selected traits; the group includes the Komodo Dragon, with its famously slow metabolism (which can be fuelled by only 12 meals a year), a longevity of up to 50 years, ambush predation and scavenging, and clutches of 20 eggs that require eight months of incubation. Snakes exhibit autarchoglossan adaptations, allowing them to hunt and capture active vertebrates, but snakes also include a range of species with strongly S-selected traits. The Common Adder (*Vipera berus*), for instance, occurs throughout a range of latitudes including those above the Arctic Circle (Willmer *et al.*, 2005). In Finland adders aggregate in dens to survive the winter and give birth to live young to provide constant temperatures for developing foetuses. Other vipers, such as the Horned Viper (*Bitis cornuta*) of South Africa and Rattlesnakes (*Crotalus* species) from the Americas, including the Sidewinder (*C. cerastes*), are desert ambush predators. The largest of snakes, such as the Green Anaconda (*Eunectes murinus*), accrue their gigantic sizes slowly over a period of decades, have long gestation periods (six months) and are ambush predators and carrion eaters rather than exhibiting the high-octane lifestyle of C-selected organisms.

However, the snakes include perhaps the best candidates for C-selected squamates: sea snakes do conduct highly active lifestyles. The Turtle-Headed Sea Snake (*Emydocephalus annulatus*) feeds in shallow-water reefs from Australia to New Caledonia. Feeding does not involve the large, infrequent meals typical of many snakes – this sea snake continuously forages for caches of small fish eggs hidden among the substrate, in a manner that has been compared to the foraging of terrestrial herbivores (Shine *et al.*, 2004). Thus a substantial proportion of this animal's efforts are directed towards actively seeking resources.

Sea snakes require a plentiful supply of resources and probably have a significant impact as predators. For example, Signal Island, New Caledonia, is a 15 ha islet that, in a recent survey (Brischoux & Bonnet, 2008), was found to host 4,087 individuals of two species of sea snake, the Blue-Lipped Sea Krait (*Laticauda laticaudata*) and Saint Girons' Sea Krait (*L. saintgironsi*). Together these two species eat around 45,000 fish (more than 1.3 tonnes) a year, foraging up to 17 km around the islet and probably coming into competition with snakes from other islets. Competition between these two species is largely avoided via *resource partitioning* – the Blue-Lipped Sea Krait forages on soft areas of seabed and the Saint Girons' Sea Krait exploits the coral matrix. In Chapter 5 we shall revisit this argument of how similar species coexist.

A similar situation has also been recorded for the Bar-Bellied Sea Snake (*Hydrophis elegans*) and the Olive-Headed Sea Snake (*Disteria major*) in Shark Bay, Western Australia (Kerford, 2000). Other marine species, such as the Yellow-Lipped Sea Krait (*Laticauda colubrina*) from Fiji, also exhibit remarkable fidelity to particular islands where they retreat for reproduction and moulting, which has been interpreted as a form of territoriality (Shetty & Shine, 2002).

Tolerances to abiotic extremes are not a prominent characteristic of sea snake ecology, and rapid development and consistent fecundity are not evident for these species. For instance, the Bar-Bellied Sea Snake is a moderately sized animal (females between 140 and 180 cm, males 20 cm smaller) that reaches sexual maturity at three to four years of age, producing between five and 19 offspring. In contrast, the Keelback (*Tropidonophis mairii*) is a semi-aquatic snake from freshwater habitats including streams, swamps and drainage ditches in northern Australia. It is unusual for a snake in that it may produce two clutches each year, and although clutch sizes are similar to the Bar-Bellied Sea Snake (5–12 eggs) the eggs are small and develop rapidly, with offspring reaching maturity a few months after hatching (Brown & Shine, 2002).

Snake reproduction is still relatively underinvestigated, but Brown & Shine (2002) suggest that natricines (grass snakes and freshwater snakes) are generally characterized by investment in multiple clutches and rapid maturity. Indeed, most tropical Natricinae mature by two years of age (Brown & Shine, 2002 and references therein); thus candidates for relatively R-selected species include the Buff-Striped Keelback (*Amphiesma stolata*) and the Red-Necked Keelback (*Rhabdophis subminiatus*) from Cambodia, and the Indonesian Garter Snake (*Xenochrophis vittata*) and the Chequered Keelback (*X. piscator*) from Java. The Chequered Keelback takes just one to three years to mature and produces an average of 38 eggs per clutch. Temperate natricines, such as the Common Garter Snake (*Thamnophis sirtalis*), have lifestyles that include a degree of tolerance of low temperatures, using denning behaviour and viviparity to maintain the temperature and development of foetuses, and thus reproductive traits differ from tropical species. However, even the Red-Bellied Snake (*Storeria occipitomaculata*) of North America is small (usually 23–30 cm in length; Blanchard, 1937), relatively fecund and ephemeral, producing young within two years and living for between two and four years, with a 'heavy investment' in more rather than larger young and extremely high mortality rates when young (Semlitsch & Moran, 1984). This, again, suggests a degree of R-selection.

However, although developmental times of the Red-Bellied Snake are undoubtedly rapid, caution should be exercised when interpreting this 'heavy investment' in large numbers of offspring, as the mean number of offspring is actually only seven (Blanchard, 1937). Other small snakes also produce small numbers of young. For example, the Eastern Worm Snake (*Carphophis amoenus*) produces just three to five eggs (Willson & Dorcas, 2004), and the smallest reptile in the world, the 10-cm long Barbados Threadsnake (*Leptotyphlops carlae*) is so tiny that there is room inside the body for only a single, elongated egg (Blair Hedges, 2008). This suggests that the form of snakes may represent a phylogenetic constraint to fecundity (a 'packaging constraint'; Brown & Shine, 2002), which is particularly relevant for smaller species.

Thus whilst developmental times and multiple clutches suggest that some natricine snakes could be relatively R-selected with respect to other squamates, it is not clear whether these snakes truly represent an R-selected extreme, or if phylogenetic constraints relating to the serpent bauplan restrict the range of adaptive strategies possible for this group. We suspect that these smaller species are intermediate SR strategists, capable of relatively rapid development and multiple clutches, but not truly equivalent to highly R-selected organisms in other groups. In conclusion, it appears that the squamates exhibit variation from S- to C-selected forms, also varying towards R-selection, but it is unclear at present whether truly R-selected squamates exist.

Other extant reptile clades, such as the Testudines (Turtles), and Crocodilians are separate lineages from the squamates, but it is perhaps worth mentioning that these too appear to exhibit phylogenetic constraint in their adaptive strategies. Some marine turtles are capable of active lifestyles, but this probably represents variability around what are largely S-selected strategies. The Eastern Box Turtle (*Terrapene carolina carolina*), the Western Box Turtle (*Terrapene ornata*) and the Painted Turtle (*Chrysemys picta*) hibernate underground for at least part of the lifecycle, and at up to 0.5 kg are the largest known animals capable of surviving freezing of the entire body during the winter (Storey & Storey, 1992). The Aldabran Giant Tortoise (*Geochelone gigantea*) can weigh up to 120 kg, has a long lifespan of up to 90 years and is an extremely late developer, with female reproductive maturity achieved at around 20 years (Swingland, 1977). Turtles are, of course, also relatively well protected. Crocodilians grow in the same manner as the Green Anaconda, slowly attaining large sizes using slow metabolisms and ambush predation. They show an additional S-selected trait not shared by anacondas – parental care.

Amphibia (amphibians)

'**The frog**' is often cited as the epitome of an *r*-selected animal because many frog species are small and fecund. Indeed, fecundity is a general feature of amphibians, most of which produce vast clutches compared to other terrestrial animals, and some frog species are small and develop rapidly – the Panamanian Golden Frog (*Atelopus zeteki*) weighs 3–15 g and lays around 370 eggs that hatch after just nine days of embryonic development (Karraker *et al.*, 2006). The smallest amphibian species do generally have the fastest developmental rates, particularly for species from temporary habitats (Morrison & Hero, 2003).

However, larger species develop slowly, and fecundity (the absolute number of eggs produced) is not a good measure of reproductive investment – large species produce more eggs, but the total mass of eggs represents a much smaller proportion of adult body mass (Morrison & Hero, 2003). This observation is also relevant to other aquatic organisms, such as fishes, that tend to be fecund but may nonetheless vary in reproductive investment relative to body mass – 'it is the reproductive investment relative to the current body size that matters' (Heino & Kaitala, 1999). We will return to this point below, when we discuss fish life-histories. First, it is evident that the larger, slowly developing amphibians are relatively S-selected.

The Goliath Frog (*Conraua goliath*; 32 cm, 3.3 kg) from western Africa inhabits fast-flowing, well-oxygenated but nutrient-poor streams, ambushing moderately large prey such as crabs. Eggs of the Goliath Frog take two-and-a-half to three months to hatch – clearly not on a par with the developmental times of small frogs such as the Panamanian Golden Frog. Large salamanders have a similar lifestyle to the Goliath Frog, living in fast-flowing, clear waters, and are ambush predators. The largest modern amphibian, the Chinese Giant Salamander (*Andrias davidianus*), weighs between 25 and 30 kg and may be up to 1.8 m long. The closely related Japanese Giant Salamander (*A. japonicus*) has an extremely slow metabolic rate and is known to be capable of living for 52 years. These species lay up to 500 eggs, but these take up to two months to hatch, with the male standing guard, and represent a relatively small proportion of the overall body mass of the adult female. Thus the developmental rates and adaptive strategies of larger amphibians are slow, conservative, and suggestive of S-selection.

Some amphibians are also capable of tolerating the most extreme of abiotic conditions. The Wood Frog (*Rana sylvatica*) is a model species for studies of stress-tolerance, capable of surviving freezing of the entire body. Willmer *et al.* (2005) state that up to 65 per cent of its body water may be frozen (i.e. 7–8 g of ice), found in its coelom and beneath the skin, dehydrating the animal. Nonetheless the nerves and reflexes function normally only 5–14 hours after thawing[9]. Freezing and desiccation tolerance is achieved by packing cellular components

---

[9]This image of a frog freezing and thawing is so delightful that it has even been used in children's books as an example of the weird and wonderful things animals get up to. Here we quote the excellently succinct description from the Usborne Beginner's *Tadpoles and Frogs* (Milbourne, 2008): 'Some frogs freeze like ice cubes in the cold winter. In spring, they thaw out and wake up'. Strange, but true.

in glucose to maintain their conformation, in essentially the same manner as the sugars that protect the cells of S-selected desiccation-tolerant plants (see Hoekstra *et al.*, 2001). Glucose is produced from glycogen stores that are built up to huge concentrations in the liver prior to hibernation. However, this ability to store, mobilize and use glucose as a cryoprotectant would not be possible without a range of other stress-tolerance systems such as cold-adapted enzymes and efficient, rapid cell-membrane glucose transport systems (Storey & Storey, 1984; Willmer *et al.*, 2005).

The Siberian Salamander (*Salamandrella keyserlingii*) also uses these mechanisms to tolerate freezing. Even species such as the Common Frog (*Rana temporaria*) hibernate, and this species may take up to six years to reach maturity and can live for up to 15 years in the wild (Miaud *et al.*, 1999). These are species that invest in the survival of the individual during abiotic extremes, rather than investing in regeneration, and there can be little doubt that these frogs and salamanders qualify as S-selected animals.

Morrison & Hero (2003) reviewed how amphibian traits change with altitude and latitude. They found that amphibians in harsher abiotic environments have shorter activity periods and hence shorter breeding seasons and longer larval periods, are larger both as larvae and adults, reach maturity later in life and have fewer clutches that are larger in absolute terms but smaller relative to adult body size. This is consistent with a gradient between small R-selected and larger S-selected amphibians towards habitats with increasingly prevalent abiotic challenges.

However, Morrison & Hero (2003) also noted that there were exceptions to these general trends, particularly in more stable habitats with greater food availability in which species could be relatively large, sometimes larger than high altitude/latitude species. This indicates a second axis of variation in amphibian traits, with larger C-selected species in productive habitats at one extreme. For example, since the introduction of the Cane Toad (*Bufo marinus*) from Hawaii to the Australian tropics in 1935 it has imposed a range of effects on native wildlife, many of which occur via competition. Although resource partitioning may limit direct competition between adult Cane Toads and native frogs (reviewed by Shine, 2010), tadpoles compete with mosquito larvae and, by forming large groups that compete for food, can reduce the survival of native frog tadpoles. Adult Cane Toads may even compete with other vertebrates for resources such as nesting burrows. It has been hypothesized (Shine, 2010) that Cane Toads may compete with a range of vertebrates for food resources (particularly in the dry season, because unlike native species the Cane Toad remains active year-round), but currently no studies have been undertaken to investigate this.

Osteichthyes (bony fishes)

We have already discussed some empirical evidence for the existence of three primary adaptive strategies in North American fishes and the parallels that have been drawn between these and plant CSR strategies (Winemiller & Rose, 1992). Winemiller (1992, 1995) then took a wider view, using a number of published studies of life-history traits to classify bony fishes worldwide according to three

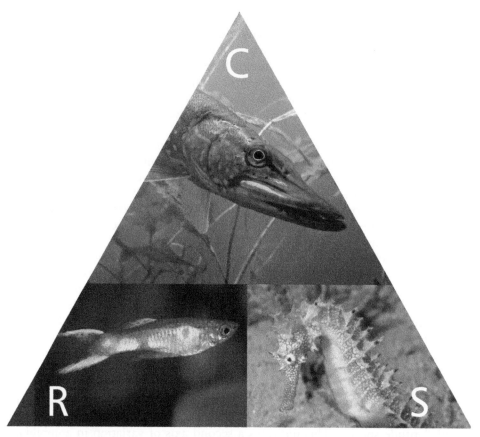

Fig. 4.4 Examples of C-, S- and R-selection for bony fishes, according to Winemiller (1995): (C) the Northern Pike, *Esox lucius* (© Krzysztof Odziomek/Shutterstock.com), (S) Jayakar's Seahorse *Hippocampus jayakari* (© Jon Milnes/Shutterstock.com), and (R) the Wild Guppy, *Poecilia reticulata* (© Dobermaraner/Shutterstock.com) (not to scale).

life-history strategies (an example of each is given in Fig. 4.4). He noted, again, that these were compatible with CSR theory, and also suggested that phylogenetic trends were apparent: 'although teleosts span all three life-history strategies, species within a given family tend to be centred around only one or two strategic end-points' (Winemiller, 1992). Families including predominantly C-selected species included, among others, herrings, pikes, jacks, snappers, sea basses, mackerels, old and new world catfishes and the larger minnows[10], which are characterized by traits such as late maturity, intermediate to large body sizes, large clutches of small eggs, annual spawning episodes, low survivorship and rapid first-year growth (Winemiller, 1995).

---

[10]See the original publication for a more extensive list of fish groups with each life-history strategy and the Latin family names.

S-selected taxa are characterized by late maturity (and may attain a range of sizes at maturity), small clutches of large eggs, 'brood guarding or maternal provisioning of nutrients to developing embryos', and high survivorship. These include cichlids, sticklebacks, salmon, trout, some tropical gobies, sea catfishes, armoured catfishes and seahorses. As we have already seen, parental care can be an important part of S-adaptation, and is epitomized by cichlids that may brood for several months. Winemiller noted that salmon invest in a different form of parental care involving the placement of eggs in 'special habitats' – referring to the effort invested in the spawning migration, which ensures that newly hatched young are not subject to the same degree of predation as found in the sea.

R-selected species (exhibiting early maturity, small size, small- to medium-sized clutches of small eggs, multiple spawning bouts each year, rapid larval growth and low survivorship) included killifishes (we will see more of these extraordinary fishes later), livebearers (guppies, platies, mollies, mosquito fishes and swordtails), small minnows, small tetras and rainbow fishes. Winemiller (1995) noted that:

> . . . by having among the smallest rather than largest clutches, opportunistic-type fishes differ markedly from the traditional model of r-strategists. Yet because of their small size, the relative reproductive effort of opportunistic strategists is actually high, despite the fact that absolute clutch size and egg size are small. In these small species, serial spawning sometimes results in an annual reproductive biomass that exceeds the female's body mass. Small fishes with early maturation and frequent spawning are well equipped to repopulate habitats following disturbances and to sustain their numbers when faced with continuously high mortality during the adult stage.

Let's examine in greater detail three examples of these contrasting adaptive strategies.

The Nile Perch (*Lates niloticus*) is a voracious predator that will eat practically anything that moves, including other Nile Perch (Katunzi *et al.*, 2006). This active fish has become a devastating invasive species, introduced for sport fishing into productive east African water bodies such as Lake Victoria and Lake Nabugabo (Johnson & Gill, 1994; Schofield & Chapman, 1999). Here it has drastically reduced the diversity and abundance of other fishes (Schofield & Chapman, 1999) and as populations of prey fish species have crashed the Nile Perch has started foraging for different prey, mainly invertebrates (Mkumbo & Ligtvoet, 1992).

Changes in fish and invertebrate populations have led to negative impacts on other animals which compete with the Nile Perch, such as the Pied Kingfisher (*Ceryle rudis*) (Wanink & Goudswaard, 1994) and, tragically, human communities that traditionally relied on tilapia (cichlid) meat, and who were not those responsible for introducing the Nile Perch. The changing economy of these regions has brought about socio-economic shifts that have been hugely detrimental to traditional economic practices, and even though the Nile Perch may itself form a resource for humans, the sustainability of perch populations is unknown (Reynolds & Greboval, 1988). The initial negative impact of the Nile Perch has earned it the epithet of 'Vile Perch' (Ogutuohwayo, 1993).

This is competition on a grand scale. In Australia, importation of *Lates niloticus* is prohibited because it is capable of out-competing its slightly smaller native cousin the Barramundi (*Lates calcarifer*; Johnson & Gill, 1994). The Nile Perch is a large fish that uses active foraging to fuel moderately rapid growth rates, with males reaching maturity at two years, or half a metre in length, with females maturing at four-years or 80 cm (Hughes, 1992). It can reach a maximum size of 2 m in length, or 200 kg, and is known to be relatively intolerant of hypoxia (reduced oxygen concentrations) and can only survive in relatively clean waters where it can live stress-free. These traits are suggestive of a highly C-selected strategy.

In contrast, the Tambaqui or Black Pacu (*Colossoma macropomum*) is a large, solitary, vegetarian piranha (maximum length 108 cm, 50 kg; IGFA, 2001; Machacek, 2008) native to river ecosystems in the Amazon Basin, where it is exposed to a range of abiotic stresses.

Tambaqui move seasonally between muddy white water rivers and acidic black waters, in which they can survive due to rapid acclimation to changes in pH, only showing signs of stress at an extremely acid pH of 3 (Wood *et al.*, 1998; Aride *et al.*, 2007). Rotting vegetation causes hypoxia and results in high dissolved CO_2 concentrations (hypercarbia), also filling the water with black humus. These waters may also contain relatively high concentrations of dissolved metal ions, such as copper and cadmium.

Elevated concentrations of humic substances and even petrochemical pollutants induce a physiological stress-response in the tambaqui, mediated by cytochrome P450 1A (Matsuo *et al.*, 2006), regulating antioxidant defences that are useful against a range of environmental stress factors (Marcon & Wilhelm, 1999). Metals may accumulate in fish gills, but the Tambaqui in particular appears to have specific ion transport systems and efficient **metal binding physiology** that protect against this (Matsuo *et al.*, 2005).

The Tambaqui is able to resist hypoxia because its gills have chemoreceptors that sense the oxygen concentration in the water and trigger **aquatic surface respiration** behaviour (the fish literally move to the surface and gulp down air; Florindo *et al.*, 2006), and red blood cell counts and haemoglobin contents acclimate rapidly, in a matter of hours (Affonso *et al.*, 2002). Similar chemoreceptors sense hypercarbia, triggering hyperventilation, during which greater volumes of water are flushed over the gills (Florindo *et al.*, 2004; Gilmour *et al.*, 2005).

The river systems inhabited by the Tambaqui are also characterized by seasonal flooding into surrounding *varzea* rainforest, where there is a greater availability of forest resources during the wet season. Seeds, fruits and zooplankton form the diet of the Tambaqui during the dry season, but during the flood pulse the diet becomes more balanced, including the leaves and roots of submerged plants (Oliveira *et al.*, 2006). Changes in the type and availability of food are reflected by the capacity of the Tambaqui to precisely regulate the production of digestive enzymes in response to food composition (De Ameida *et al.*, 2006). The Tambaqui can even regulate the transit time of food in the gut depending on its digestibility (Da Silva *et al.*, 2003). This seasonal lifestyle also depends on the storage of fat in muscle tissues and around the eye, as the fish fatten up for the lean dry season (De Almeida & Franco, 2006).

The adaptive strategy of this species optimizes physiological performance in response to both variable resource availability and injurious stresses. It is clearly S-selected.

The Annual Killifish (*Austrofundulus limnaeus*), however, differs radically from the two fishes we have discussed in detail so far. It is small, 5 cm in length, and native to shallow ephemeral ponds in Asia. Adult killifish exhibit some stress-tolerance, in this case in response to water temperatures that may fluctuate between 20°C and 37°C during the day. The higher temperatures are resisted by the protective action of heat shock proteins (Hsp70 and Hsp90), with smaller heat shock proteins expressed quickly in response to temperature fluctuations (Podrabsky & Somero, 2004). These heat shock proteins are, however, expressed by all cellular life (Klütz, 2005) and are not an adaptation unique to the Annual Killifish's way of surviving. Rather, this depends on a rapid lifecycle which takes place entirely within a single season – the adults die when the ephemeral ponds dry out. Genes survive not within individuals, but within the eggs, laying the foundation for the following season's population explosion. Within the egg, the embryo exhibits inherently depressed rates of protein synthesis and oxygen consumption, extending the viable dormancy period (Podrabsky & Hand, 1999, 2000).

Seasonal mortality due to habitat drying – a lethal disturbance event – defines the lifestyle of the Annual Killifish, the genes of which persist where individuals cannot. It is the eggs, or *propagules*, rather than adult individuals, which are truly resistant to drying out, just like the seeds of R-selected plants; probably because genetic material and metabolic machinery can be more easily protected by chaperone proteins and sugars when they are quiescent, or at least not in constant use. Examples of this kind of R-selected life-history are also evident in the Neotropics, including various species of *Cynolebias* (Arezo *et al.*, 2005) and *Austrolebias* (García *et al.*, 2008).

The case for C-, S- and R-selection in the bony fishes is particularly strong. What about their cousins, the cartilaginous fishes?

Chondrichthyes (cartilaginous fishes)

Sharks have slower metabolic rates and lower energy and oxygen consumptions than bony fishes, and can survive for weeks without eating, with 'a strong emphasis on efficient energy use' (Helfman *et al.*, 2009). However, aside from this general phylogenetic tendency towards what could be considered S-selected traits, sharks do nonetheless exhibit variation in growth rates, size, foraging activity and tolerance of abiotic stresses that suggest variation in adaptive strategies.

The Scalloped Hammerhead Shark (*Sphyrna lewini*) is a large, nocturnal hunter of fast-moving prey such as squid and teleost fishes, other sharks, rays and crustaceans. The maximum size recorded for this species is 4.3 m in length and 152 kg (Smith, 1997; IGFA, 2001), with a potential life span of 35 years (Smith *et al.*, 1998). It can form large schools during the day when migrating, but is usually solitary. This shark has acute senses that aid active foraging; the unusually shaped hammer head allows the nostrils, eyes and the ampullae of

Lorenzini (organs that sense the electrical fields of prey) to be widely spaced, allowing a relatively large area to be sampled and highly effective binocular vision that can accurately track fast-moving prey (Kajiura & Holland, 2002; McComb *et al.*, 2009). The broad, flat head also includes wide nasal passages and a highly efficient sense of smell, and the head may also alter the hydrodynamics of the body to increase agility. In sum, the rather bizarre form of this species can be interpreted as having a range of complementary functions that work together to increase the ability of the shark to acquire resources.

Reproduction occurs once a year, via internal fertilization, and 15–31 half-metre long pups are produced after a long gestation of 9–10 months (Smith, 1997) – a moderate investment in reproductive effort for a shark. Thus the lifestyle, and indeed the highly unusual form, of the Scalloped Hammerhead Shark is based on active foraging for prey, and is not characterized by particular fecundity, longevity or tolerance of harsh abiotic conditions. This is compatible with C-selection, in which energy expenditure favours further energy acquisition rather than tissue maintenance or reproduction.

Other relatively C-selected sharks include the large and extremely active Shortfin Mako (*Isurus oxyrinchus*), which eats much more food, relative to body size, and digests this more rapidly compared to sedentary sharks such as the Nurse Shark (*Ginglymostoma cirratum*) or even the moderately active Sandbar Shark (*Carcharhinus plumbeus*; Helfman *et al.*, 2009).

Large size in sharks confers the advantages of greater speed, long-distance endurance and larger mouth size (Helfman *et al.*, 2009) – traits that are important to competitive ability. Indeed, the White Shark (*Carcharodon carcharias*) even forages within the underwater equivalent of a territory, returning to the same *anchor point* from which attacks on seals are more likely to be successful; 'size dominance is one possible explanation for this pattern as larger sharks may competitively exclude smaller sharks from the most profitable hunting areas' (Martin *et al.*, 2009). (However, White Sharks may also scavenge whale carcasses. They have low oxygen consumption rates and, if the meal is sufficiently large, can feed only once every six weeks (Carey *et al.*, 1982), suggestive of luxury consumption and perhaps an intermediate SC-selected strategy.) Even the filter-feeding sharks benefit from gigantism because this allows more efficient cruising and larger mouths, which are essential to *ram feeding* (swimming with the mouth open): e.g. the 4-tonne Basking Shark (*Cetorhinus maximus*) and the largest of all fishes, the 12-tonne Whale Shark (*Rhincodon typus*). Basking Sharks can only sustain their massive bodies by continual resource acquisition, and are obliged to migrate seasonally in order to stay within zones of high plankton productivity: 'they travel long distances (390 to 460 km) to locate temporally discrete productivity "hotspots"' (Sims *et al.*, 2003). They also increase their swimming speed when zooplankton densities decline, and stop feeding at low densities (Sims, 1999). Thus large sharks invest resources and energy in behaviour that maximizes subsequent resource acquisition. The 2-tonne, 8-m wide Manta Ray (*Manta birostris*), exhibits an adaptive strategy convergent with that of these enormous sharks.

Gigantism, **for active organisms in productive habitats**, is a trait that mainly allows more efficient acquisition and exploitation of resources, as we have already discussed for the larger whales. However, it is evident from our next example that gigantism can also be achieved via the slow-but-steady route by relatively inactive animals in less productive situations.

The sluggish Greenland Shark (*Somniosus microcephalus*) has an indolent lifestyle, languishing year-round in freezing Arctic waters. It uses unusually high concentrations of trimethylamine oxide (TMAO) and glycoproteins as a kind of cellular antifreeze, protecting enzymes and other proteins against structural and functional changes induced by cold temperatures. The Greenland Shark grows extremely slowly – a rate of half a centimetre per year has been recorded for the only individual ever measured. As it may attain a length of 6.4 metres larger specimens are thought to be several hundred years old and have been estimated to require 40 to 70 years to reach maturity, for males and females, respectively (Martin, 2009a). Reproduction via internal fertilization is equally lethargic; around ten large pups are produced, each almost a metre in length at birth. The caudal fin of the Greenland Shark is broad – a feature favouring bursts of speed typical of ambush predators. Indeed, the teeth of *S. microcephalus* also suggest an opportunistic predatory or scavenger lifestyle, the upper set being simple spikes and the lower teeth forming a single serrated cutting edge. This shark is the epitome of adaptation to extreme abiotic stress and pulsed resource availability, characterized by slow growth and high investment in tissue maintenance, lack of reliance on active foraging and limited reproductive effort. However, the final size that this shark may potentially attain is not as important for survival as is the adult size of the active sharks we discussed above, for which large size is an immediate advantage for hunting/ram feeding – adaptation for the Greenland Shark is based on a sedentary lifestyle and the slow, incremental investment of energy and resources that may eventually allow some individuals to become large. Thus S-selected organisms may gradually amass giant body sizes, whereas large size must be attained rapidly for C-selected phenotypes in order to function effectively.

It is highly unlikely that any sharks are truly R-selected.

Ichthyologists tire of the misconception that the Spiny Dogfish (*Squalus acanthias*) is simply 'smallish, dirt-common . . . the underwater equivalent of a rat' (Pierce, 2009)[11]. The Spiny Dogfish may form large schools of hundreds of sharks, and adults are small for sharks, at a maximum of 160 cm long and 9 kg (Compagno, 1984; Bigelow & Schroeder, 1948), but the adaptive strategy of this species is certainly not comparable to that of a small rodent. Female Spiny Dogfish may take up to 21 years to mature and live for up to 75 years – twice as long as the Scalloped Hammerhead Shark (Cailliet *et al.*, 2001). Like many sharks Spiny Dogfish are ovoviviparous, meaning that the eggs stay in the body until the young hatch. Gestation, at up to 24 months, is the longest of any vertebrate! The culmination of this marathon pregnancy is a rather anti-climactic litter of just five small pups (a minimum of two, maximum 11).

---

[11]Note that the Simon Pierce quoted here is no relation to the author of the present book. He is, oddly enough, actually an ichthyologist who specializes in sharks. Editors of fish-related journals take heed: please refrain from sending any more manuscripts for review.

Nor do other small shark species invest heavily in reproduction – sharks such as the 22 cm-long Spined Pygmy Shark (*Squaliolus laticaudus*), the Smalleye Pygmy Shark (*Squaliolus aliae*), the Pygmy shark (*Euprotomicrus bispinatus*) and the Cookiecutter Shark (*Isistius brasiliensis*) all have reproductive features similar to the Spiny Dogfish, producing small numbers of offspring after a pregnancy more like that of elephants than rats.

The Spiny Dogfish has extremely low resting and active metabolic rates, which translate into low energetic requirements and an energy-efficient way of life (Helfman *et al.*, 2009). Tellingly, the Spiny Dogfish uses the same cellular cryo-protection systems as the Greenland Shark, and is known to have high concentrations of TMAO that protect the nervous system against hypothermia (Villalobos & Renfro, 2007). Indeed, many of these smaller sharks hunt in colder, deep waters. The Cookiecutter Shark even uses bioluminescent skin that helps it to blend in with light shining down from above, and smaller, non-luminous fish-shaped marks on its body are used as a lure when hunting. The great age of these small, deep-water hunters reflects their cold habitat (Cailliet *et al.*, 2001).

Small shark species that hunt near the surface also show stress-tolerance, but not necessarily against cold temperatures. The metre-long Brownbanded Bamboo Shark (*Chiloscyllium punctatum*) hunts in tidal pools and can survive oxygen deprivation for up to 12 hours by switching off brain functions when stranded by the tide.

Small sharks are also well defended. Spiny Dogfish produce toxins in their tails (Halstead *et al.*, 1990) and are, as the name suggests, spiny. The small, deep-water hunters mentioned above also have fin spines. For small sharks a combination of longevity, low fecundity, defensive adaptations, stress adaptations and hunting using lures and sneak attacks rather than active foraging behaviour all suggest S-selection. Their smaller size is simply an adaptation to aid their particular lifestyle, and is only one of many traits that does not in itself accurately represent the entire adaptive trait syndrome.

It is likely that the sharks as a group, and indeed other cartilaginous fishes, are restricted to the S/C side of the CSR triangle, with ruderal marine niches dominated by more prolific bony fishes. Trade-offs between *r* and *K*-selection are usually obvious in animal groups, but in the case of sharks the trade-off is essentially a *K/K* trade-off. This is not as paradoxical as it seems when we consider that *K*-selection actually encompasses both C- and S-selection.

Insecta (insects)

Greenslade (1972) originally proposed a third primary adaptive strategy for animals based on his studies of *Priochirus* beetles. The 288 known species of this genus are native to the tropics, having spread from Asia to New Guinea, the Pacific and eventually east to Central America. *Priochirus* inhabit the humic material under the bark of fallen trees, and use a chisel-shaped head and jaws to work free pieces of bark. Thus a characteristic of the genus is the ornamentation of the head, with horn-shaped protuberances or 'teeth' that aid the beetles in pushing aside loose material, like bulldozers.

Greenslade traced the evolution of this genus from continental Asia eastwards towards the Pacific islands. He found that species which are responsible for colonizing new geographic ranges have particularly sleek head morphologies, allowing the head to be thrust into the confined spaces under tightly adhering bark. These species colonize newly fallen trees at the edge of lowland forest, among temporary vegetation (i.e. disturbed sites). These are relatively small species with high rates of dispersal, which he recognized as *r*-selected (e.g. a prevalent strategy among sub-genus *Plastus*).

In continental Asia there is a greater diversity in species and phenotypes, including larger species with greater head ornamentation that inhabit relatively productive lowland forests where a greater number of fallen trees are in a later stage of decomposition, and have looser bark. Greenslade recognized these as *K*-selected – a strategy that was prevalent in the sub-genera *Syncampsochirus* and *Exochirus*.

However, he also distinguished a '**beyond** *K*' strategy for archaic species with exaggerated and bizarre head ornamentation, native to colder montane habitats (prevalent among members of the *Peucodontus* and *Stigmatochirus* sub-genera). Greenslade argued that severe habitats select for a conservative strategy involving cold-tolerance, and select against dispersal ability.

These three strategies are directly equivalent to R-, C- and S-selected adaptive strategies, respectively.

Greenslade investigated adult beetles. In contrast, Braby (2002) found that the food resources available to larvae were critical for butterflies, determining the local distribution of three related *Mycalesis* butterfly species in the Australian tropics. He found three contrasting adaptive strategies, which he interpreted using the Southwood-Greenslade *rKA* templet. This underlines a special problem in identifying the adaptive strategies of insects: the metamorphosis of holometabolous species (those with true metamorphosis in which larval and adult forms are radically different) provides an opportunity to change the distribution of resources between body parts, and thus change the phenotype and way of life. It has even been suggested that larval and adult adaptive strategies should be classified separately because feeding habits and the ecologies of the two phases may be so dissimilar (Chown & Nicolson, 2004).

The extent to which the larval or adult form dominates the life cycle is indeed a crucial consideration for the adaptive strategies of insects. However, this is not to say that the adult strategy is independent of the larval strategy – large adults can only develop from the large amounts of resources consolidated by large larvae. For example, the adult South American Longhorn Beetle (*Titanus giganteus*) is one of the longest beetles in the world, at 150 mm, but the adult form develops from an even bigger larva (250 mm long; Wootton, 1993). The relative importance of larval or adult forms for survival differs even between closely related species. Larvae of the caddis fly *Odontocerum albicorne* live in relatively productive streams, and although the larvae live for around ten months the adults cannot feed and have a brief life-span of less than two weeks. This species responds to resource availability during the larval stage by altering the size of the adult thorax and wings during metamorphosis, but the size of the abdomen (the reproductive investment) remains relatively constant (Stevens *et al.*, 2000).

In contrast, another caddis fly species, *Glyphotaelius pellucidus*, occupies ephemeral pools and has a relatively short larval stage of seven months, with adults living for four months and probably feeding on nectar. Altered resource availability during larval growth results in changes in adult abdomen size and thus reproductive investment, but has little effect on the thorax. Thus the extent of the growth phase (larval phase) or the reproductive adult phase is critical to the adaptive strategies of these caddis flies, the former species exhibiting a relatively long period of resource acquisition followed by minimal reproductive effort, and the latter strategy based on a responsive and more extensive reproductive phase in disturbed habitats.

The Common House Mosquito (*Culex pipiens*) provides an example in which the larval and adult phases have a similarly rapid developmental pace. This R-selected species can breed within 14 days of hatching, following ten days as a larva, two as a pupa, and two days as a sexually immature adult (Spielman & D'Antonio, 2001). In contrast to the caddis flies, this reflects disturbance during both phases of the life cycle, such as the ephemeral nature of the ponds and puddles inhabited by larval forms and the probability of lethal events during adult feeding (i.e. predation or simply being swatted). As an extreme example, some tiny Chalcid wasp species, which are mostly parasitoids of other insects, can complete the entire life cycle in about a week (Wootton, 1993).

These examples suggest that certain phases of the life cycle may be more exposed to selection pressures and be of critical importance for the overall adaptive strategy (e.g. the adult phase for Greenslade's beetles, the larval phase for Braby's butterflies, or both for the mosquito). Identifying the point in the life cycle at which selection pressures exert the greatest influence is probably key to understanding and classifying insect adaptive strategies.

Nonetheless, it is possible to consider the entire insect life cycle when assigning adaptive strategies. Hodgson (1993) investigated British butterflies and found three overall suites of traits: (1) common species with a long-lived adult form, a single brood per year, and larvae restricted to a single plant species which is native to productive, undisturbed habitats (C-selected); (2) rare butterflies that produce a single brood each year, with extremely long-lived and slowly developing larvae exploiting plant species typical of unproductive habitats (S-selected); and (3) common species that produce several large broods per year, eat a range of larval food plants (all of which are species of disturbed habitats), and with a short-lived adult form (R-selected). Dennis *et al.* (2004) noted similar correlations between host plant traits and the overall suite of butterfly traits, population structure and geographical range, and went on to categorize the CSR strategies of British butterflies based on those of their host plants (see Fig. 4.5). Thus species such as the Comma Butterfly (*Polygonia c-album*) are C-selected, the Cabbage White Butterfly (*Pieris brassicae*) is relatively R-selected, and the Small Pearl-Bordered Fritillary (*Boloria selene*) is S-selected. These examples illustrate that even holometabolous species can be categorized according to CSR strategies when a range of life-history traits for both larval and adult stages are considered.

The second oddity of insects is the peculiarly strict social behaviour exhibited by some species, particularly the Hymenoptera (wasps, bees and ants). The wasps

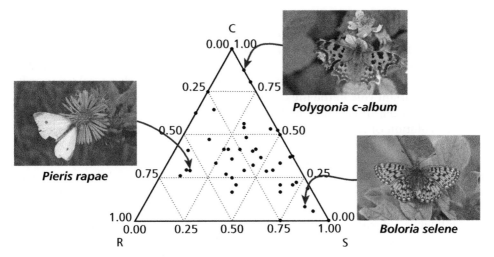

Fig. 4.5 British butterfly species, ordinated according to the CSR strategies of their larval food plants by Dennis *et al.* (2004), are mainly S- to CSR-selected but also include strongly C-selected species and some that tend towards R-selection. (Ternary plot (copyright © Royal Entomological Society of London). Comma Butterfly, *Polygonia c-album* (© Duncan Payne/Shutterstock.com), Silver-bordered Fritillary, *Boloria selene* (© Rasmus Holmboe Dahl/Shutterstock.com), Cabbage White, *Pieris rapae* (© Hway Kiong Lim/Shutterstock.com).)

are the most ancient of the Hymenoptera, and not all species are social: by comparing the behaviour of solitary, pre-social, primitively eusocial and eusocial wasps we can understand the evolution of sociality, and the role of different selection pressures in shaping the adaptive strategies of these insects.

Solitary Digger Wasps, such as *Sphex maxillosus*, exhibit primitive behaviour in which an egg is laid in a nest burrow that is stocked with paralysed crickets or grasshoppers, and then sealed up and forgotten. Thus post-hatching parental care consists simply of the larva having a meal prepared for it after emerging from the egg. The next step in the evolution of sociality is the continuous provisioning of larvae with food: for example, the Sand Wasp (*Ammophila pubescens*) behaves like a Digger Wasp, but it does not simply stock the nest burrow completely from the start: it occasionally adds caterpillars. The primitively eusocial wasps, such as the Japanese Paper Wasp (*Ropalidia fasciata*), go one step further in that the first generation of offspring do not disband and scatter, but remain to care for the second generation, and are involved in the defence of the nest against marauding ants (Suzuki & Murai, 1980; Itô *et al.*, 1985; Gadagkar, 1991). This species has a seasonal nesting cycle, with inseminated females hibernating over winter (Itô *et al.*, 1985).

Thus the first steps towards social life involve the care and investment of resources in individuals, helping them accrue sufficient resources to survive harsher periods, and defence of this investment. This is an indication that S-selection is an important factor during the evolution of insect sociality. However, sociality then facilitates resource acquisition, territoriality and the ability to compete with other insects for food resources. With the evolution of sterile castes that support a reproductive queen, the system changes gear to effectively augment the capacity of a single individual to obtain resources and reproduce (C-selection). In many social species, such as the German Wasp (*Vespula germanica*), females hibernate as adults, suggesting that S-selection still plays an important role in the life cycle. Thus we suggest that the adaptive strategies of truly social insects may be largely C-selected, but S-selection was crucial for the evolution of this situation and continues to influence the adaptive strategies of eusocial species.

Ants exhibit a range of adaptive specializations that Andersen (1995) interpreted within the context of C-, S- and R-selection for Australian species. He noted that *Iridomyrmex* species are abundant in warm, open habitats that allow for high rates of foraging, have large foraging ranges and 'exert a major competitive influence on other ants' (i.e. they are adept at finding and defending resources), suggesting that they are relatively C-selected. He also suggested that tropical rainforest canopies are a competitive environment for ant species, and that Old World *Oecophylla* and *Crematogaster* species and New World *Azteca* species are likely to be highly C-selected. Andersen (1995) considered low temperature as one of the principal abiotic stresses for ants because foraging capacity and thus resource availability are reduced in the cold, and suggested that species of cooler, shaded habitats are relatively S-selected, such as species of the genera *Heteroponera*, *Myrmechorhynchus*, *Notoncus* and *Prolasius*. He also viewed opportunistic species from warmer, disturbed habitats, such as *Ochetellus glaber*, small *Rhytidoponera* species and *Pheidole* species, as R-selected.

Aracnida (spiders, scorpions, mites and ticks)

Social spiders use **cooperative prey capture** to maximize the resources acquired by a group of closely related individuals, and are C-selected. Examples include the star of the film *Arachnophobia*, the Flat Huntsman Spider (*Delena cancerides*; Rowell & Avilés, 1995), also *Theridion nigroannulatum*, which builds colonies of several thousand individuals (Avilés *et al.*, 2006), and *Anelosimus eximius*, with up to 50,000 individuals per colony (Vollrath, 1986) – although maximum colony sizes are usually less than a thousand individuals (Avilés, 1997).

Sociality has the advantage that a joint effort can be made to construct and maintain a communal nest, which is usually draped over vegetation and includes a zone that acts as a snare. Captured prey is taken from the snare to the more protected areas of the nest where communal feeding occurs. The size of each colony is governed by competition for resources and subsequent reproductive success (Avilés, 1997) – in other words, larger societies exist in relatively productive habitats where resources are particularly abundant. Indeed, most species are native to aseasonal lowland or montane tropical rainforest, with only a few species found in warm, dry seasonal climates. Social spiders catch larger prey than solitary species (Nentwig, 1985; Uetz, 1986), with cooperation increasing the success rate of prey capture (Matsumoto, 1998; Uetz, 1989). Thus a consistent abundance of large prey is thought to be important for cooperative behaviour (Avilés, 1997). For semi-social species, sociality breaks down and colonies disperse when resource supplies dwindle (Kraft *et al.*, 1986; Ruttan, 1990). Indeed, spider colonies have been described as **foraging societies** (Whitehouse & Lubin, 2005), and have been compared to foraging flocks of birds (Rypstra, 1979).

As with the Hymenoptera, the initial step in the evolution of spider sociality is thought to have involved care of eggs and larvae. In spiders, care can range from simply relocating young to the protected part of the nest up to and including the active protection of eggs and young, and even regurgitation feeding. Semi-social species native to seasonally dry habitats invest in the defence of both young and adults. These species are active hunters that use silk only to bind leaves together to form a nest. The nest serves as a defence against ants, and is usually compact enough to be closed against the elements during the winter (e.g. *Stegodyphus* species; Avilés, 1997). Thus, like the social hymenoptera, the evolution of sociality in spiders may have depended on an initial phase of S-selection favouring parental care and defence which, for some lineages, then provided the means (teamwork) to expand resource acquisition, ultimately resulting in a strategy that favoured resource acquisition in productive aseasonal habitats (C-selection).

Intriguingly, social spider colonies are unstable, and population crashes may represent a combination of assaults by parasites and pathogens, and inbreeding (Lubin, 1991; Vollrath, 1982). Spiders do not have a winged, highly mobile phase for swapping genes between different colonies, as ants do, and must breed within the colony. This is equivalent to the susceptibility of monocultures of C-selected plant crops to disease.

Thus social spiders exhibit all the hallmarks of C-selection, but the semi-social spiders demonstrate S-selected behaviour in harsher environments. Similarly, the Mexican Redknee Tarantula (*Brachypelma smithi*) is native to desert and scrubland, usually under thorny vegetation, dry thorn forest or tropical deciduous forest (West, 2005). This tarantula is an ambush predator that feeds and moults in a burrow (Locht *et al.*, 1999) and is well defended against predation, using **urticating hairs** – the abdomen possesses a zone of highly specialized hairs that easily break off and infiltrate skin and membranes, causing irritation and allergies. Females take between nine and ten years to reach maturity, and typically live for ten more years (Locht *et al.*, 1999).

S-selection is evident in other large tropical spiders, such as *Grammostola rosea*. This species is able to live for more than two years on water alone, and extremely low respiratory rates probably contribute to this capacity to tolerate stress (Canals *et al.*, 2008). Such large spiders may also exhibit parental care (e.g. the Wolf Spider, *Brachypelma smithi*; Foelix, 1996). The Namib Dune Spider (*Leucorchestris arenicola*) even feeds its young within a nest burrow (Willmer *et al.*, 2005), the main characteristics of this species being large size, slow metabolic rates, longevity, parental care, seasonal foraging and breeding patterns, efficient excretion, discontinuous respiratory cycles and heat tolerance (Henschel, 1997). In Europe, the Purseweb Spider (*Atypus affinis*) is an ambush predator that lives in a silk-lined burrow. Although it is considerably smaller than the Wolf Spider and the Namib Dune Spider, it similarly exhibits parental care, for up to a year, and has a slow rate of development, being sexually mature at four years of age. Slow growth and development, longevity and investment in the maintenance of offspring suggest S-selection for these spiders.

The smallest spider, *Patu digua* from Colombia, is an orb-weaver (the male is the smallest, at an astounding 0.37 mm in length – so small that its distinguishing features must be viewed under an electron microscope; Forster & Platnick, 1977). Orb-weavers sit and wait for food to be snared, and can wrap insects up in silk and cache them for when prey is not available. This is particularly evident for larger species that have a relatively permanent web, such as the Golden Silk Orb-weaver (*Nephila edulis*). However, caching can only allow orb-weavers to resist a few hours or days without food – insufficient for the resistance of adverse seasons and certainly not in the same league as spiders such as *Grammostola rosea*. Smaller orb-weaver species that use ephemeral webs, such as the St Andrew's Cross Spider (*Argiope keyserlingi*), do not cache food (Champion de Crespigny *et al.*, 2001). Orb-weavers with annual lifecycles, such as *Zygiella x-notata*, are intolerant of resource-poor sites, and will relocate in order to ensure sufficient resource availability for egg production (Wherry & Elwood, 2009).

Indeed, small, ephemeral spiders are characterized by a massive pre-hatching reproductive investment. The 13 mm-long Furrow Orb-weaver (*Larinioides cornutus*) has a life cycle lasting from six months to one year, with adults either living for as little as one month and dying during the winter, or entering a resting state and dying the following spring. Up to 1000 eggs are laid, with mortality as high as 99 per cent within the first two months of hatching (Ysnel, 1991). The 17 mm-long Four Spot Orb-weaver (*Araneus quadratus*) has an annual life cycle and dies as winter closes in (Marc *et al.*, 1999), but up to 700 eggs survive the cold in a state of diapause within a protective egg sac, hatching the following spring (Burch, 1979). It is usually the propagules, rather than adults, that resist adverse conditions, as we have seen for the seeds of R-selected plants and the eggs of the Annual Killifish. Thus the smallest orb-weavers are highly R-selected.

Crustacea (crustaceans)

The Northern Crayfish (*Orconectes virilis*) is an invasive species in the Colorado River basin, USA, where it has a negative impact on populations of native fishes, such as the Flannelmouth Sucker (*Catostomus latipinnis*) (Carpenter, 2005). This crayfish is highly active, aggressive, and competes directly for food, interfering with the ability of the fish to find and utilize resources. Thus the adaptive strategy of this species appears to reflect a high degree of C-selection. However, other crayfishes are known to be even more aggressive – *O. rusticus* and *O. propinquus* compete more effectively for food and replace the Northern Crayfish when accidentally introduced into the water bodies to which it is native (Garvey *et al.*, 1994; Hill & Lodge, 1999).

In contrast, the majority of crabs are characterized by physiological flexibility and physical defence. As an extreme example, the hemocyanin of the deep-sea hydrothermal vent crab *Cyanagraea praedator* (the Smoker Crab) has a strong affinity for oxygen, resulting in efficient oxygen uptake in a potentially hypoxic environment, and a strong Bohr effect (the efficient release of oxygen from the hemolymph into respiring tissues; Chausson *et al.*, 2001). This is a physiological mechanism that maximizes metabolic performance in environments that tend to limit metabolism, and the Smoker Crab can therefore be regarded as S-selected. Physiological plasticity is similarly used by the Blue Crab (*Callinectes sapidus*), which must move between seawater and brackish water in its estuarine habitat. This species exhibits efficient regulation in the expression of the enzyme Na^+,K^+-*ATPase* in gill tissues, which is responsible for the osmoregulation of the hemo-lymph (Lovett *et al.*, 2006). This species is also well defended against predation, with large, forward-pointing spiny projections on the leading edge of the cara-pace. These adaptations are all consistent with S-selection.

Some crabs, however, have lifestyles suggestive of R-selection. The Pea Crab (*Pinnotheres maculatus*) is parasitic within the mantle cavity of the Blue Mussel (*Mytilus edulis*), causing damage to the gills of the mussel and robbing food from its host, which causes reduced mussel growth rates and shell distortion (Bierbaum & Ferson, 1986). As the name suggests, at around 6 mm Pea Crabs are extremely small, and the pea crabs in general (there are 252 species of Pinnotheridae) have rapid rates of development: e.g. *Pinnixia gracilipes* takes only 24 days to com-plete five larval moults (Lima *et al.*, 2006).

Aside from crabs, R-selection is perhaps even more apparent in crustaceans such as the Longtail Tadpole Shrimp (*Triops longicaudatus*), the eggs of which resist desiccation in arid desert soil and can hatch and complete the entire life cycle within two weeks, while the ephemeral pools persist. The eggs may remain in a state of diapause for up to 20 years and nonetheless remain viable, and are capable of surviving freezing, drought and high temperatures. Adult individuals, however, are not.

Echinodermata (sea urchins, starfish, crinoids, sea cucumbers)

Lawrence (1990) interpreted echinoderm life-histories in the context of CSR theory. He investigated a range of examples for each primary adaptive strategy, of which we shall present a small selection here. One of his general conclusions was that whilst S-selection is widespread throughout the group, C- and R-selected species are much rarer, although they do exist. He suggested that this is the result of phylogenetic constraints – only the starfish have sufficient mobility to allow active foraging, and sedentary forms do not forage in a manner analogous to C-selected plants.

Not all starfish are highly active, but the voracious Crown-of-Thorns Starfish (*Acanthaster planci*), native to Indo-Pacific coral reefs, is one such active forager that survives due to a great abundance of prey. The Sunflower Sea Star (*Pyc-nopodia helianthoides*; from North America's west coast; see Fig. 4.6) and *Meyenaster gelatinosus* (South America) have similar lifestyles. Species such as this may be the most C-selected echinoderms, according to Lawrence, but the Crown of Thorns Starfish is almost certainly not exclusively C-selected as it exhibits traits such as thorns that reduce palatability to predators and suggest a degree of S-selection. We suspect that these species may actually be intermediate SC-strategists.

Fig. 4.6 Examples of C-, S- and R-selection for echinoderms, according to Lawrence (1990): (C) the Sunflower Sea Star, *Pycnopodia helianthoides* (© Undersea Discoveries/ Shutterstock.com), (S) the Slate Pencil Urchin, *Heterocentrotus mammillatus* (© Goluba/ Shutterstock.com), and (R) the Collector Urchin, *Tripneustes gratilla* (© Cigdem Sean Cooper/Shutterstock.com) (not to scale).

Unproductive shallow habitats in the tropics host sea urchins such as the Helmet Urchin (*Colobocentrotus atratus*) and the Slate Pencil Urchin (*Hetero-centrotus mammillatus*) that invest a relatively large proportion of mass and energy into the production of spines or protective plates in comparison to other sea urchins. They are also slow growing and long-lived, typical of a conservative/ protective S-selected strategy. Lawrence also suggested that Antarctic starfish and sedentary echinoderms such as crinoids (native to nutrient-poor deep-sea habitats) could also be S-selected, although little detailed information is currently known concerning their life-histories.

Lawrence viewed tropical sea urchins such as *Lytechinus* and *Tripneustes* species as R-selected because they are found in habitats of high primary production, have rapid growth rates, high reproductive output and fecundity, short life-spans and are associated with disturbance. Many of the R-selected species suggested by Lawrence have short longevities of around one to two years, whilst the life-spans of S-selected species are measured in decades (up to 28 years for *Echinus affinis*). He concluded that echinoderm life-histories and physiologies have not been studied within the broader perspective of the range of stress and disturbance to which echinoderms are exposed, and that: 'Grime's model provides a context that can be used to interpret these characteristics' (Lawrence, 1990).

Mollusca (snails, clams, squids)

Most squid have highly active annual, live-fast, die-young lifestyles, and are a major prey item for a range of marine animals. For example, Pecl & Moltschaniwskyj (2006) investigated growth and reproductive traits of the Southern Calamary (*Sepioteuthis australis*) and also reviewed the cephalopod literature, concluding that cephalopods in general tend to have a **'live for today'** strategy, in which reproductive investment is supported by instantaneous resource availabilities rather than stored reserves. They concluded that the immediate use of resources was a logical strategy for animals with access to resources but no long-term future.

Their data revealed that the Southern Calamary has an opportunistic lifestyle, capable of varying the timing of reproduction based on environmental conditions. This is a situation similar to R-selected plants that not only complete the life cycle rapidly, but which are also capable of initiating earlier reproductive development on exposure to suboptimal conditions (Salisbury, 1942; Grime, 2001). This situation has also been described for the Common Mediterranean Squid (*Loligo vulgaris*), another annual species that produces numerous clutches of eggs (Moreno *et al.*, 2005). Squid such as these have extremely high metabolic rates (O'Dor & Webber, 1986) and use jet propulsion to move rapidly and evade predators (Cole & Gilbert, 1970).

In deeper waters the selection pressures acting on squid are very different: there is less need for speed and a greater need to conserve energy. The Vampire Squid (*Vampyroteuthis infernalis*) lurks in the cold depths, using the lowest metabolic rate of any cephalopod to survive oxygen levels of 3 per cent that would literally suffocate most animals (Seibel *et al.*, 1997). Its blood contains a form of hemocyanin that has an extremely high affinity for oxygen, allowing it to take up oxygen even in these conditions (Seibel *et al.*, 1999). It also saves energy by swimming slowly, flapping two fins on the side of its head rather than using jet propulsion (Seibel *et al.*, 1998). Its tentacles are webbed, together forming a jellyfish-like dumb-bell that can be contracted to provide an efficient burst of speed to escape predators. It can also confuse predators with bioluminescent eyespots and glowing arm tips, and as a last resort covers its tracks with a cloud of luminous mucus squirted from the arm tips (Robison *et al.*, 2003).

The 13 m-long Giant Squid (*Architeuthis dux*) is another deep-sea species, and much of what we know about this elusive animal has been inferred from analyses of tissues and stomach contents of small numbers of individuals stranded on beaches or trawled up as bycatch. The muscles of this species have low protein contents and pockets of ammonium chloride solution that potentially allow

neutral buoyancy at depth (Rosa *et al.*, 2005). Thus whilst most squid are negatively buoyant and require high protein contents and stronger muscles for constant swimming the Giant Squid is capable of maintaining a constant depth whilst
expending little energy, which could be useful for ambush predation. Oddly
enough, Giant Squid presumably become stranded on beaches because the inappropriately high buoyancy of their tissues in shallow water means that returning
to deeper water is sometimes too exhausting (Rosa *et al.*, 2005). The first
recorded observation of a Giant Squid hunting in its natural habitat shows the
squid arriving at a lure and taking bait, suggesting that it is capable of active
foraging (Kubodera & Mori, 2005). They are known to hunt at depths of around
900 m during the day, and probably 400–500 m during the night (Kubodera &
Mori, 2005); this daily travel suggests a relatively active animal. Giant Squid
may live for up to 14 years, estimated from the carbon isotope composition of
calcium carbonate *statoliths* (mineral aggregations forming part of the squid's
gravity/orientation sensing system; Landman *et al.*, 2004). This suggests that
Giant Squid achieve gigantism rapidly – a trait of C-selected organisms that
contrasts with the Vampire Squid. The even more massive Colossal Squid (*Mes-
onychoteuthis hamiltoni*), the largest invertebrate, is likely to have a similar
lifestyle but little is known of these giants.

Thus cephalopods encompass R- and S-selected species, and possibly C-selected
giants too.

R-selection is prevalent in molluscs. Carvalho & Bessa (2008) measured life-
history traits of the terrestrial Asian Trampsnail (*Bradybaena similaris*) and
concluded that it is has an '**r-strategy**'. The 10 cm-long marine nudibranch *Spurilla neapolitana* hatches three days after oviposition, matures at 67 days and
lives to an average of 157 days. An adult produces, on average, a staggering
40×10^6 eggs in this three-month period of maturity (Schlesinger *et al.*, 2009).
Xylophagid molluscs (marine wood-boring bivalves) produce up to 30,000 eggs,
colonizing logs that have fallen to the seabed. Turner (1973) observed that: 'their
high reproductive rate, high population density, rapid growth, early maturity,
and utilization of a transient habitat classify them as opportunistic species.'

Even the giant clams are relatively fecund and have fast rates of development,
although interpreting their adaptive strategy is complicated by the fact that they
are absolutely dependent on symbiotic dinoflagellate algae (*Symbiodinium*
species). These can provide, by photosynthesis, up to 100 per cent of the carbohydrates required by the clam (Fisher *et al.*, 1985). Indeed, species such as
the 1.4 m-wide True Giant Clam (*Tridacna gigas*) are restricted to shallow waters
where they have access to sunlight, and their large bodies and large mantle tissue
surface area (where the symbiotic algae reside) allow the maximum possible
interception of sunlight. In this respect they could be considered directly equivalent to C-selected plants that invest in large bodies in order to maximize exposure to sunlight. However, this is complicated by the fact that the thicker tissues
of larger clams reduce light transmission to the photosynthetic algae, meaning
that larger animals are not necessarily more productive (Fisher *et al.*, 1985).
Giant clams require clean water, cannot tolerate extremes of temperature, salinity or pH, and their shells are relatively weak compared to other molluscs (Lin
et al., 2006). This highly unusual symbiotic organism may represent a relatively

C-selected species, but perhaps the best examples of C-selected molluscs are those, such as the Hard Clam (*Mercenaria mercenaria*), that dominate the fauna of certain estuarine environments (Sma & Baggaley, 1976) and, via properties such as relatively profligate ammonia excretion, have controlling effects on the turnover of nitrogen. This species is also extremely sensitive to pollution and stress, which rapidly invoke a disease-like syndrome involving the clogging of gills with mucus followed by infection with polychaete worms, leading to declining populations (Jeffries, 1972). Thus the Hard Clam is used as an indicator species for monitoring water quality (Nasci *et al.*, 1999). It requires a clean, disturbance-free and stress-free productive habitat, where it effectively forms a monoculture.

Annelida (segmented worms)

The Common Earthworm, *Lumbricus terrestris*, undergoes competition with a range of other worm species in productive temperate grassland habitats, where the soil is rich in the organic matter on which it feeds. For instance, the growth of *L. terrestris* individuals is restricted when in contact with high population densities of the worm *Aporrectodea caliginosa*, probably due to competition for food (Eriksen-Hamel & Whalen, 2007), and *L. terrestris* competes via efficient foraging (Sheehan *et al.*, 2007). *L. terrestris* does not have rapid regeneration times, and reproductive effort is not linked to the longevity or mass of the parent – it takes up to four years to reach sexual maturity, with mating resulting in the production of an average of five cocoons over a period of up to a year (Butt & Nuutinen, 1998), which must hatch within five months or lose viability.

Other earthworm species in productive habitats can grow to enormous sizes. The Mekong Giant Earthworm (*Amynthas mekongianus*) from Laos is one of the longest of the giant earthworms, at up to 2.9 m (Blakemore *et al.*, 2007). Unlike the other giant earthworms, which usually inhabit meadows, the Mekong Giant Earthworm is, as the name suggests, native to muddy parts of the banks of the river Mekong.

Deep-sea hydrothermal vents are perhaps the most challenging habitat on Earth, but are nonetheless characterized by annelids. The Pompeii Worm (*Alvinella pompejana*) is a small annelid (a maximum of 95 mm long by 12 mm in width, 30–85 g dry mass; Desbruyères *et al.*, 1998) that colonizes the chimney walls of white smoker vents, secreting tubes that protrude outwards from the chimneys. These tubes are proteinaceous but also contain minerals, particularly iron, and protect the animal from falling mineral debris (Zbinden *et al.*, 2000). The secretion of metals and their incorporation into the tube is also a method of coping with a potentially toxic metal-laden environment (Desbruyères *et al.*, 1998).

The innate physiological capabilities of the Pompeii Worm to tolerate heat include enzymes, such as superoxide dismutase, that are more thermally stable with respect to analogues in other animals. This is due to **interaction motifs**, or regions of the molecule that form bonds between each other, reinforcing it (Shin *et al.*, 2009). The ribosomal DNA of this species is also extremely heat-stable (Dixon *et al.*, 1992). However, some of the proteins used by this species are only moderately heat stable, resulting in uncertainties as to the optimum temperature for growth and the maximum temperature that can be directly withstood by metabolism. For example, the collagens comprising the cuticle of the animal are relatively heat stable for an annelid, but maintain structure and function only up to a modest 45°C (Desbruyères *et al.*, 1998). Measurement of temperatures inside Pompeii Worm tubes is almost impossible using the bulky external tools carried by submersibles, but this annelid appears to be capable of tolerating sustained temperatures of between 45°C and 60°C (Cary *et al.*, 1998; Lee, 2003), and possibly up to 105°C for brief periods. The tube probably provides a degree of insulation during these temperature fluctuations (Desbruyères *et al.*, 1998) and the anterior of the animal protrudes into relatively cool water.

The haemoglobins used by the Pompeii Worm have high affinities for oxygen, meaning that oxygen uptake and respiration can be maintained in an extremely variable environment, coupled with a strong Bohr effect (haemoglobin loses its affinity for oxygen in the presence of CO_2 or low pH, leading to oxygen release in respiring tissues), meaning that the haemoglobin readily releases oxygen into tissues (Hourdez *et al.*, 2000)[12].

Aside from these innate traits, the Pompeii Worm is associated with symbiotic bacteria that form a biofilm over the filamentous outgrowths on the back of the worm. These bacteria – as we shall see later when we discuss the proteobacteria – use enzymes with an even higher degree of thermal stability, and the biofilm may also act to insulate the worm against pulses of high temperature (Lee *et al.*, 2008).

For Pompeii Worms, the isolation of metabolism from a harsh and variable environment, the innate stability of key metabolic components, and investment in defensive structures (tubes), are strongly suggestive of S-selection. Some annelids, however, survive solely by virtue of being prolific. The bizarre Bone-Eating Snot Flower (*Osedax frankpressi*) is one of the strangest things that have ever happened at sea. This flower-like worm colonizes the bones of dead whales that have fallen to the ocean floor, using vascularized root-like threads to penetrate the bone and invade the marrow – whalefalls are extremely productive but temporary marine microhabitats, representing as much energy and resources in one go as a thousand years of marine snow accumulation (the detritus that falls to the bottom of the ocean).

The most bizarre and macabre aspect of this bone-eating worm is its reproductive strategy, which is particularly well suited to allow dispersal and the colonization of whalefalls and other carcasses that are potentially scattered over large geographic distances (Rouse *et al.*, 2004; Vrijenhoek *et al.*, 2008). The female is extremely fecund and continuously produces offspring, but in order to do so she actually contains within her body 50 to 100 microscopic males, which provide her with sperm in return for a living. This adaptive strategy effectively converts kilograms of bone marrow into a continuous fountain of larvae, most of which will perish in the search for new opportunities. However, the

---

[12] Incidentally, the independence of hydrothermal vent ecosystems from the sun is a common misconception. The animals within these habitats require oxygen from the water. The source of this oxygen is photosynthesis.

persistence of the species as a whole requires only a small proportion to strike gold. The numbers involved are staggering – up to 900,000 females may form a population on a single carcass. It is thought that larvae that settle on bone become females, whereas larvae that settle on females become males, and literally become incorporated into the female body (Rouse *et al.*, 2004). With a highly specialized lifestyle geared up to maximize regeneration, this is clearly an R-selected organism.

Cnidaria (corals, sea anemones, jellyfish, hydras, sea pens)

Corals are symbiotic organisms composed of anthozoa, algae, bacteria and fungi that, together, form a **holobiont**. Murdoch (2007) used ten morphological, growth and reproductive traits to rank Caribbean coral holobionts according to their primary CSR strategies (see Fig. 4.7). C-selected corals are those with high

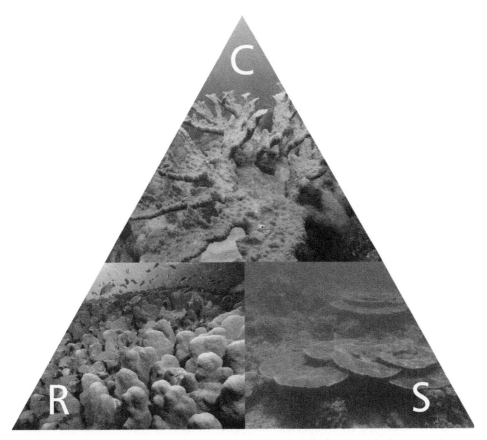

Fig. 4.7 Examples of C-, S- and R-selection for corals, according to Murdoch (2007): (C) Elkhorn Coral, *Acropora palmata* (© John A. Anderson/Shutterstock.com), (S) Table Coral, *Acropora cytherea* (© neijia/Shutterstock.com), and (R) Dome Coral, *Porites astreoides* (© Rich Carey/Shutterstock.com) (not to scale).

growth rates and large habits, persistent holobionts, late reproductive maturity (including reproduction at the end of summer after an extensive period of resource acquisition), low reproductive effort and a moderate stress response. Examples include branched, oviparous species of *Acropora* and the Ivory Bush Coral (*Oculina diffusa*). This mode of holobiont growth, for organisms that depend on photosynthesis by symbiotic algae of the genus *Symbiodinium*, is directly equivalent to C-selected plant life forms: 'rapid growth and a tall, branched structure allows these corals to overgrow all other corals under benign environmental conditions' (Murdoch, 2007). Species such as *Acropora palmata* are notoriously incapable of withstanding stress, disease and physical damage, meaning that they are the first to disappear should environmental conditions change. *A. palmata* is coated in a mucus that has antibacterial properties, but investment in this system is curtailed under stress, such as during the warm water events that cause coral bleaching, leading to increased infection by bacterial disease (Ritchie, 2006). Other C-selected corals may include *Agaricia tenuifolia*, which has a soft, fragile, branching architecture – these branches can merge with one another to provide vertical support and to occupy space, efficiently intercepting sunlight in shallow, productive reefs and allowing dominance over co-occurring hard corals such as *Millepora* species (Chornesky, 1991). *A. tenuifolia* may even dominate other branching corals such as *A. palmata*. However, at depths greater than around 6 m, where light intensities are lower, slower-growing *Millepora* species become dominant.

S-selected corals are moderately sized and have the lowest growth rates, the greatest ability to respond to stress, low palatability, low reproductive effort and the greatest longevity. They also have a limited range of dispersal. These include species with plate habits, and foliose or solitary corals such as *Agaricia fragilis* and *Scolymia cubensis*. Plate and foliose habits may be particularly prevalent in shaded sites, such as under ledges or in caves, and may function to maximize light interception where a simple, wide laminar surface is advantageous and elaborate systems of supporting stems and overtopping branches are not required.

R-selected corals have the earliest reproductive maturity, the greatest reproductive effort, high palatability and are small and relatively ephemeral. These include massive (i.e. mound-shaped), viviparous corals such as *Agaricia agaricites, Favia fragum* and *Porites astreoides*, which exhibit 'a ruderal lifestyle that is maximally adapted to frequent settlement and rapid growth within patches of marginal quality generated by disturbance'. These species invest little in defence and experience high levels of corallivory, particularly by parrotfish (Murdoch, 2007, and references therein).

Murdoch (2007) examined 200 10 m-long video transects – determining the species present in individual frames of video footage with a video camera held 40 cm away from the reef surface – to plot the distribution of over 19,000 coral holobionts belonging to 38 species along environmental gradients in the Florida Keys. He noted that coral distributions resembled 'a chaotic tangle of species varying across reefs in an idiosyncratic manner' that could only conceivably be understood in terms of a theory of life-histories. He then used models to predict how dominance should shift between C-, S- and R-selected corals along gradients of environmental stress and disturbance. He confirmed from his field data that when the prevalence of coral adaptive strategies was plotted between sites he was able to resolve a highly ordered pattern, rather than the chaotic tangle of taxa. This demonstrates that C-, S- and R-selection are useful concepts for the ecological interpretation of communities composed of these cnidarians.

Eumycota (fungi) (including notes on lichens)

A number of authors have specifically explored the adaptive strategies of fungi in terms of CSR strategies (Pugh, 1980; Cooke and Rayner, 1984; Booth *et al.*, 1998; Grime, 1988b; Nix-Stohr *et al.*, 2008). Pugh (1980) first suggested that CSR strategies could be useful for interpreting fungal life-histories, and the CSR classification of fungi is usually based on his work.

Oddly, Pugh (1980) postulated the existence of a fourth primary strategy in which high rates of stress and disturbance select for 'survivors'. He considered stressful habitats to be characterized by nutrient limitation, whereas survivors are adapted to habitats such as the phylloplane (leaf surfaces) that are characterized by 'widely fluctuating temperatures, the general level of desiccation, and the incidence of ultraviolet radiation' (Pugh, 1980). However, all of these factors are actually stresses, characterized by reduced productivity due to suboptimal metabolic performance which may be brought about either by limited resource availabilities or factors that cause injury to metabolic components such as temperature extremes, desiccation or UV light (Pierce *et al.*, 2005). Crucially, cells and tissues can potentially resist and recover from all of these stresses, and so the adaptive response involves investment in tissue maintenance and conservative growth[13]. Tellingly, Pugh's (1980) stress-tolerators and survivors both have conservative responses to abiotic environmental extremes, with survivors employing protective pigments and DNA repair mechanisms in the same manner as S-selected plants (Pierce *et al.*, 2005). Thus it appears that Pugh (1980) essentially split S-selected fungi into two groups based on the nature of the stresses they experience, but impaired metabolic performance and the conservative/ protective adaptive response common to both are typical of S-selection. Indeed, Cooke and Rayner (1984) interpreted phylloplane fungi in a slightly different light: as stress-tolerators that are adapted to a range of 'nutritive and non-nutritive' stresses. One of Pugh's (1980) main examples of a **survivor**, *Chrysosporium pannorum*, is an alpine/polar species that tolerates low temperatures and can detoxify organo-mercurial fungicides extremely effectively (i.e. it protects metabolism against disruption by heavy metals). It has a conservative, protective lifestyle directly equivalent to an S-selected plant. Thus there is no

---

[13] Disturbance, in contrast to stress, destroys cells outright, from which they cannot recover. The organism may be able to regenerate lost parts, but the cells in those lost parts are dead – disturbance is lethal to cells whereas stress is merely suppressive. Any stress may eventually kill if it is so protracted that the organism exhausts its ability to maintain metabolic performance (Larcher, 2003), but there is nonetheless a stark difference between factors such as fire, which instantly renders metabolism inoperable, and stresses that allow metabolism sufficient leeway to resist and respond. The appropriate adaptive response to each situation also differs.

need to invoke the existence of a fourth primary adaptive strategy in order to explain the adaptive response of species such as this.

Pugh (1980) suggested that R-selected fungi included soil organisms that survive as spores and 'burst into activity' when resources become available, sometimes completing the life cycle in days. Examples include species of *Fusarium* and *Penicillium*. C-selected fungi include species of rich agricultural soils, and species such as *Peniophora gigantea* that monopolize rotting wood and exclude other fungi.

In addition to the free-living saprotrophs that decompose soil organic matter a substantial component of the fungal biomass of the rhizosphere consists of mycorrhizal species that form a network or investing sheath around the roots of vascular plants. In infertile soils ecto-mycorrhizas can form the entire nutrient-absorbing surface of plant root systems. No less than six types of mycorrhizas have been recognized on the basis of the functional interactions observed between fungus and host. The majority of mycorrhizal fungi exhibit traits reflecting a strong imprint of S-selection. These are particularly evident in the ecto- and ericaceous-forms in which fungal material creates root structures with extended life-span and durability that contrast markedly with the uninfected short-lived roots of many crops and weeds of fertile arable land. S-selection is also suggested by the delayed reproduction and intermittent fruiting of many mycorrhizal fungi.

Lichens, which, of course, are symbiotic organisms involving both fungi and photosynthetic algae or cyanobacteria, have also been investigated in terms of CSR strategies. Rogers (1988) measured growth rates and the extent to which lichens of different morphologies were capable of overtopping neighbours, and concluded: 'it is possible to prepare a triangular ordination for a range of lichen growth-forms from a range of environments. In this ordination *Parmotrema austrosinense* is clearly a competitive lichen . . . *Hyperphyscia adglutinata* can be considered a stress-tolerant ruderal'.

However, these represent differences in strategy relative to other lichens, and when Rogers (1988) compared lichen growth rates and height with those of herbaceous plants he found that: 'lichens are extreme stress-tolerators as suggested by Grime (1977), but are separated from herbaceous angiosperms . . . their growth rate is less than that for the herbs'. Thus, variability in the relative extent of C-, S- and R-selection is evident for lichens, but the group as a whole is extremely S-selected in absolute terms. Rogers (1988) was able to chart the changes in lichen strategies throughout succession, and found that they 'follow the pattern of succession postulated by Grime'. Thus variation in C-, S- and R-selection is undoubtedly relevant to fungal and lichen ecology.

Archaea

Precise phylogenetic relationships towards the root of the tree of life are still uncertain, but one thing is clear: the Archaebacteria are phylogenetically closer to Eukaryotes than are the Eubacteria, or true bacteria (Madigan *et al.*, 2003).

The Archaebacteria probably diverged around 3.5 billion years ago and, as with the bacteria, the most archaic forms within the group are extremophiles, requiring extremes of temperature, pH or salinity for growth, whilst the more derived taxa do not require extreme conditions. Many extremophile taxa are thermophilic, not only surviving but actively growing at high temperatures. For example, *Pyrolobus fumarii* grows at high pressure and up to 113°C in the walls of **black smoker** hydrothermal vent chimneys. Incredibly, *P. fumarii* actually requires at least 90°C for growth. The record holders, for which viability at high temperature has been confirmed, consist of a small number of currently unnamed species of *Haloarcula* isolated from fumerole steam discharged at 180°C (Ellis *et al.*, 2008). Subsurface archaean communities appear to thrive in the highly saline zone formed where water evaporates on contact with rocks heated by magma, in a hellishly hot and salty environment. Aside from heat-loving species, many Archaebacteria are salt-loving halophiles (e.g. *Halobacterium salinarium*), acidophiles (e.g. *Picrophilus oshimae*) or alkaliphiles (e.g. *Natronobacterium gregoryi*).

The proteins and tRNA of Archaebacteria have characteristic amino acid or nucleic acid sequences that ensure that they are folded in a manner rendering them extremely stable at high temperatures (Madigan *et al.*, 2003). Heat shock proteins are also extensively used to refold partially denatured proteins, and DNA appears to be stabilized by a number of processes, one of which – coiling of DNA by **reverse DNA gyrase** – appears to be unique to Archaebacteria. *Sulfolobus acidocaldarius*, a denizen of acid thermal streams, increases the melting point of its DNA by up to 40°C by investing in a protein called **Sac7D** which binds to DNA and maintains its structure (Madigan *et al.*, 2003). Furthermore, the membranes of these organisms are comprised of durable ether-linked lipids that lack heat-labile fatty acid tails (Langworthy *et al.*, 1974; Brock, 1978). These lipids, lacking tails, are non-polar and rather than forming the bilayer typical of most cell membranes form a heat-stable lipid monolayer (Langworthy, 1977). Thus Archaebacteria invest resources in cellular protection mechanisms that conform to S-selection, in which the adaptive response protects metabolic components against injurious stresses and the organisms are characterized by relatively slow growth in unproductive habitats.

Archaebacteria also occur in colder, higher latitude habitats, but have proven extremely difficult to isolate, culture and study in the laboratory. We currently know very little about these other than the fact that we can detect their characteristic ribosomal RNA signatures in seawater and soil samples from habitats that are not characterized by the same type of environmental severity required by extremophiles (Madigan *et al.*, 2003). These organisms are likely to exhibit very different adaptive strategies from their extremophile relatives. In the words of Robertson *et al.*, (2005): 'although it is apparent that *Archaea* can be found in all environments, the chemistry of their ecological context is mostly unknown'. For now the only example we can give is that of an unnamed species of the genus *Methanosarcina* known to be capable of dominating microbial productivity in extremely fertile agricultural soils, to the detriment of other Archaeobacteria (Gattinger *et al.*, 2007). However, nothing is known of the adaptive traits that characterize this high degree of C-selection. We shall return to this example in Chapter 5, and discuss the size and dynamics of archaean communities in habitats such as arable and pastoral soils.

Proteobacteria

The proteobacteria are one of the largest and most recently diverged groups of Eubacteria, and their taxonomic diversity reflects their ecological diversity. Some Purple Sulfur Bacteria are extremophiles with similar ecologies to many Archaebacteria (Madigan *et al.*, 2003), and are thus S-selected. For example, *Halorhodospira halophila* is one of the most halophilic (salt-loving) organisms known, and is native to chronically unproductive habitats such as soda lakes and marine environments with high salinities. Proteobacteria enter into a symbiosis with the Pompeii Worm in deep-sea hydrothermal vent habitats, and have thermally stable enzymes that maintain metabolic function over a range of temperatures, some of which can sustain activity for more than ten minutes at 70°C (Lee *et al.*, 2008).

Another group of proteobacteria, the *sulfur-oxidising chemolithotrophic prokaryotes*, are found mainly in highly productive habitats. *Beggiatoa alba* is found at neutral pH and is mesothermic, inhabiting stable, eutrophic habitats such as water polluted with sewage, the rhizosphere (the biotic zone surrounding plant roots) in flooded habitats, lake sediments, and the sea surface in zones polluted by agricultural runoff. This species is relatively large (with cells 200 μm in diameter, this is a giant among prokaryotes) with many wide cells connected end-to-end to form filaments that intertwine to create tufts resembling a fungal mycelium (Madigan *et al.*, 2003). These free-floating tufts are relatively buoyant and do not settle, and biomass may build up to the extent that sewage and industrial waste systems become clogged. Population growth does not involve simply the production of new cells that are dispersed into the surrounding environment; cells are retained within the zone of resource abundance by being incorporated into the filaments – filaments that cannot survive the disturbance of high flow velocities. These bacteria exhibit the high productivity and low investment in dispersal that is characteristic of C-selected organisms.

Indeed, as a general rule, it appears that by pooling the efforts of multiple cells – be it in the form of bacterial filaments, colonies, multicellular bodies or even social groups – resources are more easily encountered, monopolized and exploited, facilitating the survival of mutually shared genes in allied cells. The largest aggregations of cells (including multicellular bodies) in productive habitats do not simply require more resources in order to function – they function in order to acquire more resources[14].

In contrast, *Pseudomonas aeruginosa* is primarily a soil organism, but can proliferate on contact with animal wounds or burns (rich but ephemeral habitats for bacteria) to become an opportunistic pathogen (Madigan *et al.*, 2003). This species is mesophilic and cannot tolerate temperatures above 43°C, nor extremes of pH. When it infects the human respiratory tract it can form biofilms that are exceedingly difficult to remove. Should this happen in the lungs then cystic fibrosis may result. Biofilms are essentially structures that allow bacteria to attach

---

[14] One implication of this is that multicellular life may have first arisen in high productivity conditions in which a joint effort to acquire resources was a distinct survival advantage.

firmly to a solid substrate, particularly where high flow velocities could dislodge cells (Madigan *et al.*, 2003), and are thus adaptations to survive potential disturbances. The closely related *P. syringae* is a similar organism: an opportunistic plant pathogen that causes necrotic lesions on foliage, also surviving by proliferation. These bacteria are the epitome of organisms that do not dominate an ecosystem until an unusually rich opportunity arises, during which they rapidly multiply to such vast numbers that the ecosystem is in danger of collapse. This can be particularly unfortunate when the ecosystem which ceases to function is that of the cells comprising a human body. Thankfully R-selected disease organisms have little stress-tolerance and a simple response such as elevated temperature (fever) is often effective in their control. The economic and sanitary impact of R-selected species is immediately evident when we think that many weeds, pests and diseases are opportunistic and ephemeral – in a word, ruderal – organisms. However, as we shall see later, not all disease organisms are R-selected – the most persistent use S-selected adaptations to evade the immune system.

Firmicutes

Cellulose, being one of the main structural polysaccharides of all plants, is an extremely common and energy-rich molecule, and the firmicutes include important bacteria involved in cellulose degradation. Cellulolytic enzymes (cellulases) break down the cellulose to release sugars, which are then taken up by the bacterium and used as an energy source. This is, of course, the basis of the symbiosis between rumen bacteria and ruminants such as cattle, which would otherwise derive much less nutritional value from the grass and other plants they ingest. In the rumen, *Ruminococcus flavifaciens* actively competes for cellulose with a chytrid fungus, *Neocallimastix frontalis*. It does so by investing in the production of extracellular proteins that inhibit the cellulolytic activity of *N. frontalis* (Bernalier *et al.*, 1992, 1993). Thus *R. flavifaciens* survives due to an adaptation allowing it to suppress, via interference competition, other organisms.

However, the rumen may be a variable environment in which resource availability depends on the diet and activity of the host animal. Energy budgets and maintenance costs of organisms are also critical to rumen community structure. For example, *Butyrivibrio fibrisolvens* has a high **maintenance energy** expenditure, investing in maintaining metabolic activity in existing cells rather than simply producing more. This results in slower, steadier population growth than *Bacteroides ruminicola*, which, in contrast, invests most of its energy and resources into population growth and has high growth rates and yields when the host animal has a rich diet. Thus *B. fibrisolvens* has a more consistent yield over a range of resource availabilities, and is favoured when cellulose is less plentiful (Russell & Balwin, 1979). This maintenance strategy is directly analogous to a strategy of stress-tolerance in plants (Pierce *et al.*, 2005), allowing dominance when resources are scarce or inconsistently available.

Whilst many rumen firmicutes compete using high substrate affinities (rapid uptake of sugars and other resources) others, such as *Streptococcus bovis*, exhibit opportunistic behaviour based on high uptake capacities (uptake rates that are highest towards higher substrate concentrations). The high substrate capacity of

S. bovis is not usually employed, and the species bides its time in the background of the microbial community. However, whenever there is a sudden increase in starch concentrations in the rumen, such as when the animal is fed on grain, *S. bovis* has the greatest capacity to capitalize on this bounty, leading to a population explosion that can involve a doubling in population size every 20 minutes (Fenchel & Finlay, 1995). This results in acidosis, and an upset stomach can cause the cow to stop eating, eventually returning to a more natural diet – which is, of course, lethal to most of the *S. bovis* population. Thus although *S. bovis* coexists within the community by occupying ephemeral niches, it is capable of dominance on the rare occasions when extremely rapid growth and proliferation are favoured, mirroring population explosions and transitory dominance in other R-selected organisms.

Cyanobacteria

Thermosynechococcus elongatus is a relatively archaic cyanobacterium that lives in hot springs and has an optimum growth temperature of 55°C. The genome of this species has recently been mapped (Nakamura *et al.*, 2002), and shows a much wider range of genes (with respect to mesophilic cyanobacteria) encoding protective heat shock proteins, many of these genes being specific to *T. elongatus*. This species undoubtedly invests a significant proportion of resources into resisting the kind of environmental pressures that could potentially disrupt metabolism.

Other species that tolerate stress include the sugar-coated desert cyanobacterium *Microcoleus vaginatus*. Exopolysaccharides (sugars excreted into the environment) mop up the reactive oxygen species that form with exposure to UV-B radiation, ultimately protecting DNA and lipid molecules (Chen *et al.*, 2009). *Chroococcidiopsis* sp. and *Cyanothece aeruginosa* are endolithic species, with communities forming a layer half a centimetre under the surface of granite rocks in the exceptionally cold, dry Taylor Valley, Antarctica. These communities eke out a living by relying on desiccation tolerance to resist drought until dew occasionally activates metabolism, resulting in short periods of photosynthetic activity (Budel *et al.*, 2008) – these species undergo extremes of both injurious stress and resource availability, and grow slowly in a chronically unproductive habitat.

Low and infrequent resource availabilities have also influenced the evolution of marine cyanobacteria. *Richelia intracellularis* survives in the ultra-oligotrophic (extremely resource-poor) waters of the eastern Mediterranean Sea using specialized heterocyst cells to fix atmospheric nitrogen, but the lack of other nutrients, particularly phosphorus, means that blooms of this species do not occur (Bar Zeev *et al.*, 2008). *Calothrix elenkinii*, in the Alberche River, Spain, is another nitrogen-fixer that survives low phosphorus concentrations using a mechanism described by Mateo *et al.* (2006) as **luxury consumption** – when pulses of high phosphorus concentrations do occasionally become available *C. elenkinii* is capable of rapid uptake and storage of this nutrient for leaner times. This ability is exhibited to a much lesser extent by *Nostoc punctiforme* in the same river system, which simply dies off during periods of phosphorus limitation. Even different cyanobacterial species within the same genus may exhibit a range of ability to cope with extreme environments, such as mesophilic and thermophilic species of *Oscillatoria* (Seckbach, 2007). These examples demonstrate that the stresses encountered by cyanobacteria encompass a range of different mechanisms, involving low resource availability and/or direct injury to metabolism, but adaptation consistently involves safeguarding metabolic performance in unproductive habitats – i.e. the nature of stress and stress adaptations are conceptually identical to those in plants (Pierce *et al.*, 2005).

In contrast, rice paddies are eutrophic (resource rich) but highly variable environments in which the cyanobacterial community changes throughout the year in response to seasonal flooding. The addition of sewage sludge can also prompt the growth of seasonal members of the cyanobacterial community (El Sharkawi *et al.*, 2006). In China, species of *Leptolyngbya* are present throughout the year, but other cyanobacteria, such as *Synechococcus* species, have an ephemeral lifestyle and are present only when the rice paddies are flooded (Song *et al.*, 2005). This genus includes a range of forms with contrasting ecologies, and the particular species of *Synechococcus* inhabiting rice paddies have so far only been identified from their DNA, and are currently unnamed. Little is known of the ecology of these species, but the genus as a whole generally requires high nutrient concentrations and high light intensities, and is usually restricted to surface waters in marine ecosystems (Ma *et al.*, 2004) (a notable exception being the S-selected *S. lividus* inhabiting hot spring effluents; Brock, 1978).

In temperate and subtropical marine ecosystems, *Synechococcus* co-occurs with another cyanobacterium, *Prochlorococcus*, but the latter is more abundant and distributed more evenly within the water column. Indeed, it has been estimated that as many as 10^{25} individual *Prochlorococcus* cells probably exist in the world's oceans (Partensky *et al.*, 1999). *Synechococcus* is not as abundant, and indeed exhibits a far greater seasonal variation in abundance and is particularly prevalent in nutrient-rich upwellings and river '**plumes**' that bring nutrients to the ocean (Paul *et al.*, 2000). Factors such as viral mortality and grazing may also be involved in fine-tuning the balance in abundance of these two cyanobacteria (Campbell & Vaulot, 1993; Hirose *et al.*, 2008). In general, *Synechococcus* exhibits a greater range of growth rates, and a greater maximum growth rate that is positively correlated with grazing intensity (Agawin & Agusti, 2005). Thus *Synechococcus* occupies a more marginal and inconsistent niche in which it is more prone to lethal events, to which it is adapted by higher rates of regeneration.

However, it would be an exaggeration to state that *Synechococcus* and *Prochlorococcus* have highly contrasting adaptive strategies. Indeed, the two genera share similar habitats and are exposed to similar selection pressures (Agawin & Agusti, 2005). They are both extremely small (in terms of both genome and physical size; Dufresne *et al.*, 2003), fast growing and taxonomically very closely related. Being at the base of most marine food chains, with the combined weight of life from viruses to albatrosses and blue whales all ultimately gaining their carbon from these primary producers, these marine cyanobacteria are clearly subject to a high frequency of mortality, and are highly R-selected. They do, however, occupy slightly different niches, usually based on subtle differences between a few traits rather than the adaptive strategy as a whole. As we shall see in the next chapter, differences in single traits have a role to play in fine-tuning the coexistence of *Synechococcus* and *Prochlorococcus*, whilst differences in strategy become apparent in particular regional conditions of resource depletion in which S-selected cyanobacteria dominate.

Primary production by cyanobacteria in freshwater habitats is dominated by species that may be functionally and ecologically very different to *Synechococcus* and *Prochlorococcus*. *Oscillatoria* species dominate stable, eutrophic lakes polluted with organic matter, using similar adaptations that we saw previously for the proteobacterium *Beggiatoa alba* – large colonies of multicellular filaments allow resources to be intercepted and monopolized. Resource acquisition is also facilitated by motility, with the filaments able to pull themselves along surfaces or use gas vesicles to modify bouyancy (Konopka, 1982), migrating upwards in the water column to absorb light, and downwards to acquire nutrients – thereby maximising resource uptake and minimizing exposure to potentially damaging high light intensities at midday (Garcia-Pichel *et al.*, 1994). This active foraging behaviour allows *Oscillatoria rubescens* to out-compete a range of other bacteria for organic nitrogen sources (Feuillade *et al.*, 1998). *Oscillatoria* is intolerant of stress: *O. salina* exhibits reduced motility in response to a range of factors, and is extremely sensitive to temperature shock, extremes of temperature and pH, light intensity and the presence of heavy metals (Gupta & Agrawal, 2006). *Oscillatoria* species are thus good candidates for C-selected cyanobacteria.

Viruses

Viruses have one important feature in common with cellular organisms that suggests that they too can be considered as life-forms: a gene-based system of heredity on which natural selection can act. Indeed, like cellular organisms an evolutionary tree of life is evident in the similarities and dissimilarities between viral genomes (Fauquet *et al.*, 2005). Genes are also encoded in DNA or RNA in essentially the same manner as cellular organisms. Raoult & Forterre (2008) have proposed the following definition of a virus as 'a capsid-encoding organism that is composed of proteins and nucleic acids, self-assembles in a nucleocapsid and uses a ribosome-encoding organism for the completion of its life cycle'. Although the origin of viruses remains unclear and is hotly debated, it is evident that the three main domains of cellular life are associated with their own characteristic viruses, suggesting that viruses exhibit an ancient divergence following the evolution of the three domains (Prangishvili *et al.*, 2006).

The minimum genome for a virus consists of just four genes (e.g. *Enterobacteria phage MS2*; Fauquet *et al.*, 2005). Small genomes can be replicated rapidly, before the immune system can respond, and so genes for defence and long-term persistence are less important to the survival of these viruses. Many R-selected multicellular parasites show a similar **hit-and-run** strategy – a massive proportion of host cell resources is invested in viral replication, and the individual virus particles are relatively ephemeral. Like many of the R-selected bacteria, R-selected viruses are opportunistic and capable of population explosions that threaten the stability of the cellular ecosystem – a mode of growth that can, of course, lead to disease or death even in large multicellular eukaryotic hosts. Viruses thus include the most extreme ruderals that can exist.

Viruses do nonetheless show a range of adaptive strategies.

Viruses of the families Fuselloviridae, Lipothrixviridae and Rudiviridae specialize in infecting extremophile Archaebacteria such as *Sulfolobus*, but in this high temperature environment most of the viral life cycle must take place in a **stable carrier state** within the safety of the bacterial cell membrane (Prangishvili & Garrett, 2004). The host membrane, as discussed above, is extremely resistant to high temperatures, as is the cellular machinery that is hijacked for replication (Brock, 1978). The fuselloviruses, which usually occur in a plastid-like state in the cytoplasm, can even integrate themselves into the host genome and, amazingly, can then separate themselves out from the bacterial genome later on (Prangishvili & Garrett, 2004). The majority of these viruses do not cause lysis and death of the bacterial cell but are released without damaging it. This method of hiding from harsh environmental conditions represents a conservative comportment, lacking the aggressive and destructive mode of replication that characterizes many viruses. The virus particles experience relatively high temperatures without undergoing catastrophic changes in conformation, and are actively involved in influencing host metabolism at high temperature. They can thus be considered equivalent to cellular organisms that grow and complete the life cycle in challenging environments.

This ability to hide without causing the cell to split appears to be a general rule for Archaebacteria-infecting viruses, because it 'reduces the possibility of direct exposure of a virus to the harsh environmental conditions' (Prangishvili *et al.*, 2006). A small number of viruses, such as the filamentous *Thermoproteus tenax virus* (TTV1; Lipothrixviridae) are even present in the environment, infect bacterial cells, replicate and cause cell lysis, all at high temperature (Prangishvili & Garrett, 2004). By these criteria the Archaebacteria-infecting viruses can be viewed as relatively S-selected. Escaping from environmental adversity by incorporation into the host cytoplasm or genome is essentially the same method that viruses such as *Herpes simplex virus 1* (HSV-1) use to hide from the immune system during chronic infections – the most persistent of diseases appear to be S-selected.

This dichotomy between acute and persistent viral infections is complicated by the fact that some viruses, such as avian influenza, cause low-intensity chronic infections in one species but can switch to acute, ephemeral infections when they jump to new host species. This led Villareal *et al.* (2000) to reject the concept of *r*/*K*-selection in viruses. However, we suggest that there are nonetheless examples of primary strategies, exhibited by viruses such as the Archaebacteria-infecting viruses, in which the virus is known to live only in one particular extreme mode. It is possible that viruses capable of shifting between either relatively acute or persistent life-histories depending on the host may have a somewhat intermediate strategy.

We also suggest that a third primary strategy is apparent in viruses: one that allows competitive dominance by relatively C-selected viruses over more ephemeral viruses. Giruses, the giant viruses (Claverie *et al.*, 2006), are large enough to be confused with bacteria when viewed under the microscope. *Acanthamoeba polyphaga mimivirus* (APMV) has the largest known viral genome[15], with 1.2 million base pairs and 911 genes that code for proteins which are used to persuade the metabolism of the host amoeba *Acanthamoeba polyphaga* to establish extensive 'virus factories' within the cytoplasm, where replication and viral assembly take place (Claverie *et al.*, 2006; Suzan-Monti *et al.*, 2007). Amoebas are extremely large cells that represent a particularly rich habitat for viruses. Indeed, the largest habitat available to viruses, as parasites of cells, consists of the largest cells, and this sets an upper limit on the absolute size of viruses, embodied in APMV. However, the C-strategy is not about being big *per se* in absolute terms, but about being able to acquire and monopolize resources within the habitat. APMV is adapted to manipulate host metabolism in a manner that, by producing virus factories, allows maximal control over resources and optimisation of viral assembly – something that other, smaller, viruses infecting the same host presumably cannot achieve on the same scale. In this sense APMV invests the resources at its disposal in a similar manner to C-selected multicellular organisms, in the context of its habitat and the potentially cohabiting organisms against which it must compete, despite being orders of magnitude smaller than a whale or a large plant. The second largest viral heavyweight, *Cafeteria roenbergensis* virus (CroV) similarly infects a relatively large single-celled organism (the extremely widespread heterotrophic flagellate from which it takes its name) and also has a relatively large genome of around 730,000 base pairs, including 544 genes that appear to code for proteins (Fischer *et al.*, 2010).

Viruses do compete with one another. In 90–99 per cent of cases in which a cell is simultaneously exposed to two viruses only one is able to replicate (Delbrück, 1945; Chase *et al.*, 1989) – a phenomenon known as 'mutual exclusion'. Mutual exclusion occurs early in infection (a 5–15-minute headstart is enough to ensure dominance by one virus) with the first virus to enter altering the permeability of the cell membrane and thus the ability of subsequent viruses to gain entry (Greiner *et al.*, 2009). However, this is unlikely to be the whole story, as even excluded viruses can have a negative impact on the replication of the initial, intracellular virus. This 'depressor effect' appears to be due to viruses at the cell surface pilfering resources; resources which Delbrück (1945) hypothesized were located just inside the membrane:

> . . . mutual exclusion is caused by impermeability of the cell membrane induced by the first virus particle which penetrates the membrane . . . The virus which is barred from entry into the cell can nevertheless compete with the intracellular virus for a common substrate. It converts the substrate in an irreversible reaction into a product which is characteristic for each virus. This competition for substrate is the cause of the depressor effect which occurs only between dissimilar viruses.

Resources are critical to virus life cycles. During virus replication the capacity of the host cell to synthesize proteins is pushed into overdrive and resources are quickly exhausted (Eigen *et al.*, 1991; Regoes *et al.*, 2005). The rate at which this occurs depends on the physical size of the cell and the pool of resources

---

[15] As the book went to press *Megavirus chilensis* was described (DOI:10.1073/pnas.1110889108) with a slightly larger genome but a similar ecology.

available to protein synthesis (Hadas *et al.*, 1997). It is thus possible that mutual exclusion involves the rapid establishment of a monopoly over host cell resources, and adaptations to aid this conceivably include traits such as virus factories that consolidate the monopoly, increasing the probability of a successful competitive outcome.

One of the most interesting aspects of research on mutual exclusion is that it has been carried out mainly for the 'giant *Chlorella* viruses' (Phycodnaviridae), so called because they infect the green alga *Chlorella*. These have relatively large genome sizes that, like APMV, encode hundreds of proteins that may influence host metabolism and produce factories that optimize the conversion of host resources into new virus particles[16]. The most widely studied of these giant viruses, *Paramecium bursaria* chlorella virus 1 (PBCV-1), has a genome size of 330,000 base pairs that encode approximately 373 proteins – this is small relative to APMV but nonetheless giant for most viruses and is equivalent to the smallest of bacteria (Van Etten, 2003). Having a large genome size also allows a degree of protection against stress, as host metabolism may be manipulated to protect PBCV-1 against the damaging effects of UV light and the virus may occasionally exist in a stable carrier state within active algal cells (Van Etten, 2003) although, unlike the S-selected Archaebacteria-infecting viruses, this does not appear to be a typical feature of the PBCV-1 life cycle.

Many unanswered questions remain with regard to the nature of competition and the function of large size in viruses. How can viruses on the cell surface compete for the same substrate as intracellular viruses and use this substrate when they lack the cellular machinery required for the expression and functioning of their genes? Is large size, by allowing protection and maintenance of virus function, actually an S-selected trait in viruses (is a large viral genome composed mainly of genes involved in maintenance or alternatively resource sequestration and optimisation of replication)? For example, many of the genes of *Cafeteria roenbergensis* virus are thought to be involved in DNA repair, although the function of a large proportion of genes is unknown (Fischer *et al.*, 2010). The natural history of giant viruses in general is also enigmatic. For example, *Chlorella* viruses only infect particular strains of *Chlorella* that exhibit mutualistic relationships with aquatic animals, but paradoxically *Chlorella* remains isolated within the animal and is defended against the virus – how do *Chlorella* viruses persist in nature (do other hosts exist, or can symbiotic *Chlorella* exist without an animal host; Van Etten, 2003)? Is there a relationship between the amount of resources provided by the *Chlorella*/animal symbiosis and the lifestyle and large size of these viruses?

In Chapter 5 we shall see that the study of microorganisms in the context of their natural habitats is a burgeoning field of research. Virus ecology is even more complex than that of cellular microorganisms, as virus/host/host-of-the-virus-host ecologies may be nested together in the style of matryoshka dolls, with the habitat of the host of great relevance to the context of the virus. Multiple scales of investigation are required. Whilst intrepid virologists sally-forth to answer these questions, we must be content to hypothesize that giant viruses

---

[16]In contrast to APMV, *Chlorella* virus factories provide solely an assembly centre for components produced in the host cell nucleus, rather than being the site of both replication and assembly.

invest resources in a manner that mainly allows further resource sequestration and territorial (host cell) control; they appear to be good candidates for relatively C-selected capsid-encoding organisms, conceptually equivalent to C-selected ribosome-encoding organisms.

Extinct groups

We can never be absolutely certain how extinct organisms lived, for the simple reason that we cannot go out into the field and observe them. A number of biologically derived molecules have been found that originally formed part of the bones of a *Tyrannosaurus rex*, the hadrosaur *Brachylophosaurus canadensis*, and the feathers of *Shuvuuia deserti* (Schweitzer *et al.*, 1997, 1999, 2009). We can even tell, by the fine structure of fossilized feathers, that the tail of the dinosaur *Sinosauropteryx prima* probably had alternating chestnut and white stripes (Zhang *et al.*, 2010). Amazing as such insights are they represent tiny windows into the biology and ecology of these extinct animals. Very often fossil evidence is based on small numbers of specimens and thus involves a large degree of uncertainty. For this reason we have decided to present this section on extinct organisms separately from extant organisms. Much of the evidence presented in the following paragraphs should be viewed with the following caveat in mind: 'this is our best estimate based on insufficient specimens to allow statistical analysis', or in other words, 'possibly'. In attempting to interpret the overall adaptive strategies of these organisms our aim is simply to determine whether or not it is realistic to do so with the available evidence.

Non-avian dinosaurs
Most evidence for the adaptive strategies of non-avian dinosaurs is, of course, derived from fossilized bones, and two aspects of bone microstructure tell us how rapidly and consistently dinosaurs grew: the degree of vascularization (rapid growth requires a good blood supply) and lines of arrested growth (LAG; equivalent in many respects to tree rings). There is some uncertainty in the use of LAGs to calculate dinosaur growth rates and ages, as different bones in the same individual may tell different stories (it is best to investigate as many as possible) and it is often assumed that each year is represented by one LAG, but this is not necessarily the case; Horner *et al.*, 1999, 2000). Furthermore, estimates of dinosaur mass used to calculate growth rates should be interpreted with caution (Packard *et al.*, 2009). However, it is clear that different dinosaur species had bones with contrasting microstructures, and that these microstructures are typical of different growth strategies in modern animals such as crocodiles, birds and mammals.

Thus Padian *et al.* (2004) cautiously used LAGs to compare the life-history strategies of different dinosaurs. They suggested that large 'typical dinosaurs' such as *Allosaurus fragilis* and the sauropods exhibited extremely rapid growth until maturity, at which point growth slowed dramatically. This contrasts with the smaller dinosaurs that grew more slowly, but these can be divided into two groups: those that continued growing for an extended period (e.g. *Scutellosaurus*), and those that quickly attained maturity, and were thus relatively small as adults. For this third strategy, reproductive maturity could be attained within as little as a single year, for example for the proto-bird *Confuciusornis sanctus*. Let's consider each of Padian *et al.*'s three strategies in turn.

LAGs in *A. fragilis* femurs of various sizes suggest that this species could grow as rapidly as 148 kg per year, probably reaching a maximum age of 22–28 years (Bybee *et al.*, 2006). A cast from the inside of a fossil *A. fragilis* skull also tells us that the brain was more like that of a crocodile than that of a bird, having a larger region for accepting sensory data than for deciding what to do with it (Rogers, 1999). Thus *A. fragilis* probably snapped at anything that moved rather than being discerning in its tastes, but unlike crocodiles it was an agile biped that was more capable of actively searching for food. The life-history traits of large size, rapid growth rate and active, frenetic feeding make *A. fragilis* a good candidate for a C-selected dinosaur. Similar life-history traits are also evident for other giant carnivorous theropods such as *Acrocanthosaurus atokensis*, *Carcharodontosaurus saharicus*, *Gorgosaurus libratus*, *Giganotosaurus carolinii*, and *T. rex* (Erickson *et al.*, 2004).

Surprisingly, evidence from LAGs suggests that the lumbering vegetarian sauropods were also good candidates for C-selected organisms. We know that they lumbered because their fossil footprints are always closely spaced, but plodding was due to their tremendous weight and not because their metabolisms were slow. Indeed, sauropod bones reveal that these animals probably had a highly efficient respiratory system like that of birds (Wedel, 2003).

In birds, a network of air sacs extends from the lungs, even occupying cavities within the bones (it is these cavities in sauropod bones which suggests they used the same system). The lungs themselves are relatively immobile; it is the air sacs that expand like bellows during inhalation, pulling air into the entire system. Air arrives at sacs behind the lungs via one system of bronchi that bypasses the gas-exchange surfaces of the lungs. During exhalation this air is forced out through the lungs, meaning that fresh air is kept flowing constantly through them, and the exchange of gases between the air and the blood is extremely efficient. Using this system sauropods may have oxygenated their blood both when they breathed in and when they breathed out, quite unlike mammals. Also, evaporation into air sacs cools organs from inside the body in a much more direct manner than is possible with mammalian sweating.

This system probably allowed sauropods to avoid the formation of respiratory dead space in their long necks, which in turn allowed a uniquely efficient method of acquiring resources: small heads cantilevered out over the vegetation on long necks, balanced by long tails, with stomach-stones (also known as '**gastroliths**') for grinding, rather than the use of muscular heads and tough teeth to grind food. This allowed them to stand on one spot and continuously rake their peg-like teeth through the surrounding vegetation without pausing, feeding incessantly to fuel the growth of gargantuan bodies. They were large for the simple reason that a massive gut allowed greater absorption of nutrients from plant matter. Sauropods probably gained few nutrients from cycads, tree ferns and most conifers, but plants of the genera *Equisetum*, *Araucaria* and *Ginkgo* are energy rich and can yield sufficiently large amounts of nutrients following a long fermentation period, requiring a protracted digestive process in a long gut, and thus a big body (Hummel *et al.*, 2008). However, these dinosaurs were particularly massive even compared to the largest of modern terrestrial herbivores, which was probably the result of higher Mesozoic atmospheric CO_2 concentrations and CO_2-fertilization of plant growth, which allowed greater productivities than those possible under current CO_2-limited conditions (Burness *et al.*, 2001).

Precisely how fast and how massive sauropods were capable of growing is a matter of debate, as contrasting methods yield estimates that differ by an order of magnitude (e.g. growth rates of between 520 kg and 5000 kg per year for *Apatosaurus ajax*; Lehman & Woodward, 2008), but even minimum estimates are rapid and leave little doubt that these animals depended on high metabolic rates. The weight of sauropods is now thought to have been overestimated by traditional statistical techniques, and although these animals were certainly massive, species such as *Apatosaurus louisae* probably weighed only around three times more than a bull African Elephant (Packard *et al.*, 2009) – a value that is more in fitting with their bird-like bone structure and probable physiology. The similarity with C-selected plants is striking, the strategy being essentially based on a large body size that allows the most efficient assimilation of resources and fast growth rates.

Sauropod reproduction was also equivalent to that of C-selected plants – the total biomass invested in each clutch represents a small proportion of the tonnes of adult biomass that laid them. Sauropod egg volume has been calculated as less than 1 litre (Chiappe *et al.*, 1998), and eggs were probably laid in earth depressions perhaps covered by soil or vegetation, with no parental care evident (Horner, 2000). Sauropods probably did not allocate a large proportion of their annual efforts to reproduction. The survival strategies of massive sauropod species – including overall form, the way they grew, moved and ate – were all based on acquiring resources and investing these in the ability to acquire more. This is highly suggestive of C-selection.

The efficient avian respiratory system, albeit in a relatively primitive form, appears to have been an early feature of Saurischians (sauropods and theropods; O'Connor & Claessens, 2005; Sereno *et al.*, 2008), but was *not* shared by the Ornithischian dinosaurs. Basal Ornithischian species such as *Scutellosaurus lawleri* grew slowly and steadily, more like crocodiles than the giant dinosaurs (Padian *et al.*, 2004). Intriguingly, a small, primitive Ornithischian, *Oryctodromeus cubicularis*, has been found fossilized within a burrow alongside two well-developed young. This behaviour has been interpreted in light of the denning behaviour of extant bird and mammal species as an adaptation for the parental care of juveniles in a harsh environment (Varricchio *et al.*, 2007) – the dinosaur equivalent of the Alpine Marmot. Similar fossilized dinosaur burrows have recently been found in what would have been a southern polar environment, suggesting that some species were capable of using these extended phenotypes to endure the cold, dark winter months during which vegetation would have been scarcer (Martin, 2009b).

The more advanced Ornithischians were characterized by a preponderance of spines, bony plates and thumb-spikes that could have been used, at least in part, for defence (although many frills, bumps, lumps and spikes are thought to have emerged mainly from sexual selection, i.e. evolution driven by intraspecific rather than interspecific interactions, and were too brittle or rich in blood vessels to afford effective defence). CSR theory predicts that S-selected organisms grow slowly and invest a substantial proportion of resources into physical or constitutive (inherently tough) defences to protect their hard-won investments. By this measure the ankylosaurs were probably the most S-selected of all dinosaurs. *Euplocephalus tutus* was typical, having such a heavily armoured skull that the brain seems to have been almost an afterthought. It even had reinforced eyelids. Many were relatively small, at 2.5–3 m in length, and had various configurations of spikes and bony osteoderms erupting from the skin, similar to the dermal scutes of crocodiles (e.g. *Gargoyleosaurus parkpinorum*, *Minmi paravertebra*). Larger ankylosaurs, at 5–10 m, possessed an additional bony club at the end of the tail (e.g. *Ankylosaurus magniventris*, *E. tutus*, *Pinacosaurus grangeri*, *Saichania chulsanensis*, and *Talarurus plicatospineus*). There is even some fossil evidence to suggest that *P. grangeri* had glands involved in salt excretion (Hill *et al.*, 2003; although this is a case in which the caveat 'possibly' must be applied). However, there can be little doubt that in these cases the animals were well defended. Many stegosaurid dinosaurs had rows of sharp spines on their backs (e.g. *Dacentrurus armatus*) or on their flanks and tails (*Chialingosaurus kuani*, *Huayangosaurus taibaii*, *Kentrosaurus aethiopicus*), although in some, such as *Stegosaurus armatus*, *S. stenops* and *Wuerhosaurus homheni*, defensive spikes were relegated to the tail alone, forming a **thagomizer**[17]. *Stegosaurus stenops* even had bony chainmail-like armour protecting the throat.

Aside from the large, fast-growing typical dinosaurs and the smaller armoured dinosaurs with crocodile-like growth Padian *et al.* (2004) also recognized the existence of small, early maturing species, which are the best candidates for R-selected dinosaurs. These include the ancient Triassic theropod *Coelophysis bauri*, and more derived Cretaceous non-avian theropods such as *Anchiornis huxleyi*, *Compsognathus longipes*, *Oviraptor philoceratops*, *Sinosauropteryx prima*, *Microraptor zhaoianus* and *Troodon formosus*, and the early avian theropods *Archaeopteryx lithographica* and *Confuciusornis sanctus*. Some of these, such as the 1.7 m-long *C. bauri* were not the smallest of creatures in absolute terms, but they had an extremely lightweight, bird-like construction. Animals such as *C. longipes*, *M. zhaoianus* and *Caudipteryx zoui* probably weighed only two or three kilograms (Padian *et al.*, 2004) and could have attained this mass extremely rapidly. *Limusaurus inextricabilis* was small and numerous enough that groups of them literally could not extract themselves from volcanic mud accumulated, probably, in the footprints of one of their giant sauropod cousins (Eberth *et al.*, 2010). The smallest non-avian therapod, *Anchiornis huxleyi*, is estimated to have been just 34 cm in length and 110 g when mature (Xu *et al.*, 2009) – this is small by vertebrate standards in general.

---

[17]The word 'thagomizer' was coined by Gary Larson in one of his *Far Side* cartoons – it was the deadly weapon encountered by the unfortunate caveman Thag Simmons. Mr Larson has, of course, used artistic licence – he is fully aware that ancient hominids and non-avian dinosaurs did not occur during the same period (Larson, 1992). The word, however, has been lovingly adopted by palaeontologists and is now a sensible palaeontological term.

It is strongly suspected that some of these smaller dinosaurs were gregarious as a defence against predation, as many had no innate defences and fossils are often found in groups (e.g. Kobayashi & Lü, 2003; Eberth *et al.*, 2010). A high proportion of fossilized juveniles compared to adults suggests that juvenile mortality was a particularly strong selection pressure for species such as *Sinornitho-mimus dongi* (Kobayashi & Lü, 2003). These smaller dinosaurs also include the most prolific egg-layers, with the greatest investment in pre-hatching reproductive behaviours: *O. philoceratops* laid between 20 and 36 eggs in neat circles or spirals, sometimes in more than one layer (Horner, 2000 and references therein). Fossils of adult *O. philoceratops* sitting on nests containing *O. philoceratops* embryos strongly suggest that this species used active incubation or brooding behaviour, like many modern birds (Norell *et al.*, 1995; Dong & Currie, 1996). *T. formosus* also invested in nest structure, clutch arrangement and size, laying between 12 and 24 eggs in a rimmed earth nest, which may have reduced the threats of flooding and predation, and even of injury to hatchlings by parental trampling. Eggs were half-buried and orientated to stand on the pointed end, and were thus held in place by the soil with the exposed portion maintaining contact with a brooding parent; fossils have been found showing parents apparently caught in the act (Varricchio *et al.*, 1997, reviewed by Horner, 2000). The simple fact that the most detailed information we have regarding dinosaur reproduction is evident for these smaller theropods suggests that these particular species invested a great deal of time and effort in regeneration – they were apparently more likely to be killed and fossilized during this phase of the life cycle.

It has been suggested (Martin, 2009b) that burrowing dinosaurs may have been able to hide and avoid the disturbances of the Cretaceous/Paleogene (K/Pg) extinction[18]. However, we only have fossil evidence for burrowing behaviour in early Ornithischians, whereas it was the smallest Saurischian species that actually survived, with their descendents then diversifying so successfully that they now flit between trees all over the planet. It is tempting to ascribe survival of the K/Pg extinction to some aspect of physical durability, such as having insulating feathers or hair and being able keep warm during a global winter, but this does not explain why the larger feathered dinosaurs (such as many dromeosaurids), hairy archosaurs (including pterosaurs) and the really tough, long-lived forms such as ankylosaurs became extinct whilst the smallest species did not. The implication is that insulation or physical toughness may not have been central to the continuation of terrestrial species during and immediately after the K/Pg extinction.

What may have really counted, after hiding from the event itself, was the ability to capitalize on the vacant niches available during the aftermath using an adaptive strategy based on opportunism and rapid repopulation. The animals best suited to do this were strongly R-selected mammals and avian

---

[18] Formerly known as the K/T, or Cretaceous/Tertiary, extinction. Incidentally, a broad range of evidence accumulated over decades now provides an extremely high degree of confidence that the K/Pg extinction was triggered by the Chicxulub meteorite impact (Schulte *et al.*, 2010).

dinosaurs. Being small, fluffy, reproducing like rabbits and equipped to eat small organisms such as detrivorous invertebrates was precisely the syndrome of traits that would have been important for hiding, repopulating and then diversifying to eventually form the dominant elements of the Paleogene fauna. Productive habitats, in which large size is an immediate advantage for acquiring resources, are a product of environmental stability. Large-scale environmental disturbances spell disaster for large C-selected organisms, but are a boon for R-selected organisms. Had any giant dinosaurs been able to survive the K/Pg event itself not only would their particular food-chains have been interrupted, they would have been over-run by population explosions of small mammal and bird species great enough to make the ten plagues of Egypt look like minor inconveniences. The ascendance of the mammals and birds was truly meteoric.

In conclusion, it is not unreasonable, based on available fossil evidence, to suggest possible adaptive strategies for non-avian dinosaurs, and these appear to be in broad agreement with C-, S- and R-selection. Our view of dinosaur adaptive strategies will undoubtedly become more detailed as more fossils are found and our knowledge of probable behaviours expands.

Pterosaurs

Padian *et al.* (2004) used similar analyses of fossilized bones to determine that pterosaurs, the Mesozoic flying reptiles, had similar adaptive strategies to dinosaurs. Small, relatively ancient species such as *Eudimorphodon* and *Rhamphorhynchus* had slow and steady crocodile-like growth. In contrast, larger pterodactyloid pterosaurs from the Late Cretaceous had highly vascularized tissues that resemble those of ducks and hawks (Riqlès *et al.*, 2000), suggesting rapid growth. A bird-like system of air sacs, and thus the potential for efficient bi-directional lung ventilation, was also more prevalent among the larger, derived pterosaurs, and appears to have evolved more than once to allow the giant but lightweight construction of the ornithocheiroids (e.g. *Anhanguera santanae*) and the azhdarchoids (e.g. *Quetzalcoatlus northropi*; Claessens *et al.*, 2009).

These growth/physiological traits could be interpreted as evidence for variability between S- and C-selection in pterosaurs. Unfortunately, there is little further evidence, particularly with regard to reproductive traits, and there is currently nothing to suggest the past existence of R-selected pterosaurs. A lack of R-selected species could explain why no pterosaurs survived the K/Pg extinction.

Universal adaptive strategy theory – the evolution of CSR and beyond *K* theories

We have just explored some of the evidence suggesting that adaptive solutions conforming to C-, S-, and R-adaptation are evident in a range of taxonomic groups throughout the tree of life. It is evident that some clades include the full range of adaptive strategies, but others contain C- and S-selected species but not R strategists, such as the universally long-lived and low fecundity sharks. Various clades have been investigated and classified within the context of a three-strategy model by ecologists that specialize in these groups: e.g. fish, corals (and dino-flagellate algae), fungi (including lichens), echinoderms and various insect groups (Rogers, 1988; Lawrence, 1990; Winemiller & Rose, 1992; Winemiller, 1992, 1995; Elliott *et al.*, 2001; Smayda & Reynolds, 2001; Regel, 2003; Murdoch, 2007; König-Rinke, 2008). Several ecologists have independently arrived at the conclusion that CSR theory can help to make sense of apparently chaotic eco-systems. Additionally, it is reasonable to interpret the adaptive strategies of extinct groups such as non-avian dinosaurs and pterosaurs in terms of C-, S- and R-selection.

Therefore, C-, S-, and R-selection appear to have been important throughout the evolution of life on Earth, and our hypothesis is broadly supported.

A **universal three-way trade-off** constrains adaptive strategies throughout the tree of life, with extreme strategies facilitating the survival of genes via: (C). the survival of the individual using traits that maximize resource **acquisition** and resource control in consistently productive niches, (S). individual survival via **maintenance** of metabolic performance in variable and unproductive niches, or (R). rapid gene propagation via rapid completion of the life cycle and **regeneration** in niches where events are frequently lethal to the individual.

Adaptations to C-, S- and R-selection facilitate three main functions: **acquisition, maintenance** and **regeneration**, respectively. The presence of these three main adaptive responses throughout the tree of life suggests that CSR theory is more than just a theory of plant strategies. Although the similarities between the *rKA*-templet and CSR theory have been pointed out by other authors, the terms *r*- and *K*- are actually population characteristics and do not describe either the selective forces influencing evolution nor the adaptive responses of organisms. In contrast, the terminology of C-, S- and R-selection is based on the evolution-ary processes that shape adaptive strategies and is more appropriate. C-, S- and R-selection thus form the basis of **universal adaptive strategy theory** – a natural extension of CSR plant strategy theory.

First steps towards a universal methodology

Are there common traits that can be used to classify all organisms according to universal adaptive strategy theory? Here we will briefly discuss the use of mul-tivariate analysis of functional traits for the classification of a range of organisms throughout Darwin's tree, and make the first tentative steps towards a **universal adaptive strategy classification**.

It is not our aim to provide, by the end of this book, a fully functioning methodology that field ecologists can employ to classify all of the organisms they are confronted with. Indeed, it took Linnaeus and all his students a lifetime of work simply to name all the organisms known at the time, and the additional work of quantifying ecologically relevant traits for these organisms would be a colossal task. The current fashion is for research groups in different countries to work together to produce databases of functional traits, which are then sub-jected to multivariate analyses in order to detect the principal directions of specialization (e.g. Díaz *et al.*, 2004; Wright *et al.*, 2004). We suggest that a universal classification system can only be achieved by such large-scale collaboration. It is certainly beyond the capabilities of a couple of botanists, and our aim is merely to get the ball rolling.

Common functional traits that have been used to compare birds and mammals include age at first reproduction, fecundity and adult life expectancy (Gaillard *et al.*, 1989). Of course, phylogenetic constraints may bias such a simple approach to using traits – most birds, for example, have a size and mass that is constrained by adaptation for flight, whereas mammals have a much greater size range (Gaillard *et al.*, 1989); whales are supported by a denser medium than air and have not encountered the same evolutionary size constraints as birds. Equally, although C-, S- and R-adaptation are apparent throughout a range of organisms, it is not immediately obvious how an R-selected rumen bacterium, for instance, can be measured and classified on an equal footing with an organism such as the European Harvest Mouse or the Southern Calamary.

We would argue that although phylogenetic constraint has undoubtedly restricted the range of adaptive strategies possible in particular clades, this should not stop us searching for trends in adaptive strategies within and between clades. Many Juncaceae are S-selected plants, but this does not in any way invalidate the fact that each particular species is adapted to survive in a harsh environment and has an S-selected function in its native ecosystem. The evolution of *Streptococcus bovis*, *Micromys minutus* and *Sepioteuthis australis* has clearly occurred within different phylogenetic constraints – one is a prokaryote, one a mammal and one a mollusc – but what matters for their ecology is not the restrictions that have bounded their past evolution but the similarities in the way in which they survive and their equivalent comportment. For comparative purposes, it is not the details of particular phenotypic adaptations that matter *per se*, but the equivalent manner in which genes are implemented and propagated, and how rapidly and economically resources are acquired, processed and invested. We have seen the evidence that *Streptococcus bovis*, *Micromys minutus* and *Sepioteuthis australis* exhibit short generation times, compared to other firmicutes, Rodentia and Cephalopoda, respectively, and reproduction is supported by instantaneous resource availabilities rather than stored resources: in this sense they have a similar general approach to passing on genes and thus an equivalent evolutionary strategy.

Allocation to reproduction can be represented as a **proportion** of the total resources acquired by the organism used for regeneration rather than maintenance or further resource acquisition, reflecting the trade-off between the three principal adaptive responses. We suggest that calculating a lifetime mass/energy budget, based on this fundamental three-way investment trade-off, may be the gold standard measure of adaptive strategies across all organisms.

Several other workers have already arrived at a similar conclusion, and have in fact measured lifetime energy budgets for certain organisms. Pecl & Moltschaniwskyj (2006) state that:

> . . . lifetime reproductive allocation, and therefore the life-history strategy adopted by an animal, needs to be understood in terms of resource allocation between reproduction and other competing needs, such as maintenance and growth.

Box 4.1: Do the sexes differ in adaptive strategy?

How can we define the adaptive strategy of a sexually dimorphic species, for which two suites of traits are evident? Pecl & Moltschaniwskyj's (2006) work on the Southern Calamary demonstrates that it is the female that incurs the main reproductive costs, and is smaller, less able to acquire resources and grow within her single year of life. This suggests that the sex with the greater total resource commitment to the continuation of genes should be considered representative of the adaptive strategy of the species (usually the female). Alternatively, the two sexes could be treated as two slightly different adaptive strategies with different ecologies.

The parallels between their *reproduction, maintenance* and *growth* and our *regeneration, maintenance* and *resource acquisition* (which is intimately involved in the growth phase of the life cycle) are clear. They also suggest that:

> . . . energy is viewed by most biologists as being the closest thing there is to a common currency of life, with lifetime patterns of energy allocation central to life-history theory.

They then went on to estimate the lifetime partitioning of energy for the Southern Calamary, based on traits such as growth rate and the timing of developmental events such as maturation. As detailed in Box 4.1, this work also highlights the need to account for the sex of individuals.

Heino & Kaitala (1999) suggest a similar approach:

> . . . lifetime reproductive allocation can only be understood in terms of resource or energy allocation between reproduction and other competing needs, such as maintenance and growth. Usually it is assumed that maintenance has priority over other needs. Energy in excess of maintenance is 'surplus' energy, which can be allocated between growth and reproduction.

Thus current resource allocation models consider only **surplus energy allocation** to either somatic growth or reproduction, whilst 'allocation to maintenance is simply ignored' (Heino & Kaitala, 1999). We suggest that resource allocation models based on the universal three-way trade-off would be a fruitful avenue for future research in preference to a dichotomous trade-off scheme that acknowledges the existence of a third allocation pathway yet ignores it.

Models of lifetime resource allocation are currently based on data such as growth rates, body size, fecundity, and total resource intake (Heino & Kaitala, 1999). Maintenance traits must also be included, such as metabolic rates or the turnover times of elements within the living system, and a multivariate approach would be useful to determine the extent to which further traits contribute towards the universal three-way trade-off. A direct measure of the extent of S-selection could involve quantifying, for instance, the proportion of mass invested in structures used to resist adverse environmental conditions, or even the expression of genes involved in DNA repair. The degree of C-selection could be expressed as the total amount of matter acquired (or biomass produced) during the lifetime of an individual, as a function of its mass. The lifetime invest-ment in reproduction (measured, perhaps, as the total mass of offspring pro-duced per mass of the mature individual) may give a reasonable estimate of the extent to which regeneration is a part of the overall adaptive strategy, or the extent of R-selection. Ultimately, if we wish to classify organisms according to universal adaptive strategy theory the key question is: what proportion of avail-able resources is allocated between acquisition, maintenance or regeneration? Measures such as these would provide ternary triplet coordinates (three percent-age values representing C-, S- and R-adaptation, respectively) that could be used to visualize the position of the adaptive strategy within the triangle of trade-offs, as is currently done for plants (Caccianiga *et al.*, 2006; Pierce *et al.*, 2007a, b). In this way the role of taxonomically diverse organisms within an ecosystem can be quantified, visualized and compared.

All organisms rely on certain elements for the production of fresh components and for growth, particularly carbon, nitrogen and phosphorus. Indeed, organic life, by definition, depends on the properties of carbon – carbon is the true common currency in the economies of all organisms, including viruses, which lack metabolism and thus innate biochemical means of storing and using energy. This should provide a satisfactory answer to J.L. Harper's question, which we encountered in Chapter 2: 'In Professor Grime's triangle, the axes are given in percentages so that any point on the triangle adds up to 100 per cent of some-thing. What is this quantity, and how can we measure it and allocate it between C, S and D?' We suggest that 100 per cent is the total amount of an essential or limiting nutrient (e.g. carbon or nitrogen) acquired by an organism during its individual lifetime, and that allocation between C, S and D can be measured by quantifying the proportion of these elements invested in traits involved in resource acquisition, maintenance or regeneration. This promises a universal comparative measure of adaptive strategies.

Summary

1   Multivariate analysis of life-history traits in mammals, birds and fishes has revealed the existence of three main directions of adaptive specialization similar to those of plants. Plant ecologists, zoologists and bacteriologists have independently developed three-way adaptive strategy theories within their own particular fields and have remarked on the similarities between them, although without attempting unification.
2   Examples of extreme adaptive strategies evident within the mammals, birds, squamates, amphibians, bony fishes, cartilaginous fishes, insects, arachnids, crustaceans, echinoderms, molluscs, segmented worms, corals, fungi (includ-ing lichens), archaeobacteria, eubacteria (proteobacteria, firmicutes and cyanobacteria) and viruses are consistent with the hypothesis that a universal three-way trade-off between resource acquisition, maintenance and regenera-tion constrains adaptive strategies throughout the tree of life.
3   Phenotypically diverse groups such as mammals, bony fishes, insects, proteo-bacteria and cyanobacteria include numerous examples of trait syndromes

consistent with the full range of C-, S- and R-selected organisms. Less diverse groups appear to be phylogenetically constrained, but trait syndromes are nonetheless consistent with C-, S- and/or R-selection. For example, sharks and squamates lack highly R-selected forms and consist of C- and S-selected taxa.

4 Fossil evidence of physiological, behavioural and reproductive traits for extinct groups such as the non-avian dinosaurs and pterosaurs suggests that it is realistic to consider these organisms in terms of C-, S- and R-selection, although determination of precise strategies is not possible.

5 Two principal routes to large size are evident throughout the tree of life: **(a)** in productive niches, rapid growth of organisms for which large size is immediately advantageous for resource acquisition and control (C-selection); and **(b)** in unproductive niches, slow, incremental growth where large size may eventually be a by-product of longevity (S-selection).

6 Organisms in disparate groups exhibit radically different phenotypes, but what may unite them as organisms with analogous adaptive strategies is equivalence in the proportion of resources invested in resource acquisition, maintenance or regeneration.

7 All organisms, even viruses which lack metabolism, are characterized by organic chemistry and face a three-way trade-off in the allocation of carbon to resource acquisition, maintenance and regeneration throughout their existence. Measuring or estimating carbon partitioning could allow comparison of adaptive strategies throughout the tree of life.

8 CSR plant strategy theory can now be regarded as a specialized sphere within the broader realm of **universal adaptive strategy theory**.

5

From Adaptive Strategies to Communities

The CSR model in particular has proved to have applicability to a wide range of situations, allowing characterisation and explanation of the relationships which we see in communities of organisms.

(Dickinson & Murphy, 2007)

Preceding chapters of this book have documented that, with increasing momentum, evidence is accumulating of the operation of the same fundamental constraints and trade-offs in the evolution of the core characteristics of organisms distributed throughout the tree of life. It is remarkable to observe that these same patterns of specialization recur in very different ecosystems on land and under water and in organisms that perform a wide range of different functions within them. The emerging challenge for ecologists is to navigate from the repeated patterns of adaptive specialization recognized throughout the tree of life to a predictive understanding of how ecosystems function and vary from place to place and in response to increasing impacts from a rapidly expanding human population. There is considerable potential to move directly to predictions of the characteristics of an ecosystem simply on the basis of the functional traits of the organisms that it contains. However, the inclusion of this chapter is intended as a signal that we do not accept that such a direct 'hop' from species to ecosystem is an adequate basis for sound ecological understanding of the majority of ecosystems.

To fully comprehend the structure and dynamics of an ecosystem it is frequently necessary that we examine its main components. These are usually described as communities and can be defined as sets of organisms that exhibit close similarities in resource capture and utilisation and perform similar functions within the ecosystem (e.g. canopy trees, shrubs, herbs, epiphytes,

The Evolutionary Strategies that Shape Ecosystems, First Edition. J. Philip Grime, Simon Pierce.
© 2012 John Wiley & Sons, Ltd. Published 2012 by John Wiley & Sons, Ltd.

herbivores, decomposers, predators, parasitoids). Why is it necessary, in our pursuit of the ecosystem, to venture into such detail? Part of the answer to this question relates to the fact that the contribution of a community of organisms to the functioning of an ecosystem depends upon the type, number and relative abundance of its component species. Moreover, in any particular ecosystem the composition of each community may be very dynamic, being determined by both the rate of extinction and by rates of immigration from the pool of candidates for membership existing in the vicinity.

An additional reason why it is necessary to devote special attention to this subject is that, to an extraordinary extent, investigations of plant, animal and microbial communities have followed different paths until quite recently. In plant community ecology we have witnessed the patient accumulation of empirical data over a period of more than a hundred years. As John Harper observed:

> Plants stand still and wait to be counted. (Harper, 1977)

In comparison with plant assemblages, animal communities have proved to be an elusive subject for investigation and have attracted a considerable volume of rather abstract theory and controversy:

> Community ecology has for too long been regarded as repugnant and intractably complex. (Pianka, 1992)

These divergent histories must be taken into account as we seek to establish the study of communities as an essential step in the ascent to the ecosystem. For this reason we will divide this topic into a number of sections, first visiting the relatively tranquil domain of plant community ecology and the general theoretical insights gained from the study of plant communities, followed by microbial and then animal community ecology.

Plant communities

The description and mapping of terrestrial vegetation began early and, especially in continental Europe, became remarkably formalized with standardized procedures for compiling lists of species and estimating their abundances in samples of uniform composition. In grasslands, heathlands and wetlands it was widely accepted that the same, or closely similar, assemblages of species recurred over large geographical areas and coincided with particular geological strata, topographies, soil conditions and land management. Such remarkable consistency in species composition led quite naturally to general acceptance of the term 'community' and many plant ecologists spent their entire working lives documenting and publishing accounts of the distributions and regional variations of communities. For some botanists there were parallels in philosophy between finding and classifying plant species and that of developing a taxonomy of plant communities. This caused some to view the community as a kind of discrete unit or 'super-organism' rather than as a mere recurring assemblage. In North America differing views about the significance of the plant community generated a debate between two leading ecologists of sufficient intensity to cause one of the protagonists to abandon ecology altogether!

It should be noted that this famous argument between Clements and Gleason was an exceptional event in the otherwise rather orderly early progression of plant community ecology. Neither of the two experts questioned the existence of predictable assemblages of plants over large areas of landscape. However, for some commentators the century-long descriptive phase in which so many plant ecologists took part was a barren exercise:

> . . . the wilderness of meticulous classification and ordination of plant communities in which plant ecology has wandered for so long began in the pursuit of answers to questions but then became an activity simply for its own sake. (May, 1985)

> This proliferation of multivariate techniques for the analysis of spatial variation in plant community structure has not, however, led to great advances in our understanding of the processes underlying these patterns. (Crawley, 1986)

These were pertinent criticisms. Description alone could not provide the basis for an understanding of variation in plant communities. It can be argued that because they steadfastly refused to follow the example of the phytosociologists of mainland Europe, British ecologists made faster progress in developing mechanistic and experimental approaches in community ecology. However, in one sense at least, this policy backfired. In the second half of the 20th century radical changes in land-use, damaging effects of pollution and climate changes began to transform British vegetation (Hodgson, 1986a, b). Because such effects were occurring in the absence of baseline information on the composition of British plant communities it was difficult to assess the scale of the problem and to devise conservation policies. Fortunately, the need for urgent action was recognized and a system of phytosociological description (Rodwell, 1991) was developed and rapidly applied. In addition, a network of permanent recording sites, known as The Countryside Survey (Bunce *et al.*, 1999) was put in place as a basis for monitoring the effects of land-use policies and environmental change on plant community composition.

Productive disturbed communities

In Chapter 3 we have described the broad relationship that exists between the adaptive strategies of individual plant species and the equilibrium between productivity, disturbance and competition in their habitats. It is now appropriate to ask: 'How far can these relationships be extended into an understanding of the composition and structure of plant communities?'

In seeking an answer to this question it is useful first to comment on the special case of communities of R-selected plants. Where productive habitats are subjected to frequent and severe disturbance we can predict that the vegetation will contain a large component of ephemerals (see Fig. 5.1). Because their populations are constantly destroyed and re-established from seed it is inevitable that

Fig. 5.1 *Baeria chrysostoma* (Asteraceae) in the Mojave Desert, California, USA, is a ruderal, R-selected species that grows following rain. The population persists, despite severe drought disturbance and the death of individuals, because these plants invest in early reproduction and seed production. (Copyright © 2003 Simon Pierce.)

such community structures will be open to invasion by resident and alien species. This results in a paradox whereby functional uniformity often coincides with extreme variability in species composition through space and time. Added to this, the widespread use of herbicides on arable fields has drastically reduced the species-richness of the communities of ephemerals in farmland and under the most intensive forms of management many species have become locally extinct. Despite the distorting effects of recent farming methods R-selected communities remain a distinctive and expanding feature of modern landscapes particularly as a consequence of urbanization and industrialization.

Productive undisturbed communities

What are the distinguishing features of plant communities that develop in productive, relatively undisturbed conditions that in theory would be expected to be conducive to C-selected adaptive strategies? In one extremely labour-intensive effort to answer this question ecologists (Al-Mufti *et al.*, 1977) conducted a programme of seasonal sampling of abandoned land in northern England. At sites where vegetation was established under fertile, relatively undisturbed

Fig. 5.2 A dense stand of tall herbs on a productive river terrace dominated by **(1)** Reed-grass (*Phalaris arundinacea*), and **(2)** Meadow-sweet (*Filipendula ulmaria*) in Lathkilldale, North Derbyshire. (Copyright © 2012 J.P. Grime.)

conditions a large peak in shoot dry matter was apparent during the summer and frequently coincided with dominance by a single species exhibiting tall stature and a capacity for vigorous clonal spread (e.g. *Urtica dioica*, *Chamerion angustifolium*; see Fig. 5.2). These sites contained few species and this appeared to be related to the ability of the dominant C-strategist to monopolize resource capture during the main growing season. Consistent with this interpretation a different vegetation structure (see Fig. 5.3) was observed at another relatively undisturbed site with a shallower soil and lower productivity. Here the biomass accumulated by the dominant C-strategist, *Filipendula ulmaria*, was lower and this had allowed the presence of an understorey consisting of the shade-tolerant herb *Mercurialis perennis*, the spring geophyte *Anemone nemorosa* and a number of large mosses.

These results confirmed the prediction that, in the absence of major impacts of environmental stress and disturbance, competitive dominance could become the over-riding determinant of community structure and in extreme cases could produce monospecific stands of vegetation. There was also evidence that debilitation of potential dominants by specific forms of stress or disturbance would allow penetration into the community by other species as forecast in the original description of the CSR model:

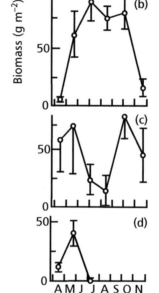

Fig. 5.3 Seasonal change in the above-ground biomass of a moderately productive tall-herb community in Lathkilldale, North Derbyshire. **(a)** *Filipendula ulmaria*; **(b)** *Mercurialis perennis*; **(c)** bryophytes; **(d)** *Anemone nemorosa*. (Re-drawn after Al-Mufti *et al.*, 1977. Copyright © Blackwell Publishing – Journals.)

'Stress and disturbance together comprise those phenomena which prevent the resolution of competition. At moderate intensities this intervention has the effect of creating spatial or temporal niches.' (Grime, 1974).

In the **centrifugal model** of Keddy (1990), which we shall see in more detail later (see Fig. 5.7 on page 121), this scenario is elaborated: here the specific characteristics of the admitted species are predicted according to the mix of factors limiting the vigour of the dominant in specific situations. The importance of this model to community assembly will become evident as we progress through this chapter.

Fig. 5.4 An ancient sheep pasture on a limestone outcrop at Cressbrookdale, North Derbyshire. The short turf is composed exclusively of stress-tolerators including **(1)** lichens; **(2)** bryophytes; **(3)** Rockrose, *Helianthemum nummularium*; **(4)** Meadow Oat, *Helictotrichon pratense*; and **(5)** Sheep's Fescue, *Festuca ovina*. (Copyright © 2012 J.P. Grime.)

Unproductive relatively undisturbed communities

The programme of seasonal sampling and sorting used by Al-Mufti *et al.* (1977) was also applied to sites in which the productivity of perennial herbaceous communities was severely reduced by impacts of low soil fertility. The most conspicuous feature of these communities (see Fig. 5.4) was the absence or low abundance of species capable of attaining biomass peaks in the summer and the presence of a variety of S-strategists, many of which had long-lived leaves forming evergreen canopies showing little change in structure throughout the year.

Plant community composition

So far in this chapter we have examined plant communities in terms of the factors governing the kind of plants that they contain. However, it is also informative to record the relative abundances of the species within a community and to attempt to identify the mechanisms controlling the hierarchy from most to least abundant. In Fig. 5.5, for example, it is apparent that at high productivity and

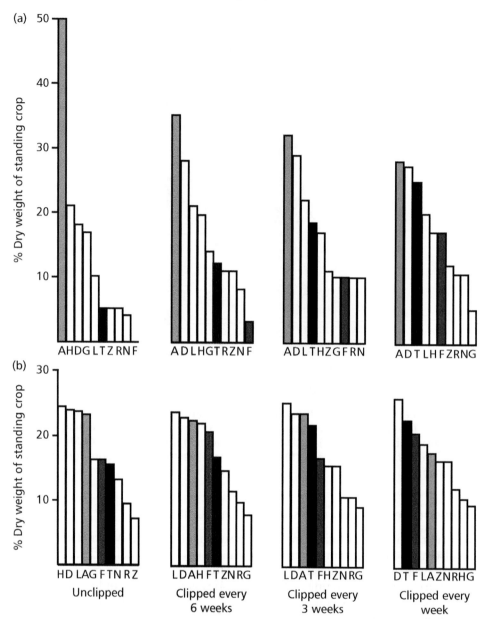

Fig. 5.5 Columns comparing the species composition of vegetation synthesized after seven months from a standard seed mixture of grasses and subjected to **(a)** high fertility (170 mg L⁻¹ N), and **(b)** low fertility (5 mg L⁻¹ N) and four intensities of defoliation. A-*Arrhenatherum elatius*, H-*Holcus lanatus*, D-*Dactylis glomerata*, G-*Elytrigia repens*, L-*Lolium perenne*, T-*Agrostis capillaris*, Z-*Bromopsis erectus*, R-*Festuca rubra*, N-*Anthoxanthum odoratum*, F-*Festuca ovina*. (Re-drawn after Mahmoud, 1973. Copyright © A. Mahmoud.)

in the absence of frequent clipping the potentially tall species *Arrhenatherum elatius* dominates the biomass, whereas the small tussock grass *Festuca ovina* never attains dominance and is prominent only under infertile, clipped conditions. This suggests that traits related to potential size can be used to predict the relative abundance of species within communities and evidence consistent with this hypothesis has been obtained (Grime 1973a, b; Al-Mufti *et al.*, 1977). Strong support for the impact of size has also been obtained in 'foraging' experiments (Campbell *et al.*, 1991) in which measurements of the scale of root and shoot exploration in resource-patchy conditions was found to accurately predict the relative abundances obtained when the same set of species was used to synthesize plant communities grown from seed for 16 weeks. It can be argued that the predictive power of plant size with respect to relative abundance may apply only in communities that allow dominance. There is a requirement for experiments across a wider range of environments and management regimes.

Recently Shipley *et al.* (2006) and Shipley (2010) have explored alternative techniques, namely a novel **'maximum entropy'** (or MaxEnt) method, to investigate relative abundance during community assembly. In essence, they provide a practical method for predicting which members of a local species pool may attain dominance, calculated from species trait values, and applied the method to a secondary succession. They employed traits that reflect investment of resources in either vegetative growth or reproductive development (seed number and maturation date, the proportion of perennial species in the community, specific leaf area and, notably, size-related traits such as the mass of organs, seeds and vegetative matter). These traits are likely to reflect CSR strategies (Shipley, 2010). Indeed, it is worth noting that the predictions made by the MaxEnt model and the shifts in trait values actually observed by Shipley *et al.* (2006) are consistent with a gradient from communities composed of many R-selected species in early succession to fewer C-selected species in late succession: the results can readily be interpreted within the theoretical and evolutionary context provided by universal adaptive strategy theory.

As with any new method the predictive power of MaxEnt has yet to be widely tested (Shipley *et al.*, 2006; Shipley 2010) and it has stirred considerable debate (e.g. Marks & Muller-Landau, 2007; Roxburgh & Mokany, 2007; Shipley *et al.*, 2007). One thing we would add to this debate is this: MaxEnt uses traits related to the primary strategy to predict the potential dominants within the species pool, but diversity of subsidiary species may depend on more subtle distinctions. In many ecosystems a large number of species may be present that, at least in terms of resource use and growth (Hubbell, 2005), survive in broadly equivalent ways, and this coexistence of similar strategies must be accounted for. In fact, later we shall see that a large proportion of this current chapter is devoted to a novel approach to this problem of species functional equivalence, based on the concept of convergence in adaptive strategy and divergence in single traits.

To return to our discussion of the biodiversity patterns evident over productivity ranges, previously we discussed a number of studies which suggested that progress in analysing the structure of plant communities could be made by recognizing that there are plants (C-strategists) that are capable of so monopolizing productive habitats that only one species occurs. These investigations also supported the hypothesis that a greater number of species could persist in conditions where C-strategists were reduced in vigour or even totally excluded by low productivity, vegetation disturbance or, more commonly, some combination of these two phenomena. These results pointed strongly to a relationship between the competition/stress/disturbance equilibrium and the species-richness of plant communities. Even more exciting, connections were becoming apparent between the CSR model of primary plant strategies and another model describing the control of species diversity published three years earlier (Grime, 1973a). The events leading to development of this earlier model will now be described and this is followed by an explanation of its connections with CSR theory.

The humped-back model

Origins

Two activities inspired the **humped-back model**. The first consisted of an experiment (Mahmoud, 1973) in which communities of grasses were synthesized from seed under controlled conditions of nitrogen supply and cutting. The results (see Fig. 5.5 above) confirmed that at high fertility and in the absence of disturbance (defoliation) the community was dominated by the tall, fast-growing species *Arrhenatherum elatius*. Although the effects of stress (nitrogen starvation) and defoliation were similar, in that dominance by *A. elatius* was prevented and species diversity was promoted, it was evident that these two treatments had different effects on the relative proportions of the species in the community. In Fig. 5.5 this difference is most evident with respect to the changes in the status of the C-strategist *A. elatius* and the S-strategist *Festuca ovina*. It was also apparent that greatest evenness across the community occurred where nitrogen stress *and* frequent cutting coincided.

The second source of information (see Fig. 5.6) influencing development of the humped-back model was an extensive survey (2748 m² samples; Grime, 1973b) of herbaceous vegetation providing measurements of species-richness in a wide range of habitats in northern England. One of the most important facts to emerge from this survey was that the capacity of species to dominate plant communities and drive down species-richness was not confined to perennial C-strategists such as the *Arrhenatherum elatius* of Fig. 5.5. Mono-specific stands of vegetation also occurred in circumstances such as riverbanks where fertile soils were disturbed by winter floods but then re-colonized each spring by dense populations of tall summer annuals (C-R strategists) such as *Impatiens glandulifera*. It was also revealed that dominance and low diversity frequently arise on less fertile soils provided that vegetation disturbance is low, allowing robust but slow-growing species (S-C strategists) to monopolize space by extensive root systems and leaf canopies (e.g. *Rubus fruticosus*) and in some instances by production of a deep layer of persistent leaf litter (e.g. *Brachypodium pinnatum*).

The survey exposed a quite different contingency in which low species-richness is frequently observed. Species-poor communities are characteristic of a wide range of habitats where productivity is severely reduced by shallow soils (rock outcrops, walls, cliffs, screes, lead-mine heaps). It was also apparent that

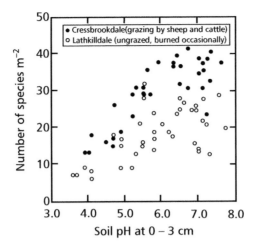

Fig. 5.6 Comparison of species-richness in two Derbyshire dales differing in management (closed circles: Cressbrookdale, grazing by sheep and cattle; open circles: Lathkilldale, ungrazed, burned occasionally). (Re-drawn with permission from Grime, 1973b. Copyright © Elsevier.)

on paths and in other habitats subjected to frequent disturbance the number of species can fall close to zero. Consistent with the experimental results of Fig. 5.5, there is strong evidence from the field survey that **the highest levels of species-richness in a wide range of habitats are associated with unfertilized soils and forms of management that involve removal of biomass by grazing, mowing or burning.** Figure 5.6 uses data from a small part of the survey to examine in more detail the patterns in species-richness associated with variation in soil pH and grazing management. In the abandoned pasture at Lathkilldale, species-richness across the entire pH range from 4.0 to 7.0 falls below that occurring at Cressbrookdale where there is a long and continuing history of grazing.

Formulation

As we have just described, experiments and field observations in the early 1970s had begun to implicate several phenomena in the mechanisms controlling species-richness in herbaceous vegetation. Where conditions allowed a high density of biomass or litter to accumulate species of robust stature were capable of exercising dominance to an extent that greatly reduced species-richness. Stress (reduced productivity) and disturbance (damage to vegetation), alone or in combination, could stimulate diversity by suppressing the vigour of potential dominants. However, the promotory effects of stress and disturbance were only sustained across a moderate range of intensities and at high severity both could eliminate vegetation completely.

These insights were brought together in the humped-back model (Grime, 1973a; Al-Mufti *et al.*, 1977) describing variation in species-richness along gradients of stress and disturbance. At this stage of development (see Fig. 5.7) the

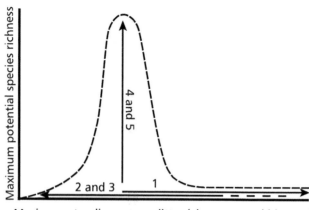

Fig. 5.7 The **humped-back model** summarizes the relationship between biodiversity and biomass, and the impact of **(1)** dominance; **(2)** stress; **(3)** disturbance or biomass destruction; **(4)** niche differentiation; and **(5)** the species pool, or ingress of suitable species or genotypes. (Re-drawn after Grime, 1973a. Copyright © Nature Publishing Group.)

main purpose of the model was **(1)** to recognize three different circumstances, (dominance, extreme stress and high disturbance) that suppress diversity; and **(2)** to define the conditions of intermediate biomass accumulation amenable to high species-richness.

One of the central tenets of the humped-back model is that in extreme environments organisms must exhibit a high degree of adaptive specialization in order to survive. This concept applies as much to high productivity environments as it does to adverse environments, because organisms that are specialized competitors are more likely to monopolize resources and exclude other species. Thus diversity in adaptive strategies should be more extensive at intermediate productivities where there is a greater range of opportunities, coincident with the greater species diversities that are evident in nature. Navas & Violle (2009) examined **functional diversity** (i.e. diversity in trait values) along the humped-back curve, using the differences in plant height within communities at contrasting productivities, and found that diversity in this trait was indeed greatest at intermediate productivities – this may be interpreted as evidence that adaptive strategy diversity underpins species diversity. However, only a single trait was investigated in this study and it is unclear how the abundance of dominant adaptive strategies affects the productivity/diversity relationship shown in the humped-back model. Presumably the incidence of generalist and subordinate adaptive strategies is greatest at intermediate productivities.

Independent confirmation and compatibility with new research

A review of the diversity/productivity relationships evident in nature (Mittelbach *et al.*, 2001) demonstrated that humped-back curves represent the most common type of relationship, although linear and U-shaped correlations, or a lack of correlation, are also apparent. However, it was also evident from this review that non-humped-back relationships were more common when investigations took place within a single community, rather than when diversity was compared between habitats – suggesting that many of these relationships were artefacts arising from the study of diversity over limited productivity ranges (Mittelbach *et al.*, 2001). This conclusion was also reached by Moore & Keddy (1989) from the comparison of diversity/productivity relationships within and between wetland plant communities. It has been recognized (Guo & Berry, 1998) that when low- to medium-productivity habitats are investigated then the diversity/productivity relationship will be positive because only the left slope of the hump comes under study, whereas investigation of medium- to high-productivity habitats yields a negative correlation because only the right flank is investigated.

The scale of investigation, or quadrat size, is a further complication, because larger quadrats will include larger numbers of species. However, Rapson *et al.* (1997) found that as long as sites of differing productivity were sampled the humped-back curve remained evident, albeit with greater species-richness and thus a more evident curve when larger quadrats were employed (see Fig. 5.8). Thus the diversity/productivity curve may be more (or less) evident depending on the methods used.

Indeed, a humped-back relationship may not be evident when biomass is collected in years that deviate markedly from climatic norms, reappearing when

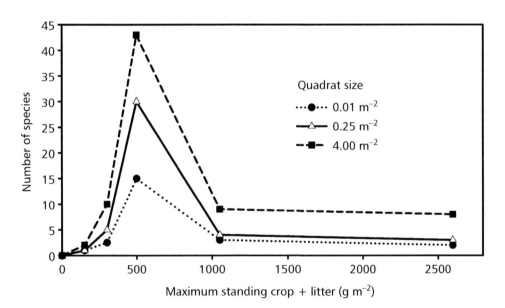

Fig. 5.8 The size of quadrat used affects the number of species that are detected, but nonetheless the humped-back curve remains evident even using the smallest of quadrats, as long as the comparison is between separate communities at different productivities. (Re-drawn from Rapson *et al.*, 1997. Copyright © Blackwell Publishing – Journals.)

conditions return to those more typical and favourable for the growth of local vegetation (Laughlin & Moore, 2009). Houseman & Gross (2006) used the experimental addition of seeds to augment the local species pool available to a mid-successional old-field site including a range of microsites of varying productivity. They found that when the species pool was restricted the diversity/ productivity relationship was negative and linear, but when seeds of greater numbers of species were added diversity increased only at intermediate productivities, resulting in a humped-back curve, 'supporting the contention that soil resource supply rates may dictate species pool–local richness patterns of diversity patterns as predicted by Grime' (Houseman & Gross, 2006). Together with the study of Laughlin & Moore (2009), this provides further evidence that the humped-back curve is essentially the maximum potential diversity relationship (Grime 1973a, b) – what Houseman & Gross (2006) term 'saturated' – and that this may not necessarily be evident if atypical conditions of stress or restricted dispersal are prevalent when productivity and diversity are measured. Indeed, Fig. 5.9 illustrates a humped-back curve coinciding with a local gradient in soil depth in a limestone grassland, and the lack of a uni-modal curve at atypical extremes of water availability.

Recent studies, published since Mittlebach *et al.*'s review, have found the signature of the humped-back curve in plant communities as diverse as tropical rainforests (Molino & Sabatier, 2001), European fens and meadows (Olde Venterink *et al.*, 2001), Australian *Eucalyptus* woodlands (Allcock & Hik, 2003), a Himalayan mountain grassland (Bhattarai *et al.*, 2004), Mediterranean wetlands (Espinar, 2006), and at the regional scale throughout Europe (Cornwell

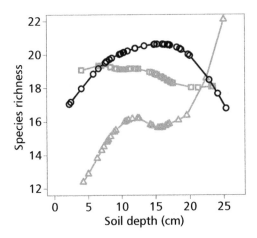

Fig. 5.9 Fitted curves describing variation in vascular plant species-richness in microsites (100 mm × 100 mm) distributed evenly across four categories of soil depth, at experimental plots at the Buxton Climate Change Impacts Laboratory in North Derbyshire, UK. A uni-modal pattern is evident under natural climatic conditions (circles) but the relationship changes with regular summer drought in July and August (triangles) and with supplementary watering, June–September (squares). Treatments applied 1993–2010. (Fridley, Askew & Grime, unpublished data.)

& Grubb, 2003). It has also been confirmed that the species evenness of communities (i.e. the extent to which dominance is shared between species) follows a humped-back relationship with productivity (Chalcraft *et al.*, 2009). Even digital organisms (self-replicating computer programs), evolving from a simple virtual 'ancestor' and exposed to different productivity levels, adapt and diverge to produce a humped-back 'biodiversity'/productivity curve (Colasanti *et al.*, 2001; Chow *et al.*, 2004), from which it was concluded that only exceptionally specialized species can evolve at extremes of productivity.

The humped-back curve is also compatible with the intriguing idea that in habitats characterized by low productivity and exposure to stress it is facilitation, rather than competition, that forms the main interaction between species shaping the left flank of the humped-back curve (the **stress-gradient hypothesis**; Michalet *et al.*, 2006; Xiao *et al.*, 2009). At the lowest productivities abiotic conditions may be so harsh that even facilitation is negligible, and only small numbers of species are appropriately adapted to survive. This view has recently gained support from transplantation experiments showing that in the most severe conditions interactions between plants are negligible (Forey *et al.*, 2010). In slightly more forgiving habitats the shelter of dominant species becomes important and an increasing range of adaptive strategies are facilitated. Indeed, alpine plant communities are characterized by positive interactions, manifest during experimental studies as the negative impacts of decreased growth and survival when neighbouring plants are removed (Choler *et al.*, 2001; Callaway *et al.*, 2002). However, as abiotic conditions improve, interactions between species shift away from facilitation and towards competition so that around the peak of the humped-back curve both types of interaction are important; competition then supersedes facilitation at moderate to high productivities.

It is evident that the humped-back model can explain another recently described natural phenomenon, which, strangely enough, has been independently christened the '**elevational humped-back curve**'. This describes the pattern of diversity often observed along elevation gradients whereby the greatest diversities are apparent at intermediate elevation, evident for both plant (Chawla *et al.*, 2008; Nogues-Bravo *et al.*, 2008) and animal (Li *et al.*, 2003; Lee *et al.*, 2004; Beck & Chey, 2008; Rowe, 2009) communities. Low diversity at high elevation is usually attributed to increasing abiotic stress, but the literature is oddly silent when it comes to a reason for low diversities at lower elevation. We suggest a simple explanation: at lower elevation a greater availability of resources and temperatures more conducive to growth favour the development of high productivity ecosystems dominated by a relatively small number of C-selected species. At high elevation abiotic stress imposes lower diversities, whereas at intermediate elevation these habitat extremes are more rarely encountered and thus organisms are adapted to a broader range of niches, encouraging diversity. In other words, the elevational humped-back curve is simply a spatial manifestation of the productivity humped-back curve.

Some studies concluding that a humped-back curve does not exist may simply rely on a rather curious interpretation of the data.

The humped-back model is inconsistent with a recent meta-analysis of 192 studies of productivity/biodiversity relationships published in a paper entitled 'The functional role of producer diversity in ecosystems' (Cardinale *et al.*, 2011). Despite this promising title the paper specifically excluded work conducted in natural ecosystems: one of the criteria for consideration in the meta-analysis was that data should be based on experimental cultures. Thus one of the key questions of the paper, 'How do diversity effects documented in experiments scale-up to "real" ecosystems?' was answered with: 'insufficient evidence to address'. To translate: all sorts of interesting things are evident when plants are cultivated, but it is not possible to tell whether these can explain natural phenomena.

Cultivation experiments do not usually emulate specific, realistic management regimes that are key forces in the creation of biodiversity.

For a field ecologist measuring diversity and biomass production in calcareous grassland with more than 40 species per square metre (e.g. at Cressbrookdale in the Derbyshire dales; see Fig. 5.6 above) the idea that a handful of species in garden plots is the best tool for the investigation of biodiversity, and the conclusion that the greatest productivity occurs with the greatest biodiversity, is naive. Ecologists have a duty to inform conservation policy and to protect the organisms under study. It is reassuring that legislation, such as the European Union habitats directive (Council Directive 92/43/EEC), recognizes the high biodiversity of meadows, and does not seek protection for stands of *Reynoutria japonica*, which have six times the biomass production of species-rich *Trifolio thalii-Festucetum puccinellii* or *Caricion austroalpinae* meadows, but just the one species (A. Verginella, S. Armiraglio, B. Cerabolini & S. Pierce, unpublished data). What is evident in natural and semi-natural ecosystems (and by semi-natural we mean long-established wild systems influenced by management) is that when species-richness and productivity of communities are compared across the full productivity range the greatest species richness is found at intermediate productivities.

Unfortunately, a re-occurring theme in modern ecology seems to be that phenomena observed in highly artificial systems or in entirely theoretical models are afforded greater credence than phenomena observed in nature, as Keddy (2005) complained in his article entitled 'Putting the plants back into plant ecology'. Scientific investigation aims to provide possible explanations for phenomena **observed in nature** – science is a method for investigating the natural world. Thus (and it is with considerable vexation that we feel the need to reiterate this rather basic philosophy) nature observation is an essential starting point for ecological investigation. Reviews and meta-analyses should focus on data obtained from natural systems, and if anything is to be excluded it should be data obtained from highly artificial experiments, not the other way around.

Recently a large-scale survey of herbaceous vegetation (Adler *et al.*, 2011) has claimed to falsify the humped-back model, describing diversity/biomass relationships as 'weak and variable'. However, the majority (91 %) of sampling points in this study exhibited biomasses of less than $500 \, g \, m^{-2}$, and the maximum biomass sampled was only $1534 \, g \, m^{-2}$. Adler's co-authors (Grace *et al.*, 2007) previously recorded biomasses exceeding $4000 \, g \, m^{-2}$ in a number of species-poor wetland communities. Despite the gaps in their dataset Adler *et al.* state that they actually found a statistically significant humped-back curve at the global scale, but were able to make it non-significant by omitting, without explanation, a wetland community and nine sites that they deemed particularly anthropogenic. This exclusion of high biomass/low diversity sites and the strong emphasis on communities of low biomass indicate that the 'Nutrient Network collaborative experiment', from which the study emerged, was not designed as an objective test of the humped-back model over the full productivity range. Ironically, in their main analysis maximum biodiversities of ~40 spp. m^{-2} were evident at biomasses of 400–500 g m^{-2}, with much lower diversities at productivity extremes, broadly agreeing with the view of the humped-back curve as an upper limit to diversity peaking at ~500 g m^{-2} (as in Fig. 5.8). We predict that completion of Adler *et al.*'s publicly available dataset will yield a humped-back maximum potential biodiversity curve.

Species-pools, filters and community composition

Habitats may, in reality, include more than one productivity gradient based on qualitatively different environmental stress or disturbance factors. This inspired Keddy (1990) to produce the centrifugal model (see Fig. 5.10), which allows the description of diversity/productivity relationships along multiple gradients (see

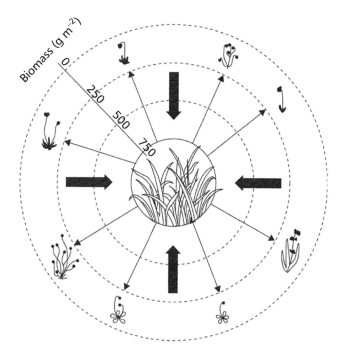

Fig. 5.10 The **centrifugal model** of Keddy (1990, 2005) acknowledges that multiple productivity gradients may exist in a habitat based on the action of more than one stress or disturbance factor, with the taxonomic identity of species along each gradient differing according to their specific adaptations. (Reproduced from Keddy, 2005. Copyright © Oxford University Press.)

also Wisheu & Keddy, 1996; Keddy, 2005). The model suggests that eutrophication (denoted by the thicker arrows in the diagram) will favour vegetation dominated by C-selected species, but that impoverishment by various factors may lead to local plant communities dominated by S-selected species of various taxonomic identities.

We can use this concept to imagine that, say, a gradient of increasing salinity may lead to species exhibiting the basic suite of conservative economics traits but with resources also invested in specific resistance traits, such as succulent tissues and extra capacity to isolate and dilute toxic sodium ions within cell vacuoles. A gradient in the intensity of another stress factor, such as low temperature, selects again for species with the basic suite of conservative traits, but this time resources are additionally invested in adaptations such as cold shock proteins or plasticity in membrane phospholipid expression (traits that specifically aid survival of chilling), rather than salt-tolerance. Both factors favour species with conservative resource economies. However, the precise manner in which resources are invested, and thus the character and taxonomic identity of the organism, depends on the nature of the stress.

This is a rather simplistic example because many S-selected species actually perform multiple resistance functions beyond the general resistance conferred by the primary strategy (e.g. the Ice Plant, *Mesembryanthemum crystalinum*, can exhibit both of the above responses; Pierce *et al.*, 2005). Also, plant responses are sometimes far from clear-cut and discrete. It is not a straightforward task, for example, to determine which plants use CAM and which use C_3 photosynthesis: plants as diverse as ferns and bromeliads may exhibit low CAM activities that can only be discerned by the combined use of multiple techniques (Holtum & Winter, 1999; Pierce *et al.*, 2002b) and some switch facultatively between CAM and C_3 in response to seasonal environmental changes (Maxwell *et al.*, 1994; Cushman, 2001; Cushman & Borland, 2002; Dodd *et al.*, 2002; Griffiths *et al.*, 2002). Each CAM species may take up a different quantity of CO_2 throughout a 24-hour period, abiding by a characteristic diel pattern of gas exchange (Pierce *et al.*, 2002b), and thus exhibiting a distinctive CO_2 uptake capacity and tempo.

In all of these examples, however, plants exhibit a suite of conservative functional traits, suggesting that although the CSD-equilibrium may decide the gross nature of adaptation and the adaptive strategies that can successfully enter the community (see Box 5.1), finer-scale adaptations may then decide the ultimate identity of the species incorporated during community assembly. This can also explain why there is one basic mechanism of stress (reduced metabolic performance limiting biomass production) and one basic conservative stress response, but a range of S-selected species with specific adaptations to particular stress factors. An analogy from the animal world is the difference between a polar bear and a dromedary – both have conservative lifestyles and rely on storing energy as fat reserves, but the former uses the insulating properties of fat to survive cold stress whereas the latter, to survive hot and arid conditions, must avoid these insulating properties by consigning fat to a hump on its back. The difference in the way in which fat is used contributes towards strikingly different phenotypes, but these animals share similarly conservative life-history traits in

Box 5.1: The CSD-equilibrium as an ecological filter

Filters are selective obstacles to the entry of species into the ecosystem, and act by excluding unsuitably adapted phenotypes. In Fig. 5.11 we can see an early example of the filter concept in which multiple filters (climate, fire regime and drought) exclude many of the species that arrive as 'seed rain' (i.e. the species comprising the local species pool; Woodward & Diament, 1991). We could similarly consider competition, stress and disturbance as separate filters in this manner, but this would imply the existence of a hierarchy in which these factors have different levels of importance. This is unlikely, because we have seen that adaptive specialization occurs in response to a three-way trade-off in which these three factors are of equal importance in shaping the evolution of primary strategies. Thus Fig. 5.12 shows a representation of the CSD-equilibrium as a single filter in which the intensity of C, S and D determine the strategies (represented by different geometric shapes) that find suitable niches (regions of the CSD-equilibrium), illustrated for habitats characterized by differing levels of productivity and disturbance. This allows entry into the community to organisms with suitable, and therefore similar, adaptive strategies.

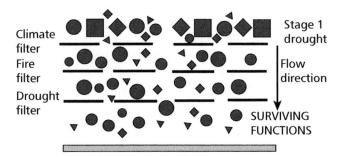

Fig. 5.11 An example of multiple filters excluding unsuitable plant functional types from a habitat characterized by seasonal drought. (Re-drawn after Woodward & Diament, 1991. Copyright © Blackwell Publishing – Journals.)

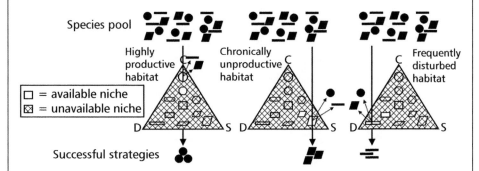

Fig. 5.12 The CSD-equilibrium is a filter that excludes adaptive strategies (represented by different geometric shapes) from niches characterized by contrasting levels of productivity and disturbance, sorting the local species pool into admissible and inadmissible strategies.

response to abiotic conditions that are potentially highly limiting to metabolic performance.

Polar bears and dromedaries clearly do not live together, and neither do salt-tolerant succulents and alpine snow-bed plants. These examples merely illustrate the point that organisms may share a generally S-selected syndrome but also rely on specific differences in the way in which certain features are employed in order to survive the local stresses found in their habitat. To see how such qualitative differences affect coexistence we must take a real-world example in which coexistence depends on slight phenotypic variations around the general theme of a shared adaptive strategy. For example, most members of the pineapple family (*Bromeliaceae*, or bromeliads) exhibit sturdy, slow-growing life-forms and are clearly S-selected (Benzing, 2000), but more than 2,500 species have evolved in the American tropics. How can we account for this biodiversity? Some species occupy exposed, arid niches and rely on sparse coverings of water-absorbing leaf hairs. In contrast, many cloud forest species are adapted to excessive rainfall that can block stomata and thus the CO_2 uptake required for photosynthesis; whilst the leaf hairs of these species are essentially the same as those of arid-zone species the hairs form a denser, contiguous leaf indumentum that repels water (see Fig. 5.13; Pierce *et al.*, 2001).

This is a clear example in which the CSD-equilibrium has selected for a common conservative bauplan, subsequently modified in a subtle but crucial manner according to the factor inducing stress. Thus survival is possible in qualitatively different types of low productivity niche by different S-selected taxa, extending bromeliad diversity.

Examples also exist of bromeliads that are almost identical in terms of growth-form and physiological traits, but different pollinators have driven evolutionary divergence in floral morphology and the daily opening times of flowers (e.g. bird-pollinated *Vriesea* species and bat-pollinated *Werauhia* species). Figure 5.14 shows

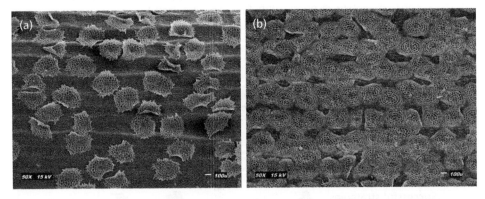

Fig. 5.13 Scanning electron micrographs of bromeliad leaf surfaces reveal that the density of leaf hairs (trichomes) is the only difference between **(a)** an absorbent leaf surface allowing survival in exposed and often arid epiphytic niches (*Aechmea dactylina*), and **(b)** a water-repellent surface allowing *Vriesea monstrum* to maintain gas exchange when wetted. (Reproduced with permission from Pierce *et al.*, 2001. Copyright © Botanical Society of America.)

Fig. 5.14 Some of the 39 species of bromeliad at the Cerro Jefe montane cloud forest, central Panama, coexisting due to differences in both overall adaptive strategy and more subtle variation in sub-sets of traits such as pollination syndromes: **(a)** *Pitcairnia arcuata* is a tall, perennial C₃ herb with broad, soft, spoon-shaped leaves almost a metre in length (C-selected); **(b)** *Guzmania calamifolia* is a terrestrial herb with tough, grass-like leaves that can dominate the understorey in these relatively open forest formations (SC-selected); **(c)** *Guzmania loraxiana* may be epiphytic or terrestrial and has thick, sturdy leaves (S/SC-selected); **(d)** *Werauhia millennia* has a practically identical vegetative form and strategy to *G. loraxiana*, but is pollinated by bats rather than insects; **(e)** *Aechmea dactylina* is an extremely S-selected epiphyte with CAM photosynthesis, thorny leaf margins and ant-symbiosis that improves nutrient status – flexible CAM metabolism allows survival of both extremely dry and wet conditions in exposed niches; **(f)** *Tillandsia bulbosa* is another S-selected CAM species but exhibits specific adaptations to wetting (Benzing, 2000), such as in-rolled leaves that shelter stomata, representing extreme variation on the basic S-selected bauplan. (Based on data in Pierce *et al.*, 2002a. Copyright © 2001 S. Pierce.)

a small selection of the bromeliad diversity in the Cerro Jefe cloud forest in central Panama. Adaptation of the overall suite of traits (the strategy) is evident alongside more subtle differences such as the flowering of *Vriesea* and *Werauhia* (and some insect-pollinated *Guzmania*) that allow coexistence for otherwise similar species. Two levels of variation are apparent: **(1)** quantitative variation of the set of traits determining the primary strategy, reflecting the universal three-way trade-off, and **(2)** variation, often qualitative, in small subsets of interlinked traits or even single traits. How do these two levels of variation operate to determine which species gain entry into the community? We suggest that variation in the adaptive strategy takes precedence over subtle variations in single traits because the strategy determines the overall partitioning of resources essential to metabolism, growth and development. Ultimately, however, the two act to determine survival and thus community structure. As two levels of variability are apparent, this problem should be approached armed with a two-tier model of community assembly.

Figure 5.15 introduces a '**twin-filter model**', which is an extremely simplified scheme that builds on the humped-back model to identify major components of the mechanism controlling the composition of a community. The initial controller is the composition of the pool of adaptive strategies and species in the vicinity of the community (Grime, 1979; Keddy, 1992); if an adaptive strategy or species is declining in the community and absent from the species pool it is unlikely to persist in the community. For example, in modern landscapes loss of traditional farming methods is leading to ineffective dispersal and local decline of species pools.

Subsequently, an ecological filter discriminates between primary adaptive strategies and is determined by the CSD-equilibrium prevailing in the habitat, limiting successful establishment to a subset of the primary strategists. The traits discriminated against by this '**primary CSD-equilibrium filter**' or '**CSD filter**' are those fundamental to the acquisition, retention and allocation of matter and energy utilized by primary metabolism, and are interdependent traits subject to the three-way trade-off (i.e. decline in one function is associated with gain in another). This filter has convergent effects because it operates on a day-to-day basis in the common circumstance where organisms occupy areas of similar habitat that subject them to common constraints with parallel and interlinked effects on metabolism and life history.

The final component of Fig. 5.15 is the '**proximal selection pressures filter**', which we will also refer to as simply the '**proximal filter**'. This determines which of the species capable of negotiating the CSD filter can actually penetrate and persist in the community. Proximal filter traits are those that affect survival but which are not integral to the CSR strategy (i.e. do not co-vary with competing primary functions) and often reflect intermittent threats to survival acting during particular phases of development or qualitative differences in how and when functions are performed *per se* (e.g. how metabolism is protected against specific stress factors, how pollination is achieved, how specific disease organisms are resisted, which particular enzyme is used for CO_2 assimilation). For convenience this can be conceptualized as a single filter, but in practice a range of 'proximal filters' may actually govern life-history divergence.

Thus the proximal filter includes a very wide range of factors that are not consistently entrained in the axes of the CSR model and admit or exclude traits such as those influencing the dispersal, dormancy and size of seeds, the

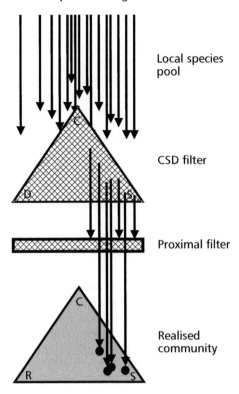

Fig. 5.15 The **twin-filter model** of community assembly. The local species pool provides a source of species that can potentially enter the community. Two filters then select against unsuitable suites of phenotypic traits, and thus against the individuals representing each species. The CSD filter represents the CSD-equilibrium (the extent of competition (C), stress (S) and disturbance (D)) of the habitat, which selects on a day-to-day basis for convergence (similarity) in the general adaptive strategies that can survive locally. The proximal filter is comprised of secondary selection pressures, such as local pollinators and the availability of animal seed dispersal vectors, which select intermittently against particular traits rather than the strategy as a whole. This filter has the potential to admit or exclude species and individuals that are well adapted to survive the CSD-equilibrium, resulting in divergence (dissimilarity) within the local subset of species.

phenology of germination and leaf growth, tolerance of climatic extremes, anaerobic conditions and soil toxins, resistance to particular pathogens and predators and dependence upon resources that are not used for growth *per se*, such as specific symbionts[1].

As we explain in Chapter 6, the distinction between CSD and proximal filters appears to have implications that extend beyond plant communities. It seems likely that this simple dichotomy can help us to identify which traits drive ecosystems and which are restricted in their role to controlling the species composition of plant communities.

---

[1]A resource may be defined as 'any external factor required to complete a developmental stage' and can include factors such as pollinators and other symbionts (Pierce *et al.*, 2005).

Evidence for the action of twin filters

The effects of the CSD-equilibrium as a filter should be particularly unsubtle, with variation in the relative importance of competition, stress or disturbance inducing major changes in the functional types comprising communities. Indeed, when CSR classification was employed to compare the structure of communities under different regimes of grazing disturbance in a relatively unproductive, nutrient-poor alpine prairie (Pierce *et al.*, 2007b), a shift towards ruderal strategies was, rather predictably, observed (see Fig. 5.16). Grazing changed the community dramatically; more than twice as many species (76 in total) were present, most of the newcomers being R- or C/R-selected, and the dominant sedge *Carex curvula* was much less frequent, with dominance being shared with a number of S-selected grasses and a legume. Thus disturbance created more opportunities by suppressing dominance by any single species. Localized nutrient-rich patches resulting from dung deposition probably also increased the number of safe sites for the establishment of faster-growing species. In doing so, biodiversity, particularly for ephemeral species, was encouraged.

This study provides strong support for the controlling effects of the CSD-equilibrium as a principal filter; the entry of a large number of broadly R-selected species into the disturbed community confirms that the CSD filter determines the general strategy, but beyond this is relatively indiscriminate about the particular taxonomic identities that can participate in the community – i.e. it leaves room for the proximal filter to operate.

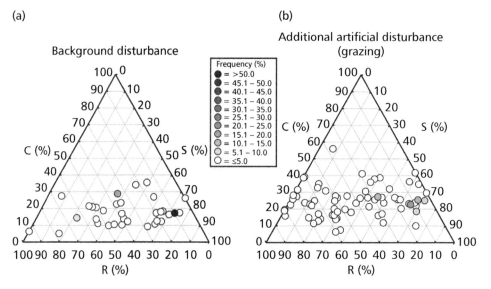

Fig. 5.16   The range and number of CSR strategies in two closely related subalpine prairies characterized by either **(a)** a background level of disturbance, and **(b)** additional disturbance by grazing cattle. The relative frequency of each species within each community is shown, with frequency categories reported in the box, inset. (Reproduced from Pierce *et al.*, 2007b. Copyright © Blackwell Publishing – Journals.)

Fig. 5.17 The chronosequence at the Cedec glacier foreland, Lombardy, Italy. Nodes 1 to 10 are sampling points running from pioneer (nodes 1–3; post 1965), early (nodes 4–5; Little Ice Age (LIA) to 1965), mid (nodes 6–7; late glacial to LIA) to late (nodes 8–10; late glacial) successional stages. (Reproduced with permission. Copyright © 2004 Fabrizio Denna.)

A similar situation can be observed when we compare changes in taxonomic identity and adaptive strategies along succession. That plant communities change progressively as they recover from disturbances is well documented, but noting the changes in taxonomic identity of plant species is not the same as understanding why species are replaced in a particular sequence. Luzzaro *et al.* (2005) examined plant community composition and adaptive strategies for ten points, or nodes, along a chronosequence at the Cedec glacier in the eastern Italian Alps, shown in Fig. 5.17. In Fig. 5.18, each number within the triangle represents a community (1=the pioneer community, 10=a late successional community), and its position represents the mean strategy at that node, adjusted to account for the relative abundance of species. These results demonstrate that in early successional stages the plant community progresses from R-strategists to S/SR-strategists with the development of the vegetation (i.e. with distance from the retreating glacier). However, by nodes 6 and 7 the succession, in terms of adaptive strategies, has stopped and the mean adaptive strategy of the community remains effectively the same until node 10. Crucially, this is not immediately obvious from species identity alone, which continues to change[2] between nodes 7 and 10. This suggests that the CSD filter is a key determinant of community composition during early and mid succession (i.e. the CSD-equilibrium is the main factor governing taxonomic identity), but as the environment becomes more stable in late succession the plant community becomes sensitive to subtle proximal filter effects that alter taxonomic identity but not the overall adaptive strategy.

Another example of the twin-filter model in action can be found on close examination of the orchid flora of semi-natural grassland communities in Europe, and here we shall take the example of the orchid diversity of one meadow in particular: Prato Olivino, at Monte Barro, near Lecco, Italy. The plant community at this site is a dry calcareous grassland dominated by the grass *Bromus erectus*, maintained by annual mowing and occupying nutrient-poor soil. Such

---

[2]Detailed phytosociological data are reported by Luzzaro (2005).

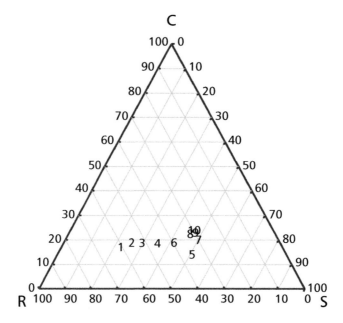

Fig. 5.18 Plant CSR strategies along the chronosequence at the Cedec glacier foreland, Italy, move from highly R-selected species towards S-selection with the development of the vegetation, but the taxonomic development of the vegetation is not reflected in the succession of CSR strategies from nodes 6–7 onwards, suggesting that proximal filter effects have come into play. (Reproduced with permission. Copyright © 2005 A. Luzzaro.)

habitats are recognized, even at the level of European legislation, as being particularly important for biodiversity. Six orchid species are prominent at Prato Olivino: *Anacamptis morio, A. pyramidalis, Ophrys benacensis, O. sphegodes, Neotinea tridentata* and *Serapias vomeracea* (see Fig. 5.19; Pierce *et al.*, 2006). Most are almost identical in terms of growth-form and are exceedingly difficult to distinguish from vegetative characteristics alone, even for a specialist with sufficient skill to discriminate orchids based on the morphology of their fruits. This suggests that the primary strategies of these species are likely to be extremely similar.

If these species exhibit essentially the same CSR-strategy (i.e. if they all have the same likelihood of passing the CSD filter), how can they coexist? In what ways have the phenotypes of these species diverged during adaptation to particular niches?

The answer is simply that whilst the primary strategies and thus general manner in which these orchids grow are convergent, divergence is evident in pollination mechanisms, the identities of the insects that pollinate them, and in slight differences in flowering times (Pierce, 2011). All except *A. pyramidalis* (which rewards butterflies with nectar) deceive naïve young native bees into visiting flowers without bequeathing nectar. However, this is achieved via different mechanisms. *Ophrys* species mislead via 'pseudocopulation' (in which

Fig. 5.19 Coexisting orchid species at Prato Olivino, Monte Barro (Lecco), Italy:
(a) *Anacamptis morio*; **(b)** *A. pyramidalis*; **(c)** *Neotinea tridentata*; **(d)** *Ophrys sphegodes*; **(e)** the Italian endemic *O. benacensis*; and **(f)** *Serapias vomeracea*.
(Copyright © 2012 S. Pierce.)

male bees are enticed into an attempted mating by flowers that mimic the form and scent of female bees) and at Prato Olivino differ in flower form and thus pollinating bee species (for details, see GIROS (2009)). *Anacamptis morio* and *N. tridentata* both use an almost identical 'food-deceptive' pollination mechanism (i.e. they mimic flowers that do provide a nectar reward), but the former flowers in April and the latter in May. Flowers of *S. vomeracea* have a distinctive pollination mechanism whereby they provide bees with shelter and warmth (the dark flowers may be 3°C warmer than air temperature during the day; Dafni *et al.*, 1981). These orchids can therefore coexist due to resource partitioning, but the resources involved are not those required for the functioning of primary metabolism and growth, but rather the pollinator resources required to complete the reproductive phase of the life cycle.

Deceptive pollination mechanisms are extremely common in the family – worldwide around 9,000 orchid species lack nectar rewards – and indeed deceptive pollination is thought to be an ancestral feature of the family. Crucially, naïve bees quickly learn to avoid deceptive flowers and this mechanism can only succeed where deceptive orchids are scattered among a profusion of species that do reward and sustain bee populations (Ayasse *et al.*, 2000; Pellegrino *et al.*, 2005). Thus the lifestyles of many meadow orchids actually depend on rarity.

There are a number of significant lessons to be learned from this example: **(1)** biodiversity is not a consequence of solely adaptation to acquire resources for vegetative growth, nor competition for these resources; **(2)** subtle differences in form and phenology beyond the primary adaptive strategy contribute to fine-scale biodiversity creation; and **(3)** species may gain a survival advantage by being subordinate rather than dominant (this is manifestly true for tropical epiphytic orchids for which survival, almost by definition, cannot depend on dominance of the plant community, nor on competition with dominant trees).

Additional mechanisms promoting diversity

So far in this review of the mechanisms affecting the species-richness of plant communities we have restricted attention to phenomena that appear to operate on a worldwide basis. It is now appropriate to refer to evidence of additional mechanisms that are suspected to play an important role on a more local scale.

Many situations have been described in which a matrix of relatively unproductive and undisturbed perennial herbaceous vegetation contains small, distinct pockets occupied by ephemerals. These may correspond to areas of shallow soil subjected to seasonal drought (Grubb, 1977) or to local disturbances such as those created by burrowing animals (Pickett & White, 1985) or ant nests (King, 1977). It is not unusual for such small-scale heterogeneity to create discontinuities of sufficient magnitude to widen the range of CSR strategists accommodated in the community. Indeed, we have already seen that Pierce *et al.* (2007) found that local disturbances by cattle were sufficient to suppress potential dominants and admit a diversity of R-selected strategists and their close allies into subalpine grassland.

Another mechanism suspected to promote species-richness in plant communities occurs in calcareous grasslands where arbuscular mycorrhizas heavily infect the root systems of the majority of the plant species (Harley & Harley, 1987) and can form extensive networks of underground connections between neighbouring species. Experiments (Grime *et al.*, 1987; van der Heijden *et al.*, 2003) synthesizing communities in the presence and absence of arbuscular mycorrhizas have revealed that they facilitate transfer of resources between individuals and in some circumstances can increase diversity by raising the biomass of seedlings and subordinate species.

Genetic diversity, intraspecific functional diversity and species diversity

The theoretical possibility that the genetic diversity of populations, or intraspecific variability, within a plant community plays an important part in sustaining species diversity was recognized in both of the following statements:

Forces maintaining species diversity and genetic diversity are similar. An understanding of community structure will come from considering how these kinds of diversity interact. (Antonovics, 1976)

Changes in genetic constitution may have important consequences at the community level. (Aarssen & Turkington, 1983)

An experimental test of this hypothesis was conducted (Booth & Grime, 2003) by synthesizing plant communities that were initially identical in species composition but had controlled differences in genetic diversity. The large stocks of genetically identical individuals required for this microcosm experiment were obtained by vegetative propagation of mature plants all selected at random from the same small (100 m^2) area of ancient pasture. After five years in which the communities were maintained under soil conditions and pasture management closely matching the parent site species, diversity was found to be greater in the communities provided with the highest level of genetic diversity (see Fig. 5.20). When molecular markers were used to compare the performance of particular genotypes in the most genotype-rich communities (Whitlock *et al.*, 2007) it was apparent (see Fig. 5.21) that under the relatively constant conditions of the experiment there were many consistent and substantial differences in the

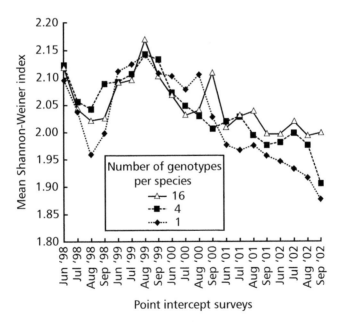

Fig. 5.20 Comparison of species diversity (Shannon-Weiner index) over five years for three communities synthesized from 11 species, with contrasted levels of genetic diversity: 16 genotypes per species (open triangles), 4 genotypes (closed squares) and 1 genotype (closed diamonds). Communities with less initial genotypic diversity exhibit greater loss of species diversity. The dip in diversity around August 1998 was an effect of transplantation from greenhouse to outdoor microcosms. (Reproduced with permission from Booth & Grime, 2003. Copyright © Blackwell Publishing – Journals.)

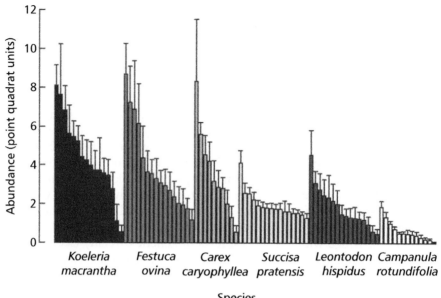

Fig. 5.21 The abundance of genotypes of six species in synthesized grassland communities maintained for five years on a natural unfertilized soil under constant conditions of simulated grazing. Each estimate is a mean from 10 replicate communities each provided initially with a random planting of the same 16 genotypes of each species. Sheep's Fescue *Festuca ovina*, Crested Hair-grass *Koeleria macrantha*, Spring Sedge *Carex caryophyllea*, Devil's-bit Scabious *Succisa pratensis*, Rough Hawkbit *Leontodon hispidus*, Harebell *Campanula rotundifolia*. Error bars indicate one standard error. (Redrawn with permission from Whitlock *et al.*, 2007. Copyright © Blackwell Publishing – Journals.)

abundance of genotypes of the same species. This strongly suggests that in this community the beneficial effect of genetic diversity on species diversity was due to the greatly reduced probability that particular species would emerge as consistent winners or losers in local neighbourhoods. Evidence supporting this explanation was obtained in another experiment (Fridley *et al.*, 2007) in which the outcomes of nine competitive pairings between three genotypes of the tussock grass *Koeleria macrantha* with three of the sedge *Carex caryophyllea* were measured. The results show that under fertile soil conditions and in the absence of simulated grazing, *K. macrantha* was a consistent winner (see Fig. 5.22a). However, when the same combinations of genotypes were maintained under infertile soil conditions and with simulated grazing, the results became genotype-dependent (see Fig. 5.22b), suggesting that under natural field conditions neither species would be vulnerable to rapid extinction. Thus genetic diversity can lead to stability.

We can hypothesize that genetic diversity leads to intraspecific diversity in functional trait values. Indeed, Bilton *et al.* (2010) investigated biomass allocation between leaves, roots and culms (flowering stems) of different genotypes of *Festuca ovina* and *K. macrantha*. The triangles in Fig. 5.23 are not CSR triangles:

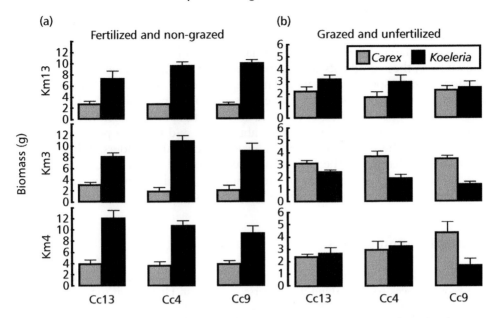

Fig. 5.22 Above-ground biomass of Spring Sedge *Carex caryophyllea* (grey bars) genotypes (Cc13, Cc4 & Cc9) and Crested Hair-grass *Koeleria macrantha* (black bars) genotypes (Km3, Km4 & Km13) in pairwise, interspecific mixtures **(a)** at high fertility in the absence of grazing, and **(b)** under low fertility and simulated grazing. Error bars indicate one standard error. (Reproduced with permission from Fridley *et al.*, 2007. Copyright © Blackwell Publishing – Journals.)

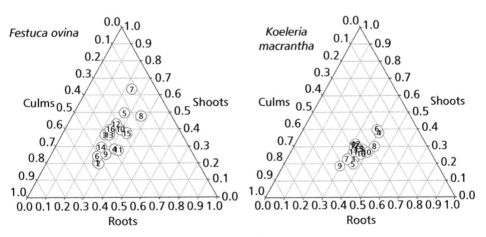

Fig. 5.23 The proportion of biomass allocated between roots, shoots and culms (flowering stems) for different genotypes (denoted by different numbers) of the grasses *Festuca ovina* and *Koeleria macrantha* from the same grassland community at Cressbrookdale, North Derbyshire, UK. (Reproduced with permission from Bilton *et al.*, 2010. Copyright © NRC Research Press.)

what they do show is a range of allocation patterns for each species grading between genotypes with smaller reproductive (culm) and larger vegetative (shoot) biomasses to larger reproductive and smaller vegetative biomasses. Although not shown here, these gradients were also reflected in characters such as the key leaf economics trait specific leaf area. This vegetative vs. reproductive trade-off is reminiscent of trade-offs we have already seen operating to shape differences in adaptive strategies between species. It is this intraspecific functional diversity that underpins variability in genotype abundance and thus the persistence of a species within the plant community. This is also crucial to understanding, as we shall see in Chapter 7, the key point that ecological filters select not against species but against some or all of the genotypes within populations of species – ecological filtering is also a process of natural selection.

Microbial communities

There are 1,000,000,000 times more bacteria in the world than stars in the sky. . . . No one would seek to understand the universe star by star. However, it sometimes seems that microbiologists are trying to understand the microbial world one cell at a time. . . . The complex nature of the microbial world may make theoretical descriptions seem impossible. But they are not. For progress will be made using simple ideas, refined iteratively and grounded in truth . . . theory will open the door to a new age in microbial ecology as we stop merely gawping at the wonder of it all, like prerenaissance peasants on a star-lit night, and start to begin to truly understand. (Curtis, 2007)

Curtis' lyrical comment reflects dissatisfaction with the modern sport of cataloguing the microbial species present at particular sites – an activity which, like plant collecting in the 19th century, unfortunately reveals little of the functions of these organisms and the processes that lead to such colossal microbial biodiversities. Thanks to the modern '**16S rRNA**' method[3] the presence of particular microbial species and groups can be detected from snippets of genetic material found in the environment (Ward *et al.*, 1990). Of course, knowing which species are present is an essential first step to understanding their ecology, and the diversity and taxonomic character of microbial communities can provide startling and useful information. Indeed, it could be argued that the 16S rRNA method has revolutionized our understanding of microbial communities simply by demonstrating the incredible breadth of microbial diversity and by revealing that there are many groups whose presence in certain habitats was effectively invisible beforehand, such as the widespread mesophilic *Archaea* (Olsen, 1994). Returning to the analogy with astronomy it seems that this method has extended our ability to investigate microbial life to the same extent that radio telescopes broadened our view of the heavens from 1933 onwards.

---

[3] The ribosome is essential to protein manufacture in all cells, and thus ribosomal RNA genes are highly conserved (relatively invariable) and so are useful for studies of broad phylogenetic relationships. The ribosome has two subunits, large and small, and it is the 16S RNA gene of the small subunit that is investigated for most prokaryotes (the subunits have different masses and shapes in eukaryotes, and so 18S rRNA is typically used for eukaryotes such as fungi and algae).

However, accurate knowledge of how communities function is unlikely to be achieved simply by knowing the generic names of the taxa involved, or that a group such as proteobacteria forms a certain percentage of the taxa present (we have already seen in Chapter 4 that extensive ecological diversity exists among the proteobacteria). As Ley *et al.* (2007) explain: 'just as pigeons and penguins are both birds but occupy very different niches, we may find it is not correct to assume that relatedness is our best guide to bacterial functions'. Indeed, it was a similar lack of correlation between taxonomic affiliation and ecological comportment that prompted ecologists to develop adaptive strategy theories, as we discussed in earlier chapters.

The comparative approach epitomized by Brock (1978) placed emphasis on the physiological ecology and life-history traits of particular species, and inspired generations of microbiologists. This approach is particularly useful for comparing the abilities of different species to compete with one another and to withstand environmental insults. It is, however, as Curtis complains, another way of trying to understand the microbial world one cell at a time. Furthermore, many of the novel organisms that have now been detected in the environment, such as the mesophilic *Archaea*, have evaded laboratory culture and are thus not amenable to detailed physiological studies of this kind. These shortcomings prompted Comte & del Giorgio (2009) to lament:

> . . . the limitation of current approaches is that they do not provide an integrative view of the overall response of bacterial [community] structure to environmental forcing, and how its different components interact to shape the resulting bacterial processes at the ecosystem level.

Clearly a comparative approach is needed that employs the 16S rRNA method to detect the species that survive or perish along environmental gradients. With such an approach we can ask, moving from one set of environmental conditions to another, who are the winners and losers? A small number of studies, reviewed by Green *et al.* (2008), have taken a functional approach to microbial biogeography, and have identified particular traits and life-histories that 'may indicate the spatial patterning of ecological strategies'. In what we are sure will become a classic example, DeLong *et al.* (2006) described changes in **functional gene repertoires** with depth in the North Pacific Subtropical Gyre. They found, along a gradient from surface waters to a depth of 4 km, a shift in microbial function away from highly active metabolisms (characterized by motility, foraging and high rates of viral infection) towards slow growth (low productivity, greater antibiotic expression and cell surface characteristics indicative of stronger defence against predation). Crucially, they were also able to detail the taxa involved at seven depths along this gradient.

However, a general theory of microbial adaptive strategies is currently lacking. In a review essay in the journal *Nature*, entitled 'The role of ecological theory in microbial ecology' (Prosser *et al.*, 2007), it was suggested that CSR strategies could provide useful macro-ecological analogues for comparative microbial ecology (this has been suggested before; Wardle & Giller, 1997), and Green *et al.* (2008) similarly drew inspiration from community processes and methods currently used in plant ecology. Universal adaptive strategy theory and the humped-back model predict that species with certain modes of investing resources will be consistently associated with particular conditions of productivity, disturbance and competition and, as we shall now discuss, there are some tantalizing glimpses that this is indeed the case for microbial life.

Comprehensive investigations of microbial diversity over the full productivity range or along changing gradients of stress or disturbance are extremely rare, although some do exist and show how intricate and responsive microbial communities can be. Indeed, Horner-Devine *et al.* (2003) found that, in experimental ponds in which fertility was manipulated, algae exhibited a humped-back relationship, as did many bacteria. However, they also found that the diversity of β-proteobacteria in the same experimental ponds showed no significant relationship with fertility and that α-proteobacteria exhibited a U-shaped relationship. This was interpreted as a reaction to the fact that the diversities of other bacteria and algae were so great at intermediate productivities that, by a mechanism that was not identified, they limited α-proteobacterial diversity in the same habitat – i.e. that the U-shaped relationship for one taxon was imposed by the humped-back relationship in a range of other taxa. However, U-shaped relationships are rare and extremely enigmatic, potentially representing special conditions of an unknown nature. A review of microbial diversity/productivity relationships in aquatic systems (Smith, 2007) found the same patterns evident for microbial as for macrobial systems.

For soil microorganisms, no single study currently exists that tests diversity over the full productivity range, although soil bacterial diversity is known to exhibit a humped-back relationship with the extent of heavy metal stress. At low concentrations of cadmium, copper, mercury or zinc competitive exclusion reduces diversity; at the highest concentrations only extremely stress-tolerant species survive; the greatest diversities are found in between these extremes (Davis *et al.*, 2004; Ranjard *et al.*, 2006). However, by comparing different soil communities at contrasting points along the productivity gradient it is evident that humped-back relationships may exist based on soil fertility. Desert soil crusts are characterized by low microbial diversities, and Yeager *et al.* (2004) found that young crusts – essentially a pioneer stage in a soil-forming succession – are formed by a single nitrogen-fixing cyanobacteria species, *Microcoleus vaginatus*, which literally prepares the ground for other cyanobacteria, lichens and mosses later in the succession. Similarly, Antarctic soils are characterized by a range of stresses including low and variable temperatures, freeze-thaw events and low nutrient availabilities. Yergeau *et al.* (2007) demonstrated that soil bacterial diversity generally decreases with increasing latitude in response to these abiotic stresses. Thus bacterial diversity declines with the harshness of the habitat, in a similar manner to the low plant diversity evident at low-productivity sites characterized by stress or disturbance, thus conforming to the left-hand side of the humped-back model.

At lower latitudes and higher soil fertilities (the right-hand side of the humped-back model) Girvan *et al.* (2004) found that arable soil had an extremely diverse microbial community. Adding fertilizer increased the number of bacterial and fungal cells and microbial biomass but, crucially, created a more homogeneous community comprised of fewer species, each with disproportionately large populations. Similarly, Gattinger *et al.* (2007) examined the abundance and diversity of methanogenic Archaebacteria in arable soils with moderate to high fertilities, manipulated with the addition of cattle manure. This was part of a long-term experiment on soil fertility at Bad Lauchstädt, Germany, which at the time of the study had been running for 103 years. They found that manure addition increased archaean biomass but decreased archaean diversity – at the highest rates of soil fertility and biomass production just one species dominated (a previously unknown species of *Methanosarcina*).

In these examples extremely high fertility leads to dominance by small numbers of species that are very efficient at producing biomass, suggesting C-selection, whilst extremely low fertility/high stress can be tolerated only by a minority of species exhibiting S-selection. Thus when a range of productivities are compared a humped-back curve does appear to be present for soil communities.

The effects of plant strategies on soil microbial communities

A number of studies confirms that soil microbial communities change not just with resource availability or injurious stress but also with vegetation type and even the identities of the species that form the plant community (e.g. Rich *et al.*, 2003; Boyle-Yarwood *et al.*, 2008). Bardgett & Walker (2004) noted that the development of microbial communities on relatively sterile deglaciated terrain at Glacier Bay, Alaska, depended first on the establishment of bacterial communities associated with *Alnus sinuata* and the moss *Rhacomitrium canescens* (species with nitrogen-rich litter) and the subsequent arrival of plant species with poorer quality litter and associated fungi. They concluded that plant traits have profound controlling effects on the character of the soil community. A recent literature review (Berg & Smalla, 2009) concluded that both plant species and soil type have controlling effects on rhizosphere microbial communities, but that 'there is no general decision about the key player: both factors can dominate depending on biotic and abiotic conditions'.

Microbial diversity may thus be heavily influenced by the plant species present, but are characteristic soil microbial communities associated with particular plant adaptive strategies? Bremer *et al.* (2007) cultivated eight species of herbaceous plants on meadow soil for three years, and found that each plant species was associated with characteristic rhizosphere communities of denitrifier bacteria. They also classified the plants into two extremely broad 'functional groups', grasses and forbs, but found no clear differences in the microbial community between these. In a more complex story, Le Roux *et al.* (2007) found strong differences in the microbial communities (ammonia-oxidizing and nitrite-reducing bacteria) associated with legumes, grasses, small herbs and tall herbs. Sugiyama *et al.* (2008) also found differences between microbial groups, in that fungal diversity, but not bacterial diversity, was correlated with 'functional type composition' (grasses, forbs and legumes) in semi-natural grasslands in northern Japan. Bezemer *et al.* (2006) found a similar effect of grasses or forbs, but that effects could be highly species-specific (i.e. different grasses may have contrasting effects). Thus different plant growth-forms may (or may not) have contrasting effects on the microbial community, but we are still unable to predict which growth-forms will have which effects and have little idea of how consistent these effects are.

Why the confusion? At this juncture we should reiterate a point we made at the start of Chapter 3: the use of taxa or broad life-form categories to delimit 'functional groups' is fraught with problems. It is an enduring myth that 'since form and function have an evolutionary origin, many functional groups are closely linked to their phylogeny, and hence can be delineated taxonomically' (Sieben *et al.*, 2010). What these authors are saying is that taxonomic groups such as grasses (the family Poaceae) essentially form a functional group and have the same general ecology, as do all legumes (Fabaceae). Species sharing the same life form, such as bushes, trees, forbs or graminoids are often assumed to function in the same manner and, by extension, have the same plant functional type (e.g. Urcelay *et al.*, 2009). However, in practice it is sometimes difficult to use such broad groups to investigate communities in an ecologically meaningful way (e.g. Zhu *et al.*, 2009).

The reason for this is simply that these categories are extremely inclusive. We have already seen (Box 3.3) that the grasses include the full gamut of CSR strategies, which are evident even within local sub-groups such as temperate lowland species. Pierce *et al.* (2007a) point out that the morphological traits which define grasses as a taxon are symptomatic of the ecological conditions in which the family first arose, but grasses have since radiated into a range of strongly contrasting habitats. In other words, we should actually expect higher level taxa such as families to be characterized by a degree of functional diversity rather than representing a coherent functional type, because taxonomic diversity represents adaptation in response to a range of ecological situations. Indeed, the grass species investigated above by Bremer *et al.* (2007) are known to exhibit contrasting CSR strategies including C-, CR-, CSR- and S/CSR-selected forms (Grime *et al.*, 2007). Thus grouping species according to family-level taxonomic affiliations or by general life-form categories provides a blunt tool for dissecting the fine anatomy of communities.

All very well and good, but what does this have to do with microbial communities? Previously we posed the question: 'Are characteristic soil microbial communities associated with different plant adaptive strategies?' We would argue that most studies of the effects of plant adaptive strategies on soil microbial diversity have not really investigated plant adaptive strategies at all, but rather general life-forms with only broad relevance to how species function and survive. Let's be clear about one thing: by **adaptive strategies** we mean precise suites of traits that we can quantify and compare between different species, even between different species of grass, and thus within communities.

Patra *et al.* (2006) investigated the effect of three grass species on the soil microbial community and rhizosphere activity in a permanent pasture in Theix, France, which had experienced the same mowing and sheep grazing regime for at least 35 years. Two of these species, *Holcus lanatus* and *Dactylis glomerata*, have previously been classified as CSR-selected (i.e. an intermediate adaptive strategy; Grime *et al.*, 2007) whereas *Arrhenatherum elatius* is strongly C-selected. Patra *et al.* found that the C-selected species was associated with genetically distinct communities of nitrate reducing and ammonia oxidizing bacteria, with respect to the two CSR-selected species, but the N_2-fixing bacteria were the same for all three plant species. This suggests that different fractions of the microbial community may respond in different ways to contrasting plant CSR strategies. Le Roux *et al.* (2007) arrived at a similar conclusion.

An extremely small number of systems have so far been investigated in such a way that they can be interpreted in terms of CSR strategies, and these include only a small range of plant species that are not widely divergent in adaptive strategy. For example, although Patra *et al.* (2006) and Bremer *et al.* (2007) found different microbial communities associated with herbaceous species known to have contrasting CSR strategies, only one or two plant species of each strategy were investigated; whether particular plant CSR strategies always select for characteristic soil communities and particular microbial CSR strategies remains far from clear. A pertinent recent development is a paper (Orwin *et al.*, 2010) comparing the microbial communities and microbially mediated ecosystem effects associated with contrasted CSR plant strategies cultivated as monocultures for seven years on a standard soil of intermediate fertility. This study concluded that 'many of the soil properties measured showed strong correlations with co-varying suites of plant traits linked to plant growth strategies' (Orwin *et al.*, 2010). It is apparent from these studies that vegetation type, the identity of plants and the character of plant adaptive strategies/suites of plant traits may be associated with the development of markedly different soil communities and properties, suggesting that there is a potentially strong link between plant-soil feedbacks and the CSR strategies that make up the plant community.

We suspect that the situation may be highly complex and modulated by the effects seen in the twin-filter model (i.e. plant CSR strategies may affect the overall character of the soil microbial community, but more subtle phenotypic differences may inflate microbial biodiversity). Complicating factors include variability in specific abiotic pressures, such as soil pH or salinity – 'complicating' in the sense that a more complicated situation provides additional niches and may support greater microbial diversities.

Facilitation in bacterial communities

Facilitation is a further complicating factor for microbial community structure and some of the best examples come not from the soil but from the rumen. Competition is prevalent in the rumen, but situations exist in which competition cannot explain niche segregation. For example, *Clostridium* strain C7 is cellulolytic and provides sugars for *Klebsiella* strain W1, which in turn fixes nitrogen and provides nitrogen and vitamins for *Clostridium* (Leschine *et al.*, 1988; Cavedon & Canale-Parola, 1992). In this way competition between the two is minimized, and would indeed be catastrophic, but facilitation favours biodiversity.

Microbiologists refer to this kind of facilitation as syntrophy, which literally means **feeding together** and signifies that the organisms provide resources for one another. It can thus be seen as a reciprocal, mutualistic symbiosis. However, facilitation in microbial communities does not necessarily involve mutualism or even resources, but can also function indirectly via the amelioration of abiotic stresses. For example, bacteria that have high respiratory rates may reduce local oxygen contents sufficiently to provide anoxic microhabitats for anaerobic bacteria in an otherwise aerobic setting (Fenchel & Finlay, 1995). This is a similar situation to facilitation in plants, whereby facilitation is more of a one-sided relationship in which the presence of one species provides a suitable abiotic environment for a second. This does not necessarily involve trophic interactions *per se*. Thus, like plants, a balance between competition and facilitation probably influences microbial community structure.

Coexistence in marine surface waters

In Chapter 4 we provided evidence supporting the view that the cyanobacteria *Synechococcus* and *Prochlorococcus*, which dominate marine primary productivity on a global scale, are extremely R-selected. If they both inhabit essentially the same surface water habitat and share the same adaptive strategy, how can they coexist?

First, it is worth mentioning that the taxonomy of these genera is extremely chaotic, with *Prochlorococcus marinus* divided into different strains depending on specific ecological characteristics, particularly optimal growth at high- or low-light intensities (Dufresne *et al.*, 2003; DeLong *et al.*, 2006; Kettler *et al.*, 2007). Crucially, if we wish to know what the underlying differences between these R-selected organisms are we must look at the fine details that distinguish strains – at the differences between genomes, and at single genes and what they do. Kettler *et al.* (2007) were able to show genetic differences between different ecotypes of *Prochlorococcus*, but these differences were so slight that they concluded: 'much of the variability exists at the leaves of the tree'. *Synechococcus* and *Prochlorococcus* are so similar that their genomes only differ by 13 genes (although some *Prochlorococcus* strains differ from one *Synechococcus* strain by as many as 33 genes; Kettler *et al.*, 2007), most of which are involved in differences in the photosynthetic machinery, sodium transport or proteins involved in iron metabolism. It is only in habitats in which environmental conditions differ drastically from those suitable for *Synechococcus* and *Prochlorococcus* that organisms with radically different genomes and adaptive strategies dominate, such as the S-selected *Richelia intracellularis* in extremely nutrient-poor waters. Thus coexistence among *Synechococcus* and *Prochlorococcus*, organisms that are extremely similar in terms of genome, phenotype and overall adaptive strategy, appears to be encouraged by the subtle effects of the proximal filter, whilst it is likely that the CSD filter governs general shifts in microbial function and biodiversity over regional scales, or with depth (e.g. DeLong *et al.*, 2006).

Novel techniques for investigating microbial adaptive strategies

We have seen that microbial diversities change in response to stress and productivity and that plants, by determining many of the niches available, may further affect soil microbial community structure. Could these patterns hinge on the adaptive strategies of microorganisms? How can we quantify microbial adaptive strategies in natural communities, in order to test this hypothesis?

Despite our cleverness as a species, having progressed through the Atomic Age, via the Space Age to the Information Age in the space of just the 20th century, it is only in the opening years of the 21st century that we have started to develop the techniques required to investigate the ecophysiology of microbes in natural settings. In this section we take a look at the dawning 'Microbial Ecology Age' that these techniques are now starting to usher in.

Orphan *et al.* (2009), for example, studied the physiology of Archaebacteria in marine sediments using a technique called **FISH-SIMS** ('fluorescence *in situ* hybridization-secondary ion mass spectrometry'). FISH is essentially a method of tagging particular organisms using a probe strand of DNA associated with dyes that fluoresce when exposed to light of particular wavelengths. This allows the target cell to be visualized under the microscope. SIMS then zaps tagged cells with a fine beam of ions, knocking off particles that can be analysed using a mass spectrometer. Orphan *et al.* provided the Archaebacteria with ammonium labelled with a heavy isotope of nitrogen ($^{15}NH_4$), and the FISH-SIMS technique allowed them to track this nitrogen and determine how quickly ammonium was taken up and incorporated into proteins. They were then able to investigate a syntrophic relationship whereby Archaebacteria supplied proteins for associated sulfate-reducing bacteria, and could even determine the impact of population sizes and the physical distances between syntrophic partners on the strength of these processes.

This elegant technique can essentially show us who does what, how quickly and where, within natural microbial communities. These kind of measurements of metabolic heterogeneity and interspecific interactions can provide precisely the kind of information that will be required to examine wild microbial communities in terms of the universal adaptive strategies of component species.

In 2007 the journal *Environmental Microbiology* asked prominent researchers to imagine what they would see in a crystal ball for the future of microbial ecology. Below are some prescient replies.

Kuypers & Jørgensen (2007) predicted that analysis of individual cells *in situ* will provide a clearer picture of 'who is there?' and 'what are they doing?' in the emergent field of **single-cell environmental biology**. This is essentially the approach of the FISH-SIMS method applied by Orphan *et al.* (2009), suggesting that we are starting to see the expansion of single-cell environmental biology as a research field, although this remains prohibitively expensive for most research groups. A range of new techniques could potentially be developed for this field: 'we may need to think of new -ISH names in parallel to FISH, like 'RISH' (radio-isotope *in situ* hybridization), 'SISH' (stable isotope *in situ* hybridization) or 'HISH' (halogen *in situ* hybridization)' (Kuypers & Jørgensen, 2007).

Rohwer's (2007) crystal ball showed that advances in computing power will ultimately lead to **real-time microbial ecology** in which imaging techniques are used to follow the dynamics of microbial activity and community composition. Advances in our capacity to sequence genetic material and store and analyse large amounts of data will also lead to the field provisionally titled 'environmental –omics' by Hugenholtz (2007) in which:

. . . viral, bacterial, archaeal and eukaryotic fractions will be routinely sampled and sequenced in parallel so that different trophic levels can be analysed in conjunction. Expressed mRNAs and proteins will be routinely obtained and analysed from the same environmental samples to get a window on community function instead of just metabolic potential. Fractionating individual populations and cells from the community for independent sequencing will become commonplace and greatly facilitate dissection and interpretation of the community data. But moreover, population genomics will mature in its own right and sampling (sequence coverage) of naturally occurring populations will go much, much deeper bringing the evolutionary processes that drive and shape populations into sharp focus . . . the structure and dynamics of microbial populations will be placed convincingly into their many and varied ecological contexts. . . . we would do well to use macro-ecological theory as a guide in this endeavour.

When the technologies foreseen in the crystal ball have been developed we will have a clearer knowledge of the adaptive responses of different species and the relative abundance of viable cells in natural communities and along environmental gradients. At that point, perhaps universal adaptive strategy theory and the humped-back and twin-filter models will provide a framework that helps us understand how galaxies of microbes form, and why every time we look through the lens of 16S rRNA we see a bustling microbial cosmos.

Animal communities

Most animals, unlike plants, do not stand still and wait to be counted, and many, particularly those that fly, migrate or control large territories, are able to range between sites and gain access to numerous plant communities. This renders precise delimitation of animal communities problematic. Nonetheless plant and animal communities may be strongly interdependent, particularly with regard to symbiotic relationships such as those between most flowering plants and insect pollinators, and animal community structure may be closely related to the character of the plant community.

Plants affect animals not only because they are a source of food, but also by determining the physical structure of the habitat and by limiting the amount of information an animal can glean about the state of resources and companions (e.g. potential mates, rivals or offspring) in its surroundings (Schulze *et al.*, 2005). Bird species-richness, for example, is related to the extent and complexity of the vegetation canopy, and thus the number and type of microsites where prey may be found (MacArthur & MacArthur, 1961; MacArthur, 1964; Recher, 1969). Similarly, the diversity of small carnivores (various cats, mongooses, foxes and genets) and their prey in the southern Kalahari Desert is most closely correlated with the extent of shrub cover. This acts to enrich the structural diversity of the habitat to essentially provide predators with a greater range of hunting options (Blaum *et al.*, 2007). This lead Blaum *et al.* to the surprising conclusion that it is the plant community, rather than the prey community *per se*, that has the greatest direct impact on these particular carnivore communities.

Whilst there can be little doubt that plant community structure may contribute to animal diversity what interests us is the identity of the adaptive strategies involved: how does the character and diversity of adaptive strategies among primary producers impact on the diversity, abundance and adaptive strategies of dependent animals?

Primary producers delimit animal diversity/productivity relationships

We have just seen that shrub cover is the most important factor determining carnivore diversity in the Kalahari Desert because it determines the physical structure of the habitat. Similarly, **live coral cover** affects reef topography and in doing so is the single most important habitat variable associated with the diversity of fish assemblages in the Mafia Island marine park, Tanzania, affecting the species-richness and abundance of corallivore fishes, invertivores and planktivores (Garpe & Öhman, 2004). In reef communities the corals, via their algal symbionts, and also free-floating phytoplankton are the primary producers. However, even the diversity of planktivore fishes changes in association with coral cover, reflecting the fact that corals mainly govern the structure of fish assemblages not via trophic relationships but by providing retreats from predators. Branched corals, considered by Murdoch (2007) to be C-selected, provide greater numbers of hiding places and are known to support greater fish diversities than the less complex, massive (mound-shaped) R-selected corals (Hiatt & Strasburg, 1960). Fishes of contrasting lifestyles all exhibit greater diversities and abundance in the vicinity of C-selected corals, suggesting a broader range of adaptive strategies in the fish assemblage. However, the effect of single coral strategies is complicated by coral biodiversity: greater diversity 'provides more ecological niches than are available in the single species stand' (Risk, 1972). Coral diversity is greatest in habitats with intermediate coral cover/disturbance (Aronson & Precht, 1995). Thus fish diversity – even the diversity of fishes dependent on a food chain in which phytoplankton are the primary producers – is highest in reefs with diverse coral assemblages and moderate to high cover (Risk, 1972; Friedlander & Parrish, 1998) where C-selected corals mix with other strategies.

Aronson & Precht (1995) examined one hectare of the Belizean Barrier Reef and defined disturbance intensities based on the extent of storm-generated coral debris. At intermediate disturbance/coral cover, coral communities exhibited not only the greatest species diversities, but also the greatest species richness and evenness (see Fig. 5.24) – i.e. rather than a few species dominating many subordinate species, communities were characterized by a larger number of more equally prevalent species. Parallels with the plant humped-back diversity/productivity relationship are immediately evident.

Cornell & Karlson (2000) suggest that coral humped-back relationships reflect, in part, a gradient of light availability, with low diversities at greater depths due to low-light stress. Thus 'species-richness of corals along depth gradients shows a unimodal, hump-shaped curve that peaks at intermediate depths'. However, they also suggest that low diversities in shallow habitats may be explained by two alternative situations: (1) high productivity encouraging rapid growth and interspecific competition, or (2) high extinction rates due to frequent disturbance. Thus two gradients may be involved: a true humped-back gradient running from chronically unproductive deeps to highly productive shallows, or

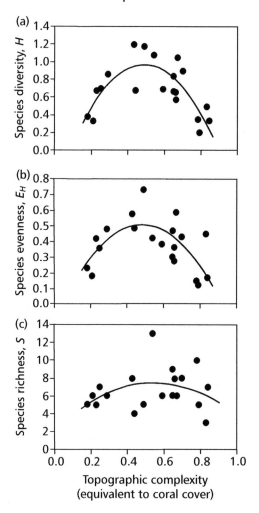

Fig. 5.24 Different measures of coral biodiversity (Shannon's species diversity index (H), Shannon's species evenness (E_H) and species-richness (S)) in a Belizean reef all show a humped-back relationship with the extent of coral cover. (Redrawn with permssion from Aronson & Precht, 1995. Copyright © Elsevier GmbH.)

alternatively what amounts to an **apparent** humped-back curve that actually follows a sequence of low→intermediate→low productivity characterized by regression into disturbance (if we imagine the centrifugal model, in Fig. 5.10 above, this would be equivalent to moving inwards along one stress gradient and, half-way through, hopping onto an adjacent spoke of the model and moving outwards again to a different situation of low diversity). Humped-back relationships are not always evident for reef communities, but Jackson (1991) suggests that this may be because of the restriction of many studies to small areas of reef. In the words of Aronson & Precht (1995): 'in many reef habitats, coral colonies are larger than the scale of observation'.

In sum, coral adaptive strategies, by determining the topography and niches available within the habitat, appear to be the most important factor determining the character and adaptations of local fish and invertebrate communities. The diversity/productivity relationship seen for corals appears to be directly equivalent to the humped-back curve evident for terrestrial plants, and can similarly be interpreted as emerging from diversity in CSR strategies – i.e. due to the ability of a greater range of functionally diverse species to coexist where biotic and abiotic pressures are least extreme and do not select for only the most extreme of strategies.

Corals are of course not the only primary producers in marine ecosystems. Ocean surface productivity is mainly determined by the activity of cyanobacteria such as *Prochlorococcus* and *Synechococcus*, which we have argued (Chapter 4) are both highly R-selected. We have seen that many strains of these species exist, as do other marine cyanobacteria, but it remains to be seen whether intermediate ocean surface productivities are associated with the greatest diversity in phytoplankton traits and adaptive strategies. We do know that phytoplankton may be classified in terms of CSR strategies based on simple morphological traits (Reynolds, 1984, 1991; Elliott *et al.*, 1999; Bonilla *et al.*, 2005), that phytoplankton CSR strategies are most diverse at intermediate disturbance intensities (Weithoff *et al.*, 2001), and that freshwater phytoplankton diversity exhibits a humped-back relationship with productivity (Leibold, 1999; Dodson *et al.*, 2000). Thus it should be relatively easy to test the hypothesis that the diversity of marine phytoplankton adaptive traits and strategies exhibit a humped-back relationship with ocean surface productivity. Could humped-back curves exist between ocean surface productivity and animal biodiversity?

Fock (2009) investigated the diversity of bathypelagic fishes (i.e. free-swimming species found below 1 km in depth) and surface productivity at 66 sampling stations throughout the Atlantic Ocean and part of the Southern Ocean, along what was essentially a planetary-scale transect covering a latitudinal range of 65° N to 57° S. In Fig. 5.25a & b we can see that surface water chlorophyll contents

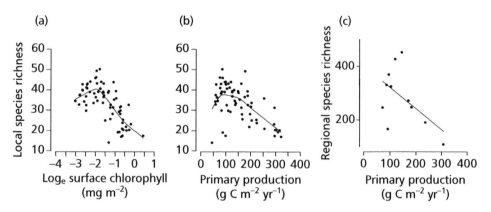

Fig. 5.25 The relationship between deep-sea fish diversity and productivity at local (a and b) and regional (c) scales in the Atlantic Ocean. (Redrawn with permission from Fock, 2009. Copyright © Blackwell Publishing Ltd.)

and productivity both exhibit humped-back relationships with the local species richness of bathypelagic fishes.

For some reason Fock then attempted to fit a linear regression to the relationship between regional species-richness and primary productivity (see Fig. 5.25c), but concluded only that this relationship 'does not fit well with the linear model' (indeed, the relationship was statistically non-significant). However, it is evident that these data are far from linearly distributed, and when we plotted the data and searched for a 'best-fit' relationship (see Fig. 5.26), we found a statistically significant correlation between diversity and productivity only when we fitted a peaked curve to the data, confirming that the relationship is most probably humped-backed (i.e. diversity of deep-sea fishes over regional scales is apparently greatest at intermediate ocean surface productivities)[4].

In Fock's study, as with many studies of marine diversity/productivity relationships, it is implicit that the activity of primary producer organisms has downstream effects on the diversity of dependent animals. It is also evident that effects on diversity cascade across communities that may be only indirectly dependent on the primary producers. After all, Fock showed that the diversity of deep-sea

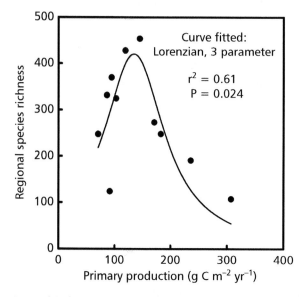

Fig. 5.26 The relationship between regional-scale diversity and productivity for bathypelagic fishes became evident (statistically significant) only when we fitted a peaked curve to the original data from Fock (2009) (i.e. the data shown in Fig. 5.25c).

---

[4]It is clear that Fock made an oversight during the interpretation of a vast and authoritative database of ocean biodiversity painstakingly acquired over 18 years. It is possible that many similar conclusions in the literature represent misinterpretations of data. This may seem like methodological nitpicking but is central to the debate surrounding diversity/productivity relationships because the definitive final statements published by the authors sometimes do not correspond with their data. In these cases the original data must be re-analysed in a statistically appropriate manner (Mittelbach *et al.*, 2001).

fishes is dependent on phytoplankton productivity occurring more than a kilometre above the zone where the fishes feed and go about the daily business of
survival. Similarly, the species richness of Procellariiform birds (a group of sea-
birds including albatrosses and petrels, which predate fish) follows a humped-
back relationship with phytoplankton primary productivity (Chown & Gaston,
1999). When these studies are considered collectively it is evident that the adaptive strategies of primary producers at the ocean surface may have knock-on
effects on the diversity of communities throughout the water column, and even
above and beyond it.

Is there similar evidence from terrestrial habitats for downstream effects of
plant CSR strategies on animal communities, particularly at higher trophic levels?
We saw in Chapter 4 that host-plant CSR strategies determine the strategies of
British butterfly species (Dennis *et al.*, 2004), and thus it is highly likely that insect
communities depend on the strategies apparent within the plant community.
Indeed, Gobbi *et al.* (2010) investigated life-history traits of both plants and carnivorous carabid beetles forming communities along a primary succession at the
Cedec glacier foreland in northern Italy. We have already described the succession
of plant CSR strategies along precisely this chronosequence (see Figs. 5.17 and
5.18 above); R-selected pioneer plants give way to S-selected species as the physical environment becomes more stable and the soil depleted of nutrients. Gobbi
and co-workers found that beetle functional traits varied alongside changes in
plant traits, and that the succession of both plants and animals could effectively
be divided into two stages: **(1)** a true succession characterized by gradually increasing vegetation cover, mainly R-selected plants and beetles with rapid developmental rates and long wings suggesting greater mobility, and **(2)** a 'climax' community
of S-selected plants supporting beetles with S-selected traits such as protracted
larval development and restricted mobility. They concluded: 'different plant
strategies may ultimately have different ways of affecting carabid assemblages via
changes in the quantity and quality of vegetation' (Gobbi *et al.*, 2010). These
changes occur over a distance of a few hundred metres and thus cannot be attributed to climate, with the main abiotic change being soil impoverishment towards
late succession. The simplest explanation is that soil nutrient status determines
the strategies and productivity of the plant community, which in turn govern the
character of adaptive traits for dependent animals and ultimately for the carnivorous beetles. Microclimatic conditions along the succession were not recorded,
but these too would presumably depend on the structure of the vegetation and
would therefore reflect the adaptive strategies within the plant community.

Thus there are strong grounds for suspecting that the adaptive strategies of
primary producers in both marine and terrestrial habitats affect the relationship
between diversity and productivity not just for the primary producers themselves
but also for a range of directly and indirectly dependent organisms. Indeed, a
review of the literature (Mittlebach *et al.*, 2001) found that humped-back curves
are common for animals and that, as with plants, this becomes particularly
evident when diversity is compared across, rather than within, community types.
This suggests that the various ambiguous or linear relationships that are
in greater evidence when this distinction is not made may simply be due to
inappropriate observation across restricted productivity ranges.

A further parallel between animal and plant biodiversity/productivity relationships arises from findings (reviewed by Bulleri 2009) that facilitation may be an important factor affecting communities of marine invertebrates at low productivities (i.e. to the left of the humped-back curve).

Twin filters and animal community assembly

The humped-back model can explain why the CSR strategies of animals change along productivity gradients, but cannot explain all of the species diversity within communities. The twin-filter model, to recap, predicts this, in that the CSD-equilibrium imposes convergence in the traits that affect ecosystem processes, but differences in non-CSR related traits may allow divergence within this subset of the species pool to allow more than one species of each strategy to coexist in the local community. Thus two species with similar strategies are likely to be segregated into different niches and coexist if they diverge in specific traits such as the precision of foraging, the expression of particular stress proteins, the type of symbiont upon which they depend, or even subtle differences in traits such as hairiness (as we saw for the bromeliads). In contrast, if two species in the local species pool have identical CSR strategies and are also highly similar in all other aspects then it is more probable that one of them will not pass the proximal filter – i.e. will not find and establish in a suitable niche to form part of the community. This is essentially an extension of the humped-back model that adds a layer of finer detail and which can explain the coexistence of similar strategies within the same habitat.

Could this also apply to animal communities?

We have seen, in Chapter 4, the example of two putatively C-selected sea snakes that coexist due to slight differences in the zones in which they forage (the Blue-Lipped Sea Krait and Saint Girons' Sea Krait; Brischoux & Bonnet, 2008). Both snakes have the same fast-paced acquisitive method of fuelling active metabolisms, and thus the CSD filter can explain why snakes in this habitat are restricted to C-selected strategies. The proximal filter has selected for slight differences in the sites where resource acquisition occurs, but not the pace or means of acquisition. Thus more than one C-selected species can coexist because a slight difference in behavioural ecology minimizes competition. We also saw the case of the European Harvest Mouse, which may share habitat with the Bank Vole and has a similar R-selected strategy. A trait not entrained in the CSR axis (a prehensile tail and thus the ability of the mouse to climb grass stems) is thought to minimize competition between these species (Ylonen, 1990) and thus allows diversity among R-selected mammals in disturbed habitats. This can similarly be interpreted as an example of the proximal filter in action; two R-selected mammals can coexist because they are sufficiently different in one key trait. Species must be suitably adapted to survive in the habitat, but not so similar that niche overlap and excessive competition prevent coexistence.

MacArthur's (1958) classic study of coexistence in warblers provides a more complex example in which five species coexist despite similar physical characteristics:

These species are congeneric, have roughly similar sizes and shapes, and all are mainly insectivorous. They are so similar in general ecological preference, at least during years of abundant food supply, that ecologists studying them have concluded that any differences in the species' requirements must be quite obscure.

MacArthur found that although these species ate essentially the same insects they foraged for them in different zones of the canopy. Thus the particular locations where food may be found, rather than the type of food *per se*, can be considered as the main difference in resource requirement between species. Additionally, MacArthur's statement that the warblers have a similar ecology 'at least during years of abundant food supply' reveals another, deeper distinction between these species:

> . . . of the five species, Cape May warblers and to a lesser degree bay-breasted warblers are dependent upon periods of superabundant food, while the remaining species maintain populations roughly proportional to the volume of foliage of the type in which they normally feed. There are differences of feeding position, behaviour, and nesting date which reduce competition. (MacArthur, 1958)

Thus MacArthur essentially saw two levels of adaptation that allowed coexistence: adaptation in general lifestyle (reliance on consistent but relatively scarce food vs. infrequently abundant food) and more particular adaptations involving specific foraging and reproductive behaviours. We would argue that the first reflects the adaptive strategy (a response to the CSD filter), whilst the second are subtle differences in individual traits that allow further niche segregation (the proximal filter).

It is worth noting that where proximal filter effects involve the differential exploitation of resources this has traditionally been labelled as **resource partitioning** (e.g. Martin, 1996). However, proximal traits actually encompass a broader range of possibilities and allow niche segregation by the differential spatial and temporal exploitation of **opportunities for survival**, not simply resource availabilities.

Being a process of survival, perhaps it would be valuable to view coexistence from the perspective of an evolutionary biologist. Here we present an example which is remarkably similar to that of MacArthur's warblers, but is undoubtedly far more famous as a classic study of speciation and coexistence: **Darwin's finches**.

Darwin collected a number of finch specimens during his stay at the Galápagos Islands. In the second edition of his travel journal, commonly known as the *Voyage of the Beagle*, Darwin (1845) highlighted the differences, particularly in beak morphology, between finch species. He also noted the restriction of species to particular islands and even suggested, briefly, that it was conceivable that they were descended from the same ancestral stock – an idea that he had developed during the seven-year interval following the publication of the first edition. However, the Galápagos finches were not thoroughly investigated as an example of evolution until David Lack, inspired by Darwin's collections, visited the islands and charted the distribution of all 14 species throughout the archipelago

(Lack, 1947). One of the main findings of his study was that a number of species with contrasting body sizes, plumage and beak shapes adapted to gather contrasting foods coexisted on certain islands. This Lack eventually explained as the result of an initial period of geographic isolation and speciation that, by creating sufficient differences between species that they no longer competed for food, then made coexistence possible.

Crucially, Darwin's finches[5] illustrate that coexistence, in this case, is not the result of the 'ghost of competition past', with the present community the winners of some long-concluded battle between species jammed together in the same place. Quite the opposite; the coexistence of Darwin's finches relies on a period of isolation, of innovation, specialization and speciation (i.e. reproductive isolation of groups previously belonging to a continuous population). This kind of speciation occurring in isolation is known as '**allopatric speciation**'. Species arising in this manner can subsequently migrate and re-encounter the preceding habitat of their evolutionary lineage, where sister species have followed their own evolutionary paths. In such cases species are less likely to compete because they are able to take advantage of different resources. (Or, as we saw with the sea snakes and warblers, exploit different physical locations where the same resource may be found.) Evolutionary biologists have a name for this process of speciation in response to changing ecological circumstance: **adaptive radiation**.

Adaptive radiation has been defined as 'the evolution of ecological and phenotypic diversity within a rapidly multiplying lineage' (Schluter, 2000). Thus the term is usually applied to localized 'bursts' of evolutionary change occurring as a founder population encounters a range of ecological opportunities that open up divergent evolutionary paths. However, it is also possible to use the term in the context of more general events such as the adaptive radiation of the birds or of the flowering plants, whales or tetrapods, for instance. One of the characteristics of adaptive radiation is that it involves speciation. Thus whereas the theory of natural selection can explain how species evolve, it is the theory of adaptive radiation that can actually explain how new species arise and diverge from one another (truly the **Origin of Species**; Coyne, 2009) – and it does so in an ecological context. Crucially, the segregation of species into contrasting niches can be understood as a process whereby adaptive radiation makes previously unexploited resources available to offshoot lineages of a parent species. In other words ecological specialization, speciation and – whenever divergent species re-encounter one another – coexistence can occur without competition as the main driving force.

Competition can however be an important selection pressure allowing evolutionary lineages to diverge in the same habitat. This '**sympatric speciation**', a form of adaptive radiation, has been demonstrated experimentally. The Three-spine Stickleback (*Gasterosteus aculeatus* complex) is not so much of a species

---

[5] 'Darwin's finches' is a term coined by Lack, and was in fact the title of his book, but perhaps one should say 'Darwin & Lack's finches' or, as Schluter (2000) suggests, simply 'Lack's finches'. If we choose to be pedantic, perhaps 'Darwin, Lack, Grant & Grant's finches' would be most appropriate – or we could be really pedantic and acknowledge that the finches belong to nobody.

as a broad taxonomic group encompassing a highly variable range of phenotypes that are classified together in the same 'complex'. Notably, individuals with similar phenotypes tend to mate with each other (Schluter, 2000). Schluter (1994) investigated a large, benthic (bottom-dwelling) form and a smaller limnetic (surface-water) form that coexist naturally, and an intermediate form occupying an intermediate feeding zone that lives in the absence of the other forms. He found that when this solitary form was placed in company with the limnetic form, the phenotypic traits of daughter generations of the two forms started to diverge in a process known as 'character displacement', which is known to reduce competition (Grant, 1972). Schluter was able to calculate that it would only take around 500 years for competition to drive the evolution of distinct benthic and limnetic forms, starting from an ancestor with the intermediate phenotype. This process could, over geological timescales, rapidly produce two lineages that are incapable of interbreeding (i.e. two species) and which have distinct ecological requirements and phenotypes. This is a clear case in which adaptive radiation and ecological specialization can occur without physical isolation, with competition as the main selection pressure.

There now exist a number of examples of such **ecological speciation**, which have recently been reviewed by Schluter (2009), but a study published following Schluter's review demonstrates how rapid ecotypic divergence may be. The Midas Cichlid (*Amphilophus* cf. *citrinellus*) colonized the Apoyeque crater lake in Nicaragua with a single founding event approximately a century ago, and has since evolved two distinct ecotypes that differ significantly in diet, body shape, head width, jaw size and lip morphology, which Elmer *et al.* (2010) suggest represents the initial stages of speciation. We will return to these examples of rapid evolution in fishes in the next chapter, as they form the basis of some of the few experimental investigations into how ecological speciation during adaptive radiation can affect ecosystem functioning, during which we will also investigate the concept that evolutionary changes may be so rapid as to occupy the same timescales as ecological processes.

These different ways in which adaptive radiation may occur illustrate a key point with regard to the messiness of ecology – although we can produce a theory, such as universal adaptive strategy theory, to describe the general constraints to evolution, it is likely that the only theories that will ever be able to explain the specific adaptations of organisms, on a case-by-case basis, are those of natural selection and adaptive radiation. This is because we can never be sure whether adaptations within a community all arose sympatrically, as a result of pressures operating within that particular community alone (*sensu* the Threespine Stickleback), or whether species evolved in allopatry and have since come together, as in the case of Darwin's finches.

For example, it is uncertain which form of adaptive radiation could be invoked to explain the coexistence of MacArthur's warblers. The different warbler species have contrasting geographic distributions in Ohio, Vermont and Maine (MacArthur, 1958), and the study was conducted where the distributions of all five species happen to overlap and they are 'sometimes found together'. Thus they do not actually form a discrete community, although they can exploit the same habitat. Does the current 'community' represent species that have evolved in isolation, via allopatry, and now occasionally coexist, or has competition at a single ancestral site resulted in sympatric speciation and ecological specialization followed by migration into ecologically diverse neighbouring sites?

This kind of question is difficult to answer, and requires detailed knowledge of the evolutionary histories of the species involved. Indeed, Grant & Grant's (2008) more detailed 30-year study of Darwin's finches has revealed that the evolutionary and ecological relationships between finches are actually much more complex than the simplified version which we presented above. Although an allopatric phase of adaptive radiation (without competition) does indeed appear to represent a starting point for speciation, populations that subsequently encounter one another can compete. This has resulted – for particular populations of particular species on particular islands – in a secondary, sympatric phase of adaptive radiation during which character displacement has further altered the phenotypes of the competing species, in a similar manner to the Threespine Sticklebacks[6]. This can occur because Darwin's finches tend to select phenotypically similar mates, again like the Threespine Sticklebacks, using physical traits such as body size, shape and, in this case, songs (Ratcliffe & Grant, 1983) meaning that reproductive isolation can occur even between sympatric groups. Indeed, mates actually prefer ecologically appropriate phenotypes, meaning that sexual selection reinforces ecological character divergence; positive feedbacks between sexual and natural selection encourage sympatric speciation (van Doorn *et al.*, 2009). Nonetheless, hybridization between species has also occurred for Darwin's finches, complicating matters further.

Adaptive radiation and community assembly

Previously in this chapter we have noted the similarities between communities of plants, microorganisms and animals, such as the prevalence of humped-back species-richness/biomass relationships. Furthermore, animal communities have provided the all-important clue that adaptive radiation and ecological speciation are key to understanding community assembly and coexistence.

It is possible that some kind of interplay between allopatric and sympatric speciation is the norm during the creation of the species pool and community assembly. This is illustrated by the bromeliad species we saw earlier in this chapter. Some of the bromeliads at Cerro Jefe, Panama, are *stenoendemics*, meaning that not only are they **endemic** (restricted to a particular region) but also to a specific habitat type. In this case several species have only ever been found in this particular montane cloud forest – on a single hilltop (Pierce *et al.*, 2002a). Notably, these species are distinguished mainly by floral and reproductive traits, and are reproductively isolated from one another by simple virtue of having different times of the day, or night, when flowers open, exploiting

---

[6]We won't present the precise details of which populations of which species have been modified by an additional bout of sympatric adaptation; these are available in Grant & Grant's (2008) authoritative text.

different pollinators. We can be reasonably sure that sympatric speciation has occurred (particularly for species within the same genus, such as *Guzmania*, that differ in having either insect or hummingbird pollinators) and can explain the coexistence of these particular species. In this case, character displacement acts to minimize competition for pollinators. In contrast, several species have a much broader geographic range and a type of extremely flexible photosynthesis (crassulacean acid metabolism) typical of climatically variable arid lowland habitats that just so happens to also be advantageous for growth in variable but wetter cloud forest habitats. These traits have allowed a small number of lowland species to invade the montane cloud forest where they now coexist with the stenoendemics (Pierce *et al.*, 2002a).

In this case the local species pool has probably been wrought by both sympatric and allopatric adaptive radiations. However, once again this is an example of a single group of organisms at one discrete site. Imagine the kind of supercomputer we would need to model the adaptive radiations of multiple taxonomic groups over entire continents, or even entire oceans, and over geological timescales! This grandeur is daunting enough to make any ecologist quake in their hiking boots. Indeed, if decades of patient field study are required to understand the ecology of a single geographically limited adaptive radiation such as that of Darwin's finches, it is little wonder that the role of multiple adaptive radiations during the creation of biodiversity over larger scales has proven so bewildering and difficult for ecologists to comprehend.

The point is that multiple adaptive radiations of different types are undoubtedly involved in creating communities, and the current confusion in ecology can be ascribed to the fact that we ecologists usually attempt to interpret snapshots of the current state of communities without understanding the waves of adaptive radiation that have washed repeatedly over the landscape.

Let's expand on this analogy. When we view the composition of a community we could be forgiven for drawing an analogy with the composition of sand grains, pebbles or silt visible in a still photograph of a beach (e.g. see Fig. 5.27), in which different adaptive strategies are represented by differently sized grains. The actual beach, of course, is not a static structure like that depicted by the still image, but the result of past waves that have continually swamped the beach to produce novel assortments of grains. The character of the grains determines the overall structure of the beach, and the precise identities of grains may vary slightly with each wave. For communities it is waves of adaptive radiation that re-assort adaptive strategies and may substitute species. Local species pools are continually renewed by adaptive radiations, with novel species then running the gauntlet of selective ecological filters to enter the community (or not). Sand grains or pebbles can be buried or washed away entirely; the shadowy twin of speciation – extinction – also shapes the species pool by decreasing biodiversity. Communities are assembled in four-dimensions as the '**ghost of adaptive radiations past**', and it is perhaps more useful to conceive of the community as the latest frame in a slow-motion movie.

We predict that detailed knowledge of community assembly rules may benefit from molecular-biogeographical accounts of species' distributions and ecological preferences in both space and time. It is not sufficient to look at the photograph

Fig. 5.27 The communities we are presented with in nature are comparable to frames from a slow-motion movie of a beach: they are composed of different 'grains' with contrasting characteristics (strategies) that are periodically re-assorted as waves of adaptive radiation wash over the landscape. Species can also be swept into the abyss of extinction. Photo shows Spotorno beach, Ligura, Italy. (Copyright © 2010 S. Pierce.)

of a community; we must watch the whole movie, perhaps by deciphering the waves of adaptive radiation recorded in species' genomes or in fossil evidence[7].

Speciation and extinction clearly underpin biodiversity but are rarely a feature of ecological theories. One exception is the *Unified Neutral Theory of Biodiversity and Biogeography* (Hubbell, 2001), and the fact that it recognizes the importance of speciation as a factor influencing community assembly is a conceptual advance. This theory was originally borne from the need to explain how the rich diversity of tropical trees Hubbell observed in Panama can arise in the same habitat, with the same resources and thus presumably with similar niches available. It is one of the most controversial of recent ecological theories because being 'neutral' it specifically demotes the importance of life-history and phenotypic differences as a factor in the formation of biodiversity, suggesting that niches are not an *a priori* necessity for the biodiversity patterns evident in nature.

---

[7]For example, a detailed fossil timeline has demonstrated the adaptive radiation from a common ancestor of subsequently co-habiting species of planktonic radiolarian in the genus *Eucyrtidium* (Kellogg & Hays, 1975).

However, it is debatable whether the modes of speciation considered by the theory are those seen in nature. These modes may be seen as approximately equivalent to allopatric and sympatric speciation, but are depicted as random events rather than as responses to environmental pressures – essentially speciation without adaptation. Hubbell decided on this view 'in the absence of a generally accepted, quantitative, genetical or ecological theory of speciation'. This has recently gained some support from the suggestion (Venditti *et al.*, 2010) that phylogenetic patterns for a range of animal, plant and fungal groups may exhibit the statistical hallmarks of random speciation events, leading to the conclusion that 'speciation is freed from the gradual tug of natural selection, there need not be an "arms race" between the species and its environment, nor even any biotic effects'.

This should be taken with a generous pinch of salt.

The statistical distributions (curves describing the rate at which taxa arise) interpreted by Venditti *et al.* (2010) as indicative of random speciation are exponential curves, suggesting that a single event kick-starts speciation followed by negligible subsequent evolution. Curves suggestive of adaptive radiation are 'a variant of the exponential model' in which secondary factors exert some influence on evolution following the initial burst of speciation. Thus both modes of speciation produce extremely similar exponential signatures. In fact, four of the five statistical models investigated by Venditti *et al.* (2010) could produce exponential curves, the difference being the mathematical assumptions underlying their calculation and thus how the curves were interpreted: 'these statistical models can produce almost indistinguishable densities, but imply different modes of causation'. Thus the difference is not in the shape of the curves, but how they are assumed to have come about.

When faced with empirical evidence that speciation occurs as an ongoing process of phenotypic adaptation to ecological conditions, supporting a modern **ecological theory of adaptive radiation** (Schluter 1994, 2000, 2001, 2009), one cannot help but suspect that subtle differences between types of exponential curve have been misinterpreted based on flawed assumptions. Perhaps more importantly, if speciation occurs with a bang or a bang followed by a fizzle, this can tell us nothing of the factors that lit the fuse. We cannot tell from phylogenetic trees alone the situations in which speciation occurred. Were populations isolated and, most importantly, how did they differentiate? Presumably differentiation relied on adaptation, with ecological and phenotypic diversity arising rapidly; but surely this is adaptive radiation? We would argue that despite the existence of alternative theories of speciation the ecological theory of adaptive radiation provides the most parsimonious and empirically well-supported explanation for the origin of species. The implication is that by deliberately ignoring the ecological theory of adaptive radiation (a theory 'widely accepted by the middle of the twentieth century' and 'regarded as one of the most highly successful theories of evolution ever advanced . . . none of its most significant claims have yet been overthrown'; Schluter, 2000), Hubbell (2001) has relied upon unrealistic assumptions about the mechanics of speciation.

Furthermore, empirical evidence demonstrates the signature of adaptive radiation in Hubbell's tropical forests: trees in Panamanian lowland rainforest exhibit a range of leaf functional traits conforming to the worldwide leaf economics spectrum (Santiago, 2007, Santiago & Wright, 2007) indicating that adaptation obeys the same economics trade-offs exhibited by plants in other habitats. Panamanian rainforest epiphytes exhibit striking ecophysiological differences that only become evident when compared throughout 24-hour periods or between seasons (e.g. Pierce *et al.*, 2002a, b) – adaptive differences that may underpin niche segregation for this component of the plant community. As we pointed out earlier, epiphytes and the trees on which they live both rely on obvious adaptive differences in order to coexist and form plant communities.

Hubbell (2005) points out that despite the existence of trait variation large numbers of species with similar traits coexist in niches such as the shaded understorey, and it is difficult to conceive how taxonomic diversity could arise in such cases: 'the literature on functional groups has been remarkably silent about the assumption of species equivalence within functional groups'.

We suggest that the **twin-filter model** and adaptive radiation provide a possible solution to this conundrum: coexisting species may exhibit convergence in traits that are crucial to the acquisition and investment of the resources required by metabolism, in response to the CSD filter, and divergence in traits that affect survival but do not have controlling effects on metabolism, promoting taxonomic-richness. Indeed, returning once again to our example of Panamanian bromeliads, we have seen that a number of species coexist that are practically indistinguishable in terms of physiology and growth but which diverge in traits such as pollination syndrome. Has Hubbell (2005), by comparing only relative growth rates and adaptations to shade, truly ruled out all phenotypic characters that might affect survival? Do understorey trees in Panamanian lowland forest share the same pollinators, drop fruit at the same time of year, respond the same way to herbivory, etc. etc.? The simple answer is 'no'. Working in the same Panamanian rainforests, Leigh (1999) pointed out that tree species may coexist because, depending on subtle inter-annual climatic variability, reproduction is more effective for certain species in certain years and the forest is assembled by bursts of success for different species. Thus species that have similar growth rates in the same habitat may indeed grow in an equivalent way but survival and persistence do not depend solely on growth. Subtle differences in single traits, particularly those involved in reproduction, may foster coexistence.

We can look to the fossil record for confirmation that adaptive radiations driven by reproductive innovations have indeed had significant impacts on forest biodiversity. In contrast to the innovations in physiology and growth-form characterising the early evolution of land plants, later plant adaptive radiations were driven by reproductive traits such as those of the seed plants, in the late Devonian, and the flowering plants, in the late Triassic (Smith *et al.*, 2010). The main impact of these adaptations was to allow novel tree species admission into existing forest communities in which advanced growth-forms and sophisticated physiologies were already dominant (Pierce *et al.*, 2005) (the earliest angiosperms were probably slow-growing, evergreen, shade-tolerant understorey shrubs; Feild *et al.*, 2004, 2009). Indeed, whenever niches are already occupied the possibilities for adaptive radiation are limited (Brockhurst *et al.*, 2007). Such difficulty in entering established communities can explain why it took as long as 75 million years for flowering plants to become widespread after their initial evolution (Smith *et al.*, 2010). Their eventual spread and rise to dominance was probably initiated by the evolution of faster growing strategies in riparian habitats, which Royer *et al.* (2010) describe as 'ruderal'. It is likely that smaller-scale adaptive radiations based on refinements of reproductive development, pollination symbioses and seed dispersal mechanisms continue to influence rainforest biodiversity. Indeed, the taxonomic diversity of complex communities is unlikely to arise *en masse*, but involve the continual interlacing of novel species throughout the existing matrix of the community with each new wave of adaptive radiation.

In order to understand the current community we must chart these past events, but we are currently extremely ignorant of the functional differences between species within the context of the adaptive radiations that have thrown them together. A corollary of this is that it is difficult to produce simple, meaningful mathematical expressions of diversity or coexistence based on the traits of organisms currently found together in a given community. This has, understandably, been the goal of many ecologists admirably seeking a quantitative, predictive understanding of community assembly rules. In most cases the peripheral messiness of communities – the precise taxonomic identities of the particular organisms that coexist – can only be understood in terms of adaptive radiation and, for ecologists, this is currently something that must simply be endured.

Hubbell (2005) offers an easy way out, suggesting that when faced with such dilemmas it may be more practical to describe community processes in terms of a neutral model.

We counter that the aim of science is to acquire a greater depth of knowledge, and that a model of biodiversity that is incompatible with empirical insight into the processes of phenotypic adaptation and ecological speciation that underpin biodiversity will not bring us any closer to the truth. In fact, if we are to view the *Unified Neutral Theory of Biodiversity and Biogeography* as a scientific theory then it is difficult to avoid the conclusion that by starting from a demonstrably false premise of neutrality it enjoys the curious and paradoxical status of a self-falsifying theory.

The work of evolutionary ecologists demonstrates that adaptive radiation and extinction are key to understanding the formation of the species pool and community assembly rules. However our integrated knowledge of evolution and ecology is currently too rudimentary to allow us to define these rules. At least by appreciating the convoluted ecological impact of adaptive radiation we can finally understand why ecology has remained so incredibly difficult to encapsulate in general theory.

We suggest that to truly understand the underlying assembly rules we must be capable of an almost impossible feat – we will need to be both precise and general. Ideally (and in science the ideal method is rarely feasible), we must understand the precise evolutionary outcomes of particular traits and events, but follow these incredibly complex interactions over landscape scales for millions of years. For real communities this could be attempted using detailed molecular phylogenies coupled with knowledge of each species' ecophysiology and functional traits. Perhaps a more achievable approach would be to run experiments in vast yet highly controlled virtual environments where both precision and generality can be attained (Hunt & Colasanti, 2007; Bornhofen *et al.*, 2011). Information technology skills, a sound understanding of the ecological theory of adaptive radiation, of universal adaptive strategy theory and of species relative abundance – and a really powerful computer system – will all be essential to ecologists pursuing community assembly rules.

In the next chapter we shall waste no more time agonising about the question of fine-scale community structure, set it firmly to one side, and focus on the ecosystem. Despite the uncertainties surrounding community assembly rules, we shall argue that adaptive strategies, reflecting a trade-off between how rapidly or conservatively resources are used, strongly affect the gross character of the community and ecosystem processes. We shall also see that whilst adaptive strategies shape the ecosystem, ecosystem functions can in turn feed back to affect microevolution and adaptation.

Summary

1   A long tradition of vegetation description and classification, particularly in Europe, has verified the widespread existence of plant communities each associated with particular conditions of soil, climate and management.

2   The structure and dynamics of a plant community are largely determined by the extent to which potentially robust adaptive strategists (C, C-R and C-S) are allowed to exercise dominance. In conditions of high productivity and little disturbance, all three, whether represented as herbs, shrubs or trees, are capable of reducing the community to a mono-specific stand.

3   The effect of moderate intensities of stress (reduced resource supply) and/ or disturbance (damage to the vegetation) is to restrict the vigour of potential dominants and to allow incursions by plants that play a subordinate role in the community.

4   Where either stress or disturbance or some combination of the two is severe, potential monopolists are excluded and, according to the balance between stress and disturbance, the community is likely to be made up of S, S-R or R strategists.

5   Points 2, 3 and 4 above, provide an essential account of how CSR theory can generate predictions about the **kinds** of plants that may be expected to occupy communities under defined conditions. However to recognize the mechanisms that control the taxonomic identity and number of the species that represent the adaptive strategies admitted into particular communities at specific locations we need to refer to two additional models (see 6 and 7 below).

6   The humped-back model provides a compact summary of the factors affecting species-richness along gradients of productivity and disturbance. Low diversity arises not only through dominance but also occurs under extreme

resource shortage and severe disturbance. The model also recognizes that the number of species attained 'within the hump' is affected by the niches available in the community and the pool of species in the surrounding landscape.

7   The twin-filter model attempts to be specific concerning the rather different traits that determine how adaptive strategies and particular species are admitted or excluded from communities. An initial '**primary CSD-equilibrium filter**' selects against traits that have major controlling effects on the movement (acquisition, retention and investment) of matter and energy used by primary metabolism, and thus has key effects with respect to ecosystem functioning. The **proximal selection pressures filter** selects against traits that affect survival, often during particular phases in the life cycle such as reproduction or juvenile development, without strong controlling effects on metabolic function, and is more exclusively concerned with admission into the community.

8   It is also apparent that there are additional mechanisms that influence community composition and diversity in specific circumstances. Arbuscular mycorrhizal networks may increase species-richness by subsidizing the resource supply to seedlings and subordinate members of communities. Genetic diversity in coexisting populations appears to be capable of sustaining species-richness by diversifying the outcome of neighbourhood interactions.

9   Technical developments have only recently allowed us to glimpse the staggering diversity of microbial communities. One of the most important findings emerging from these studies is that the high soil microbial biomass evident in the most productive terrestrial habitats is accrued by the action of small numbers of species that are able to competitively exclude others. Low microbial diversity characterized by taxa adapted to resist extremes of abiotic conditions is evident in the lowest productivity soils. This provides strong support for the humped-back model and suggests that microbial adaptive strategies may play a central role in determining community structure in the same manner as those of plants.

10   In future, macro-ecological theories such as universal adaptive strategy theory and the humped-back and twin-filter models may provide useful contexts for understanding micro-ecological processes, and may help to link microbial ecology as a discipline with that of macro-ecology. However, it is not currently possible to determine precise CSR strategies for microorganisms; this will probably depend on investigating the changes in species along environmental gradients using modern 16S rRNA techniques and on other exciting methods that are currently being developed for determining 'who does what, where' in natural settings.

11   The CSR strategies of primary producers have either been shown (for terrestrial plants and corals) or are strongly suspected (phytoplankton) to affect the character and diversity of animal communities that are both directly and indirectly dependent upon them. These effects may be based on trophic relationships, but often also on the fact that contrasting life-forms of different plant or coral adaptive strategies determine the physical

topography of the habitat, governing the availability of hiding places and the quality of information animals can obtain about the state of resources and companions. Thus the diversity of primary producers may translate into diversity within the animal community.

12 Competition is an important ecological process, but is not ubiquitous and cannot be invoked as **the** fundamental process that creates biodiversity. This role falls to **adaptive radiation**, because speciation reflects ecological diversification. Adaptive radiation, either occurring in allopatry (isolated populations) or sympatry (divergence in the same habitat), builds an ecologically diverse group of organisms (the **species pool**), subsets of which can then survive together and coexist in the same habitat by occupying contrasting niches, thereby minimising competitive interactions. Extinction, by detracting from the species pool, is also a key concept for biodiversity.

13 Examples including **Darwin's finches** demonstrate that ecological concepts such as coexistence and diversity make sense only in the light of the evolutionary histories of organisms. Communities are four-dimensional assemblages. Whatever is currently observable in three dimensions represents the 'ghost of adaptive radiations past'.

14 Understanding community assembly rules will involve the almost impossible feat of following the precise evolutionary events and interactions of organisms over millions of years and over landscape scales – something that may only be achievable with experiments conducted in virtual environments.

6

From Strategies to Ecosystems

Though the organisms may claim our primary interest, when we are trying to think fundamentally we cannot separate them from their special environment, with which they form one physical system. It is the systems so formed which, from the point of view of the ecologist, are the basic units of nature on the face of the earth . . . there is constant interchange of the most various kinds within each system, not only between the organisms but between the organic and the inorganic. These ecosystems, as we may call them, are of the most various kinds and sizes.

(Tansley, 1935)

In the first five chapters we pursued an uncompromising agenda in asserting that natural selection is more than a universal process by which a great diversity of life has evolved on Earth. We argued, with supporting evidence, that within all branches of the tree of life, constraints of habitat interacting with the limited potentiality of the organisms themselves have restricted the outcomes of natural selection to a rather narrow range of basic alternatives in life-history, resource allocation and physiology. To the detriment of ecology as a predictive science, this evidence of a consistent predisposition in the fundamental architecture of evolution has been obscured by a focus on micro-evolution and, until recently, a tendency to ignore widespread evidence of repeated patterns in macro-evolution. The purpose of this book, and Chapter 6 in particular, is to review the opportunities for ecological interpretation and prediction that arise at the level of the ecosystem from recognition of the occurrence of repeated patterns of adaptive specialization in component organisms. The exciting potential of this approach to ecosystems arises from two specific proposals that emerge from the evidence presented in earlier chapters. The first consists of the triangular model of primary adaptive strategies, defining universal patterns of functional

The Evolutionary Strategies that Shape Ecosystems, First Edition. J. Philip Grime, Simon Pierce.
© 2012 John Wiley & Sons, Ltd. Published 2012 by John Wiley & Sons, Ltd.

specialization in organisms and (we now add) in the ecosystems they occupy. The second is the twin-filter model that usefully differentiates between traits that drive ecosystem functioning and those that determine which among many candidate species actually gain admittance to particular ecosystems at specific locations. In this chapter we explain how these two models can clarify our understanding of how ecosystem functioning varies from place to place and under changing conditions of environment and management. We also identify areas of ignorance and uncertainty that remain barriers to further progress. First, however, we pay homage to ecologists who were among the first to begin to establish linkages between the functioning of ecosystems and the characteristics of component organisms.

Back to Bayreuth

> We are seeing massive changes in landscape use that are creating even more abundant successional patches, reductions in population sizes, and in the worst cases, losses of species. . . . How are the many services that ecosystems provide to humanity altered by modifications of ecosystem composition? . . . what is the role of individual species in ecosystem function? (Schulze & Mooney, 1993)

From this excerpt from the editorial statement it is evident that the Bayreuth Symposium, with contributions from a wide spectrum of ecologists from around the world, was ground-breaking in the sense that it was the first comprehensive attempt to make explicit connections between declining biodiversity and ecosystem functioning. As we might expect in view of the novel agenda, a great diversity of information of varying relevance to the main theme was presented and it is fascinating to see in some contributions early exposure of ideas that were to become research initiatives in the following decade. Many contributors recognized the need to recognize functional groups as a mechanism for reducing the problems that arise where ecosystems contain an intractably large number of species. In most instances the basis for recognising the functional types of greatest informational value with respect to ecosystems appear to have relied at this early stage on taxonomic and subjective judgements. This impression is confirmed by the following conclusion drawn by the editors in the closing section of their book under a subheading entitled 'Are there functional groups?':

> The unsolved problem is whether functional groups, which consider species collectively, are sufficient to describe ecosystem processes. Any grouping of species will depend on the objectives. Therefore, the result of defining functional groups will be quite different, depending on the aim. . . . There is no universal classification of functional groups, because the traits that are important in predicting effects on ecosystem processes differ strongly among ecosystems as well as within ecosystems for different processes . . . . (Schulze & Mooney, 1993)

We agree that different sets of organismal traits are often associated with different aspects of ecosystem functioning. In particular we recognize the usefulness of defining functional types in relation to different scales of investigation. However, there is no way to sidestep the outright conflict between 'There is no universal classification of functional groups' (Schulze & Mooney, 1993) and the very different view expounded earlier in this book and henceforth in this chapter. It is necessary that we obtain a clear understanding of how such divergent opinions came to exist with respect to the nature, diversity, scope and utility of functional groups. We believe that an important clue to this controversy is to be found in the form of the 'Rivet Popper' analogy of ecosystem deterioration enunciated by Paul Ehrlich in his Foreword to the Bayreuth Symposium. In Box 6.1 we comment on the relevance of this analysis of the impact of declining biodiversity on ecosystem functioning and examine its influence on a decade of theoretical and experimental research.

As the end of the millennium approached, some major changes in philosophy and procedure occurred. A protocol involving standardized, multi-trait screening and multivariate analysis was proposed (Díaz & Cabido, 1997) and implemented for plants (Grime *et al.*, 1997; Díaz *et al.*, 2004) in an attempt to provide a more objective basis for recognising functional groups. It has also been possible to perform comparisons of trait variation in plants at a regional (Kleyer, 1999) and world scale (Díaz *et al.*, 2004; Wright *et al.*, 2004). These large-scale collaborative projects are slowly changing the balance between theory and empirical evidence. To utilize this increasing capacity to recognize functional types of organisms it will be necessary to discover their potential role (if any) in the functioning of ecosystems. Equally important, however, it will be essential that we explain in terms of natural selection the mechanisms by which particular functional types of known influence on ecosystem functioning succeed or fail in specified instances of ecosystem assembly.

Box 6.1: The rivet-popper analogy

The Bayreuth Symposium (Schulze & Mooney 1993) begins with an extraordinary vision of how losses in biodiversity are taking place across the world and are destined to impact on ecosystems. In his Foreword, Paul Ehrlich states that:

'Ecologists generally accept the viewpoint expressed in the "rivet popper" analogy that a policy of continually exterminating populations and species eventually will dramatically compromise ecosystem services' (Ehrlich, 1993).

He then provides an explanatory quotation:

'Ecosystems, like well-made airplanes, tend to have redundant subsystems and other 'design' features that permit them to continue functioning after absorbing a certain amount of abuse. A dozen rivets or a dozen species, might never be missed. On the other hand, a thirteenth rivet popped from a wing flap, or the extinction of a key species involved in the cycling of nitrogen, could lead to a serious accident.' (Ehrlich & Ehrlich, 1981)

(Continued)

Ehrlich then used this perspective to argue that we cannot predict the critical points at which attrition will cause serious disablement of ecosystems and he advocated new research to address this problem. The ecological journals bear witness to a swift response to this rather specific call to action. The published record (see for example Karieva, 1994; Naeem *et al.*, 1994; Karieva, 1996; Tilman *et al.*, 1997; Tilman, 1999; Hector *et al.*, 1999; Kinzig *et al.*, 2001) reveals a large volume of theoretical and experimental studies in which ecologists sought to explain how losses of species might be impacting on ecosystem functioning. There is also available a substantial critique (Huston, 1997; Huston *et al.*, 2000; Wardle *et al.*, 1997; Wardle *et al.*, 2000; Smith & Knapp, 2003; Thompson *et al.*, 2005) that will be essential reading when the history of this episode is recounted.

Why did the rivet–popping analogy of Ehrlich & Ehrlich (1981) prove to be such a contentious approach? Perhaps the most serious objection to this model was that it did not capture the commonest scenario by which organisms are lost and ecosystems deteriorate and disappear. In particular, the random deletions of species applied in many experimental simulations of species loss did not accurately reproduce the predictable functional shifts resulting from changes in land use. Burning, ploughing and fertilizing of ancient woodlands and grasslands, for example, are consistently identified as the most potent drivers of species extinctions, but these occur as predictable functional shifts changing the whole ecosystem, as sets of species with particular traits are replaced, via dispersal and immigration, by other sets with different traits. This concept of species replacement has not been considered in plant cultivation experiments that purport to support the rivet-popper analogy (e.g. Cardinale *et al.*, 2011), in which ecosystem degradation is viewed as a steady decline in species numbers culminating in zero – i.e. species removal without consideration of the dynamics of species entry into the community.

'Species are not becoming rare or extinct in random fashion like bulbs popping on a Christmas tree. Species are eliminated by the same processes that are destroying or transforming their host ecosystems. . . . it is not easy to see the relevance of simulations of species loss that depend upon random deletions set against constant environmental conditions.' (Grime, 2002)

An additional limitation of many of the experiments that sought to detect benefits of species diversity upon ecosystem functioning was their scant attention to insights available from direct study of natural communities and ecosystems in the field. Even quite old textbooks recognized the principle that, in the majority of ecosystems, mass flow processes are responsible for phenomena such as primary production, nutrient cycling and carbon storage and are likely to be controlled by the dominant organisms in each community. Species-rich communities do occur in nature and it remains likely that benefits of biodiversity to ecosystems, many involving facilitation, will be demonstrated eventually in some natural habitats. However, as explained in Chapter 5, it is only in the low biomass communities associated with the left-hand side of the **humped-back model** that rising diversity and increasing biomass usually coincide. We urge caution in studies where benefits of high diversity to ecosystem productivity are reported to occur at high biomass. In immature, synthesized, perennial, communities (e.g. Hector *et al.*, 1999) it is essential to establish that such relationships have not resulted from the coincidence of a small number of high-yielding dominants with a large number of surviving but suppressed species with little impact on the community. We concur with Thompson (2010) that there has been much wasted effort (and money) in the search for benefits to ecosystem functioning that supposedly arise from high species-richness.

The Darwinian basis of ecosystem assembly

In Chapter 5 we introduced a twin-filter model that attempts to describe how admission and persistence of adaptive strategies and species are controlled at the level of each community. Here, as a preliminary to considering the significance of the twin-filter model in the larger context of the ecosystem we will reprise the key features outlined in Chapter 5.

Evidence supporting the twin-filter model is much more readily available for plants and for particular communities of animals that have been the subjects of intensive study. As its name suggests, two mechanisms form the essence of the twin-filter model. First, it is proposed that the equilibrium between competition, stress and disturbance prevailing in the habitat (the CSD filter) selects core traits such as those having controlling effects on the movement of matter and energy utilized by primary metabolism; traits embroiled in the three-way trade-off between resource acquisition, retention and investment. With the exception of finely partitioned, heterogeneous habitats such as limestone pavement, some coral reefs, deep-sea vents or seasonally flooded forests, it is not unusual for quite extensive areas of habitat to experience a similar CSD-equilibrium. It is therefore predicted that the most common effect of the CSD filter will be to exert a strongly convergent effect on the core traits of the functional types and species recruited within each of the various component communities of the ecosystem. As we explain in more detail later in this chapter, empirical studies such as the Integrated Screening Programme confirm that the core traits filtered by the CSD-equilibrium are very directly related to ecosystem functioning, and as these traits govern the movement of matter and energy we can draw the conclusion that the CSD filter is the main controller of major ecosystem processes.

The effect proposed for the CSD filter immediately poses the question: 'How do species remain in stable coexistence within communities and ecosystems if they have gained entry through a filter that encourages trait similarity?' Our response to this question is to acknowledge that frequently such coexistence does not occur. Often the general effect of the CSD filter will be to resist entry and to continuously expel many of the individuals that have been admitted. The severity of this effect will increase in circumstances where low stress and disturbance permit dominance by C-selected strategists. As we explained in Chapter 5 circumstances can arise, for example, in productive herbaceous vegetation and in marine mollusc beds, where the equilibrium of the CSD filter is so dominated by competition that monospecific communities are maintained. In the majority of communities and ecosystems, however, the levels of stress and disturbance are sufficient to restrict competition and to bring the proximal filter into play. We suggest that the mechanism by which this occurs is through the creation of spatial and/or temporal niches capable of exploitation by species that, whilst still all attuned to the same CSD-equilibrium, are specifically adapted to particular stress or disturbance factors. An alternative way in which to describe the diversifying role of the proximal filter is to suggest that, with the reduced vigour and abundance of monopolists, the fine temporal and spatial structure of the environment becomes more exposed and influential, permitting incursions by species and genotypes that enjoy fitness benefits on a local or temporary scale within their community but without necessarily exhibiting major changes in the activity and pace of primary metabolism and thus ecosystem functioning.

A further comment is appropriate concerning the identity of the traits selected by the proximal filter. These may be predicted to vary to a considerable extent in relation to traits affecting the overall adaptive strategy. In particular, for animal, plant and microbial species we would expect that the comparatively long life-spans of S-selected individuals would be associated with physiological specializations diversifying mechanisms by which individuals tolerate limited food supplies and challenging physical and chemical conditions. By contrast, in the transient habitats where R-selection prevails we suspect that the major avenues of adaptive specialization will be concentrated in the regenerative phase of the life-history with coexistence mechanisms strongly linked to different mechanisms of juvenile dormancy, rapid maturation, reproduction and dispersal.

How do primary adaptive strategies drive ecosystem functioning?

It is now appropriate to shift attention from the processes by which organisms are recruited into ecosystems to the way in which variation in their traits affects the way ecosystems function. More specifically we will test the hypothesis that the three-way trade-off of the universal adaptive strategy theory developed in earlier chapters provides the basis for a broad understanding of variation in the core structure and functioning of ecosystems.

Plants are the dominant contributors to terrestrial ecosystems and for this reason we will allow them first place in this account. In particular processes such as litter decomposition and mineral nutrient cycling there is no doubt that animals and microorganisms contribute significantly as ecosystem drivers and there is an urgent need to define their many different roles. However, we will argue here that the best prospects for rapidly creating a functional classification of ecosystems for the purposes of basic science, conservation and management is to develop a framework based initially on the functional characteristics of the primary producers. In this chapter we focus mainly on terrestrial ecosystems, where it is relatively easy to collect and identify the main contributors to biomass and, as explained in Chapters 3 and 5, progress has been made in identifying the traits that are most informative with respect to ecosystem functioning. However, as we saw in the previous chapter, not only is a similar relationship apparent between biodiversity and productivity in aquatic ecosystems but adaptive strategies appear to be subject to the same trade-offs. Thus the processes that we shall go on to describe here for terrestrial ecosystems may also prove to be of relevance to aquatic ecosystems when more information becomes available.

We will also argue that considerable progress may be achieved even if attention is confined to the species that are the dominant contributors to the biomass. The case for using dominant plant species in this way relies on a wealth of evidence, some quite old, that the character and viability of many ecosystems depends on the dominant plants. Excellent examples here are *Calluna vulgaris* in heathland (Gimingham, 1972), *Ammophila arenaria* in coastal dunes (van der Putten *et al.*, 1993), redwood forests (Stone & Vasey, 1968), mangroves (Smith, 1987) and seagrasses (Silander & Antonovics, 1982).

The plant traits that drive ecosystems

Table 6.1 lists nine plant traits that are identified in Chapter 3 as useful in defining adaptive extremes associated with CSR theory. The table also summarizes the way in which the traits vary in relation to the three extremes of specialization. Effects of this trait variation on five aspects of ecosystem functioning are predicted in the lower part of the table. The essential basis of these predictions will now be summarized briefly for each aspect of ecosystem functioning.

Primary production
It is predicted that in the low stress/low disturbance conditions associated with dominance by C-selected plants, production will be sustained at a high level and this will allow the accumulation of a large mass of vegetation in circumstances where predators, parasitoids or chemical intervention by farmers provide effective 'top-down' protection against herbivores. Where high potential productivity coincides with frequent disturbance and biomass removal, production is likely to be reduced as the vegetation composition shifts to vegetatively resprouting perennials and ephemerals with seed banks.

In habitats characterized by chronic shortages of one or more resources, dominance by S-selected plants will coincide with continuously low production. Over many centuries this low production in so-called 'marginal habitats'

Table 6.1 Key traits of different plant adaptive strategies and some of their effects on ecosystem functioning (Grime, 2003). Note that statistically significant correlations between traits (i.e. trait syndromes representing adaptive strategies) were derived from the ISP, as presented in Table 3.2 and Grime *et al.* (1997)

	Primary strategy		
Plant traits	C-selected	S-selected	R-selected
Life-history	long	very long	very short
Life-span of leaves and roots	short	long	short
Potential growth rate	rapid	slow	very rapid
Concentration of nutrients in leaves	high	low	high
Concentration of carbon in leaves	low	high	low
Leaf toughness	low	high	low
Palatability	high	low	high
Leaf decomposition rate	rapid	slow	very rapid
Seed or spore production	delayed	very delayed	early
SOME EFFECTS ON ECOSYSTEMS			
Primary production	high	low	moderate
Carbon concentration in vegetation and soil	moderate	high	low
Retention of nutrients and pollutants	weak	strong	very weak
Resistance to physical damage	low	high	low
Recovery from damage	rapid	very slow	very rapid

provided a respected limit guiding conservative exploitation, protection of soils and ecosystem services and the survival of 'low-density-low – intensity' human cultures. Increasing human population densities, deforestation and overgrazing, allied to climate change, threaten widespread collapse of ecosystems dominated by S-selected vegetation.

Herbivory

In landscapes with long and complex histories of farming it is often extremely difficult to analyse the past and present interactions between plant adaptive strategies and mammalian herbivores of various kinds. Even in semi-natural terrain, vegetation-herbivore relationships may be affected when herbivores or their predators experience land enclosures or hunting. One consequence of these problems is that it has been easier to investigate plant–herbivore interactions by concentrating attention on feeding by invertebrates. Boxes 6.2, 6.3 and 6.4 provide examples in which the role of particular invertebrates within the ecosystem can be interpreted as a direct consequence of the adaptive strategy of the dominant plants exploited. First, in Box 6.2, we see a simple case in which two closely adjacent sites differ radically in soil fertility and the associated adaptive strategies of the dominant plants which, in turn determine that in one site the dominant snail is a herbivore but in the other the commonest snail functions as a decomposer. In Box 6.3 there is a more extreme comparison involving tropical trees. The third example (Box 6.4) concerns a two-link chain of herbivory that relies rather precariously upon exploiting the immature phase in the development of otherwise effectively-defended leaves. Later in this chapter we comment on the extent to which case studies of this kind can be assimilated into a common framework.

Carbon concentration in the vegetation and soil

Although the rate of carbon fixation in stress-tolerant vegetation is slow, a high proportion of the photosynthate is predicted to be incorporated into recalcitrant organic forms that resist herbivory and slow the subsequent processes of decay. Under cool, wet, anaerobic soil conditions, residues from S-selected trees and their mycorrhizal symbionts are associated with massive depositions of soil carbon that in the case of the boreal forests are a very significant contributor to the global stock of terrestrial carbon. In mature tropical forests mineral nutrient constraints frequently coincide with S-selected leaf canopies and extensive mycorrhizas but here, under continuously warmer conditions, decomposition restricts storage in the soil and the vital contribution of these forests to the world carbon budget is heavily dependent on the stocks maintained in the trees themselves. Vegetation composed of R- or C-selected species has a capacity to capture carbon at a rapid rate, not least because they frequently occupy fertile soils where production is not seriously constrained by limitations of water or nutrients. The downside here is that the same conditions are conducive to high rates of release of carbon dioxide back to the atmosphere through plant and microbial activity. On first inspection, C-selected herbs, shrubs and trees offer an attractive option for remediation of the continuing release of carbon from fossil reserves to the atmosphere. It is clear however that on the

---

Box 6.2: Snails at the Winnats Pass

The Winnats Pass in North Derbyshire is a spectacular steep-sided limestone gorge in which highly contrasted ecosystems face each other on north- and south-facing slopes. In addition to the very contrasted climates that arise from the different exposure to the sun the bedding planes of the limestone on the north-facing slope are such that drainage water tends to seep outwards into the soil and springs occur in several places. The strata dip into the south-facing slope allowing rapid drainage into the underlying limestone. These differences are associated with a well-defined contrast in flora and fauna between opposing sides of the gorge.

A consistent feature of the Winnats Pass is the difference in distribution and ecology between two large snails, *Cepaea nemoralis* and *Arianta arbustorum*, both of which are abundant at the site and are approximately the same size at maturity. *C. nemoralis* is confined to the south-facing slope over which it occurs uniformly and at low density in a short turf dominated by slow-growing, tussock grasses such as *Festuca ovina* and *Koeleria macrantha*. In marked contrast, *A. arbustorum* is aggregated at high population densities in areas of deeper fertile soil on the north-facing slope dominated by the tall herb *Urtica dioica*, with an understory of another broad-leaved but shade-tolerant herb, *Mercurialis perennis*.

In a series of field investigations (Grime & Blythe, 1968) and laboratory feeding experiments (Grime *et al.*, 1968, 1970) it became apparent that, despite similarities in size and close taxonomic affiliation the two snails play radically different roles in their neighbouring but contrasted ecosystems. Analysis of the gut contents of individuals (evacuated by gorging the snails on filter paper immediately after their removal from the field) revealed that, whereas *C. nemoralis* was feeding mainly on the dead leaves of tussock grasses, *A. arbustorum* was exploiting the living foliage of *Urtica dioica*. The investigation also exposed differences between the two snails in life-history, feeding behaviour and desiccation tolerance that appeared to be closely attuned to their habitats. However, for the specific purposes of this chapter, the key facts are that:

- on the south-facing slope, dominance of the vegetation by S-selected, unpalatable tussock grasses has resulted in dominance of the initial step in the above-ground food chain by a mollusc capable of exploiting low quality, relatively indigestible food;
- on the north-facing slope, the presence of the C-selected, nutritious leaf canopy of *Urtica dioica* provides a concentrated source of high-quality, weakly-defended food sufficient to maintain the energetic foraging and high reproductive rate of an R-selected snail.

The message emerging from this study is that the adaptive strategy of a dominant plant may dictate the biology of the animals (and, as we shall see later, also the microorganisms) that exploit it for food. The evidence from the Winnats Pass further suggests that plant adaptive strategy will exert controls on the relative volumes of materials passing through the herbivore and decomposer food chains.

Box 6.3: The ruler from Madagascar

Office clerks in the era of Charles Dickens were equipped not only with high desks and quill pens to produce their 'copper-plate' hand-written documents. Each was also supplied with a remarkably heavy object resembling a small, black rolling-pin so smooth and heavy that it was easy to imagine that it was made of iron rather than wood. These rulers provided the straight edge against which to ink across the page and they were made from a tropical tree known as Gaboon Ebony (*Diospyros crassiflora*) that grows extremely slowly and produces wood of exceptional density. Figure 6.1a illustrates a leaf of ebony photographed in Madagascar in its natural habitat, an extremely nutrient-deficient bog. A second picture (Fig. 6.1b) was taken at a site about 50 metres from the first and shows the mixed tree canopy on a highly productive river terrace. We have in these two photographs further evidence to implicate the primary adaptive strategies of plants in the relative importance of the herbivore and decomposer food chains they support. In the ecosystem dominated by ebony we see an S-selected leaf canopy so long-lived and resistant to herbivory that its leaf surfaces are colonized by lichens. On the nearby river terrace, defence expenditure against herbivory is so ineffectual and secondary to competitive foraging for light that many leaves are full of holes even before they are fully-expanded.

Fig. 6.1 Leaves with differing degrees of inherent defence: **(a)** the S-selected Gaboon Ebony tree *Diospyros crassiflora*, exhibits long-lived leaves highly colonized by epiphytes, showing no signs of herbivory, and **(b)** extensive herbivory evident in the canopy over a nearby highly productive river terrace. (Copyright © 2001 J.P. Grime.)

Box 6.4: A chink in the armour: Oaks, Great-tits and caterpillars at Wytham Woods

It is now appropriate to examine conditions in which the controlling effects of plant adaptive strategies on ecosystem structure and functioning are more complicated. In the study reported by Charmantier *et al.* (2008) we encounter a food chain in which a relatively stress-tolerant tree species with substantial investment in anti-herbivore defence is nevertheless exploited by a small, short-lived (R-selected) bird at the University of Oxford Field Station at Wytham Woods. Oaks (*Quercus robur*) are the dominants of the tree canopy at Wytham and here, as at many other sites in Britain, they are exploited by a large number of herbivorous insects. The majority of these insects cause only slight damage to the leaves, which have a hard exterior and are produced in a single cohort in the spring (Jones, 1959; Shaw, 1974). However, two species of Lepidoptera may cause severe defoliation of mature trees and seedlings (Newbould & Goldsmith, 1981) and, at Wytham, caterpillars of one of these, the Winter Moth (*Operophtera brumata*) are a food source of critical importance to the local population of Great-tits (*Parus major*). Long-term monitoring of the breeding success of this bird species at Wytham has revealed a critical chain of events each spring that depends upon spring temperatures, leaf expansion, caterpillar development, and food supply to the Great-tit hatchlings. In a remarkable investigation extending over five decades Charmantier *et al.* (2008) show that the population of Great-tits at Wytham has displayed remarkable plasticity allowing close tracking of variation in spring temperatures in date of breeding.

This investigation adds a cautionary note to any overly simplistic attempt to use **universal adaptive strategy theory** to predict the relative importance of herbivore and decomposer food chains. British Oaks are classified as SC-strategists and their leaves would be predicted to enter the decomposer food chain. However, in common with many other SC-strategists, the young leaves have a window of vulnerability in the phase of expansion before the hard exterior of the mature leaf is fully developed. This means that, according to the climate prevailing in each spring, the capacity of the caterpillars and Great-tits to exploit the immature leaves will vary to some extent. This phenomenon has consequences for Oaks: if defoliation is severe, trees may be stimulated to produce a second cohort of leaves.

---

scale required this enterprise would become a direct competitor for the environments, water, nutrients, transport and storage facilities increasingly necessary to support conventional agriculture.

Retention of mineral nutrients and pollutants

As in the case of carbon, the residence time of mineral nutrients and pollutants in the ecosystem are predicted to be much longer where the vegetation is dominated by stress-tolerators. This is because the durability of the living plant parts above and below ground and their dead residues is likely to be much higher. Evidence consistent with this prediction is presented for the persistence of radio-isotopes from the Chernobyl Incident (see Box 6.5).

Resistance and resilience

The leaves and roots of S-selected plants are predicted to have experienced intensive selection promoting low palatability and toughness sufficient to deter

Box 6.5: Persisting pollutants and the Chernobyl Incident

Already in several sections of this book reference has been made to the slow-dynamics of stress-tolerant plants and the long residence time of carbon and mineral elements assimilated into their tissues. The Chernobyl incident provides a striking illustration of the wider implications of this phenomenon. On 26 April 1986, at 1:23 a.m. local time, reactor number four of the Chernobyl Nuclear Power Plant, near Pripyat in the Ukrainian Soviet Socialist Republic, was undergoing an experimental test of emergency cooling systems when a series of human errors led to a chain reaction. The resulting spike of power, estimated at around ten times the typical heat output of the reactor, converted coolant water into superheated steam that became trapped inside the reactor building. The resulting steam pressure explosion ripped the roof off the reactor, and the outside air rushed in to help ignite 1,700 tonnes of graphite reactor components, sending a plume of radioactive isotopes, chiefly cesium-137, billowing into the night sky. Thanks to the spin of the planet and the Coriolis force, this lethal cloud moved westward to bathe much of northern Europe in ^{137}Cs-contaminated rainfall. The tragedy killed 56 people directly due to radiation poisoning and may kill an estimated 4,000 more as cancers develop later in life, but the aspect of this pernicious legacy relevant to our discussion is the fate of the radioactive rain that fell on upland ecosystems in Britain.

In Wales and northern England upland areas were particularly exposed, and the detection of high levels of ^{137}Cs in pasture grasses and sheep essentially called a halt to agriculture. In response to such a severe impact on local economies it was essential to estimate the period for which the contamination would remain at unsafe levels. However, after two years it was apparent that these estimates were hopelessly optimistic. Contributing to a British Government Inquiry as to why ^{137}Cs had proven so persistent, Grime (1988c) suggested that estimates of residence times had been based on data from productive pastures on fertile soil, which failed to capture the slow rates of turnover and tight sequestration of ^{137}Cs into tough, long-lived tissues and persistent, slowly decomposing litter of S-selected upland plants. Furthermore, it was suggested that as S-selected species retain leaves and roots for the entire year the assimilation of ^{137}Cs could occur over a relatively long season. Interactions with the soil microbial community may also have contributed, as plants of infertile soils rely extensively on mycorrhizal fungi for mineral nutrition and these have a high affinity for heavy metals such as cesium.

Until recently these ideas had not been borne out by evidence beyond the observation of ^{137}Cs remaining in the environment and the predictions arising from CSR theory. However, at least one key conjecture – that CSR strategies govern the extent of ^{137}Cs sequestration in plant tissues – has now come under scrutiny. Willey *et al.* (2005) investigated the retention of ^{137}Cs in 61 plant species for which Grime *et al.* (1988) had previously calculated CSR strategies. They noted the accumulation of ^{137}Cs in slower-growing strategies and concluded simply that, 'Grime's plant growth strategy theory . . . might be of general use for predicting the behaviour of 134/137Cs in ecosystems'. For the record, and although not strictly part of the Chernobyl case study, Willey & Wilkins (2008) then went on to show that S-selected plant species are particularly retentive of cobalt, further suggesting that the conservative nature of S-selected species can have controlling impacts on the residence times of elements within ecosystems.

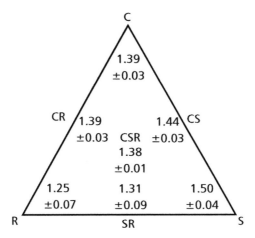

Fig. 6.2 The construction cost (g glucose g^{-1} LDW) of the leaves of boreal plant species differing in CSR strategy. C, competitors (12 species); R, ruderals (8 species); S, stress-tolerators (7 species); CS, CR, SR, CSR, secondary strategies (12, 6, 5 and 23 species, respectively). The figure shows mean values for species groups and standard errors. (Redrawn with permission from P'yankov *et al.*, 2001. Copyright © Springer.)

herbivores and to resist other forms of physical damage. This is because the fitness penalties arising from weak defence are predicted to be so much greater in slow-growers with their low potential to replace lost parts and lost individuals. Such investment in defence is not expected in ruderals or competitors where selection is more likely to have involved, respectively, heavy and early seed production and a capacity for rapid vegetative re-growth.

These predictions have been confirmed by the calculation of 'construction costs' (specifically, the number of grams of glucose required to produce one gram of leaf dry weight) for 73 species of contrasting CSR strategy (P'yankov *et al.*, 2001; see Fig. 6.2). From this study it was concluded that leaves of S-selected plants were the most costly to produce and contained the highest contents of compounds, such as phenolics, lignin, waxes and pigments, used in defence against herbivores and environmental pressures. R-selected species had the least costly leaves and invested the majority of resources in 'functional substances' such as carbohydrates and proteins. C-selected species exhibited intermediate construction costs. From a biochemical perspective, Harborne (1997) noted the existence of three plant defence strategies: **(1)** perennial 'growth-dominated' plants with rapid growth, poor inherent chemical and physical defences but with a highly inducible facultative defence system based on the expression of costly nitrogenous compounds; **(2)** perennial plants with slow growth rates and physically tough leaves that are inherently well defended (and more likely to include physical defences such as spines) but with comparatively unresponsive chemical defences; and **(3)** annual plants for which defence is relatively limited. We would argue that these reflect C-, S- and R-selection, respectively.

Where the resistance of stress-tolerators is exceeded by severe damage consequences may be lethal to the whole ecosystem due to the low capacity for re-growth and failure of a majority of S-selected species to develop persistent seed-banks in the soil. This is in marked contrast to the resilience of R-selected communities associated with frequently disturbed ecosystems of agricultural land and human settlements where vegetation remains in an early successional state characterized by short life-histories, early reproduction and long-term persistence of seed and openness to colonisation by widely dispersed seeds and spores.

Some of the most compelling evidence that the sets of plant traits listed in Table 6.1 allow prediction of ecosystem characteristics relate to measurements of resistance and resilience following extreme events. One of the earliest investigations is that of Leps *et al.* (1982) who compared the effects of the 1976 drought on two neighbouring grasslands similar in species richness but differing in productivity in the Czech Republic. One grassland was of recent origin on fertile soil and was dominated by fast-growing competitive and ruderal strategists. The second was older, had lower productivity and was occupied by slow-growing, stress-tolerant species. Reactions of the two ecosystems to the drought were in agreement with the differences predicted in Table 6.1. Resistance to drought, measured as the ability to sustain above-ground production, was much higher in the ecosystem containing stress-tolerant vegetation whereas resilience (recovery of yield in the year following the drought) was greater in the ecosystem with R- and C-selected species.

Results similar to those reported by Leps *et al.* (1982) were obtained by Tilman & Downing (1994) following exposure of fertilized and low productivity control plots to an unusually severe summer drought. Further support for predictions based on the plant traits listed in Table 6.1 is also available from investigations in which ecosystems of contrasted productivity and vegetation composition have been subjected to experimental manipulation of climate (MacGillivray *et al.*, 1995; Grime *et al.*, 2000).

Studies evaluating plant strategies and strategy-related traits as predictors of ecosystem responses to climatic events and impacts of management are relatively scarce and most of them concern investigations of drought. However, there are a few instances where experimental data provide a reminder of the unrealized potential of predictions based on strategy-related traits. An example is provided in Fig. 6.3 where estimates of potential relative growth rate measured in the laboratory are shown to be equally effective in predicting resistance to drought and to late frost in a comparison conducted across five neighbouring ecosystems.

The propagation of trait influences through food chains

Ideally, tests of the use of the adaptive strategies concept to analyse and predict aspects of ecosystem structure and functioning should involve plant, animal and microbial components. Functional classification for the purposes of ecology remains in its infancy. Arguments continue about the criteria and methods to be applied, and in earlier chapters we have commented on the particular difficulties in groups such as insects and bacteria where functional classification awaits resolution of taxonomic uncertainties. Progress is being made but there are likely to

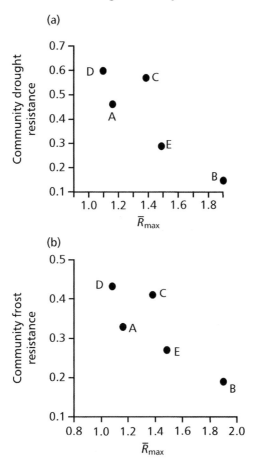

Fig. 6.3 Use of the mean maximum potential growth rate of component species to predict **(a)** community drought resistance, and **(b)** community frost resistance in five contrasted plant communities at the Buxton Climate Change Impacts Laboratory. Community resistances are calculated as the ratio of the estimated total above-ground biomass in the perturbed community to that in control communities examined at the same time. Arranged in order of increasing productivity the five communities form the sequence D, C, A, E and B. (Re-drawn with permission after MacGillivray *et al.*, 1995. Copyright © Blackwell Publishing – Journals.)

be delays before we can be sure of the precise roles of many organisms in particular ecosystems.

These difficulties place a particular responsibility upon the shoulders of plant ecologists. Plants have a foundational role in most ecosystems; they occupy most of the biomass and are relatively easy to identify and to characterize as ecological players. They also provide the substrates (particularly carbon) and shelter for many other organisms. This leads to vital questions concerning the potential of plant traits and, in particular, dominant plant traits, to predict the characteristics of associated animals and microorganisms. If animals and other heterotrophic organisms are so dependent upon the plants they exploit why has this not allowed a relatively straightforward approach to understanding ecosystems? Does the simple principle of 'you are what you eat' apply widely in ecology and allow us to predict what kind of organisms will be associated with each kind of food plant? Can this logic support a Darwinian theory of ecosystem assembly and functioning? More specifically, in view of the evidence presented in Chapters 3, 4 and 5, **does ecosystem functioning vary in relation to the strategies of the dominant primary producers?**

Efforts to address this question have been limited by shortage of datasets allowing comparison of the strategies of plants with the traits of the organisms that exploit them. However, British butterflies and their food plants provide a spectacular exception to this general paucity of information and in two publications (Hodgson, 1993; Dennis *et al.*, 2004) convincing support has been presented linking the strategies of the plants exploited by the caterpillars to the traits of each butterfly. The conclusions of the second of these studies are as follows:

> Butterfly biology is linked to host-plant strategies. An increasing tendency of a butterfly's host plants to a particular strategy biases that butterfly to functionally-linked life-history attributes and resource breadth and type. In turn population attributes and geography are significantly and substantially affected by host choice and the strategies of these host plants. . . . high host plant C and R strategy scores bias butterflies to rapid development, short early stages, multivoltinism, long flight periods, early seasonal emergence, higher mobility, polyphagy, wide resource availability and biotope occupancy. . . . Increasing host-plant S-strategy scores have reversed tendencies, biasing those butterfly species to extended development times, fewer broods, short flight periods, smaller wing expanse and lower mobility, monophagy, restricted resource exploitation and biotope occupancy, closed, aerially limited populations with typical meta population structures, sparse distributions, and limited geographical ranges, range restrictions, and increased rarity. (Dennis *et al.*, 2004)

This unusually comprehensive and definitive study strongly supports the hypothesis that similarities in functional traits are frequently propagated through food chains. Many more studies of this kind are urgently needed to examine the possibility that provisional classification of ecosystem types can be based on the functional characteristics of dominant plants.

Complicating factors

The predictions developed under the preceding heading and in Table 6.1 are deliberately naïve in the sense that they are derived exclusively as implications of a plant-centred application of CSR theory. We think they are valuable as broad indicators of the direction of ecosystem specialization under particular conditions of environment and management. However, to progress nearer to the full realities of ecosystem functioning it is necessary to consider how far the adaptive strategies of the plants correspond to those of the animals and microorganisms that coexist and interact with them. Indeed, later we discuss the ecosystem effects of CSR strategies, rather than individual traits, in all organisms, but before examining detailed evidence on this point it is necessary to recognize that in natural environments complications frequently arise. These can obscure the tendencies of plants and their associated organisms to aggregate into groups with strong functional affinities. These masking effects can be summarized as follows:

1 Where relatively pristine ecosystems occupy extensive areas of uniform landscape it is usually possible to recognize the dominant populations of plants and other organisms, to examine interactions between them and to analyse ecosystem functioning. It is much more difficult, especially in heavily impacted urban and agricultural landscapes, to disentangle the influences of organisms from those caused by human interventions. Even where the imprint of man is slight, landscape complexity can impede research. This is especially true where small parcels of contrasted landscape exist in close proximity and exchange resources and organisms on a constant or intermittent basis. Here it may be difficult to satisfactorily delineate the physical extent of each ecosystem. Similar complications can arise where ecosystems experience traumatic events such as floods and fires or visits from insect herbivores or migrating ungulates. In such cases an adequate understanding of ecosystem functioning may require continuous investigation over a number of years.

2 Ecosystem processes such as carbon, energy and mineral nutrient capture, storage and loss are mainly characterized in terms of the magnitude and annual time-course of the mass-flow processes between trophic components. It is relatively easy to make an inventory of the plants, herbivores and other organisms present in an ecosystem. However to analyse ecosystem functioning it is necessary to conduct a 'who consumes who' operation to identify the plants, animals and microorganisms that are the key drivers of these processes. Unfortunately, the detective work cannot end at this point. In order to understand ecosystem functioning we must identify the consumers of the actual plant parts (most often the leaves or fine roots of dominant plants) that are of sufficient bulk to control the mass-flow processes of the ecosystem. Only by establishing 'who consumes particular parts of who' therefore, do we begin to trace the ecosystem pathways in sufficient detail.

3 In the relatively simplistic set of predictions outlined in Table 6.1 it is forecast that in low stress/low disturbance conditions dominance will be possible by robust perennials that monopolize resource capture and it is argued that their success is achieved by allocating a high proportion of captured resources to construction of new leaves and roots. Although this remains a valid interpretation it is an incomplete explanation of competitor dominance. We need to explain why plants with low expenditure on defence are not subject to devastating attack by herbivores. In the majority of ecosystems the food chains extend beyond the herbivores to include organisms that capture and consume herbivores (carnivores) or deplete their numbers by parasitizing them. If we are to really get to grips with how these enemies of the herbivores influence ecosystem functioning it is essential to determine the extent to which their presence reduces the consumption of the vegetation by the herbivores. Towards the end of this chapter we shall describe the Fretwell/Oksanen Hypothesis (Fretwell, 1977, 1987; Oksanen, 1981), which seeks to explain how the 'top-down' protection of vegetation by the enemies of herbivores varies according to habitat productivity.

Ecosystem processes

So far we have presented arguments based on first principles and some case studies and experimental evidence suggesting how CSR strategies can be useful in predicting ecosystem properties, but which also warn that there is unlikely to be a simple, direct path from adaptive strategies to ecosystems. With this caveat in mind we now turn to the effect of adaptive strategies on specific ecosystem processes and examine the physical nature of the fluxes of matter and energy, and feedbacks operating between different communities of plants, animals and microorganisms within the ecosystem.

Dominance and mass ratio effects

Dominant species have the greatest influence on ecosystem processes via the '**mass ratio effect**' – in other words, they contribute most of the biomass that actively controls fluxes of energy and matter through the ecosystem (Grime, 1998; Smith & Knapp, 2003). In the words of Hillebrand *et al.* (2008): 'as species differ in important traits the traits of dominant species contribute more to aggregate processes in communities and ecosystems than the traits of rare species'. A recent review (de Bello *et al.*, 2010) points out that only a small number of studies currently compare the ecosystem impact of dominant plant species against that of the entire community, but that the majority of evidence nonetheless favours the mass ratio hypothesis over rival ideas such as that of niche complementarity (the hypothesis that ecosystem processes depend directly on the number of species simply because more species provide greater trait variability, exploiting more niches and performing more work; e.g. Tilman *et al.*, 1997).

Mass ratio effects are not only evident for plants; the same rules also appear to apply to soil microbial communities. Wertz *et al.* (2007) found that sterilized soil achieved the same activities (rates of denitrification and nitrite oxidation) typical of non-sterilized soil when inoculated with bacteria isolated from the original non-sterile soil. Crucially, activity remained unaltered even when the inoculate was diluted, thereby excluding the rarer, subordinate members of the bacterial community and leaving only the most abundant species. This indicates that a small number of dominant species are responsible for carrying out the work, and that highly diverse but numerically subordinate species are effectively irrelevant to bulk ecosystem processes. In other words, much of biodiversity is characterized by '**hangers on**' occupying niches peripheral to those of particular species that are truly important to how the ecosystem functions (we stress that this does not invalidate the fact that subordinate species may have particular economic, medicinal or cultural worth – we certainly do not imply that unimportance to overall ecosystem properties automatically translates into a lack of merit).

Griffiths *et al.* (2000) also found that soil functions were generally maintained when they experimentally reduced species richness. However, in this study species richness was reduced not by dilution but by fumigating soil with chloroform, which had the effect of killing certain species, including the most abundant. The fact that soil functions were able to resume following fumigation suggests that previously subordinate species were able to proliferate and effectively take the place of the previous dominants. Thus Griffiths *et al.* concluded that 'there was no direct relationship between biodiversity and function', and that diversity matters because **functional redundancy** (i.e. the presence of different taxa that may perform similar roles and substitute for each other) increases the resilience or stability of the ecosystem in the face of environmental perturbations. In a review of similar studies, Bardgett (2002) arrived at the conclusion that 'most evidence that is available suggests that there is no predictable relationship between diversity and function in soils, and that ecosystem properties are governed more by individual traits of dominant species'.

Indeed, the functional identity of plants is more important a determinant of ecosystem processes in temperate grassland than species richness (Mokany *et al.*, 2008). This suggests that ultimately it could be the adaptive strategies of dominant species that determines the magnitude of ecosystem processes. Cerabolini *et al.* (2010b) found that in two neighbouring and highly contrasting plant communities, for which species-richness was equal, greater diversity in the CSR strategies of dominant species was associated with greater biomass production. This study, although limited with regard to the number of plant communities investigated, demonstrated that a greater number of species *per se* is not a prerequisite for more active or effectual ecosystem processes: two communities with the same species richness differed in resource availabilities, species identities, resource-use efficiencies, the richness of adaptive strategies present and the total amount of biomass produced. Species-richness has no power to explain these differences. Dominant species determine the extent of bulk ecosystem processes, and what matters most for the fluxes of matter and energy in ecosystems is the diversity in how dominant species survive. Biodiversity is relevant to ecosystem function, but not according to the simplistic '**more species = more effective**' formula of popular mythology.

Fluxes and feedbacks between communities

Communities, of course, do not operate independently and form active, conjoined components of the ecosystem. Understanding the extent to which different adaptive strategies use or mobilize resources is essential to our ability to understand the fluxes and feedbacks that operate between communities of plants, animals and microorganisms.

We have already seen (Table 6.1) a summary of the effects of key plant traits on ecosystem functioning. Aside from these primary producers, it is also important to recognize that the suites of traits forming the CSR strategies of heterotrophic organisms have similar effects on the movement of matter and energy, albeit in less significant quantities. In general, S-selected plants, animals and microorganisms are characterized by longevity, persistence and resource retention, and thus consolidate or concentrate resources within their bodies/cells over relatively long periods. They may not be particularly effective at bringing in resources from the surrounding area, but once a unit of matter or energy is captured it tends to stay in that part of the ecosystem for the longer term. Thus S-selected organisms represent a relatively conservative compartment within the ecosystem, which has the effect of stockpiling matter. At the opposite extreme,

R-selected organisms are ephemeral, spending resources almost as soon as they are intercepted, and the R-compartment is likely to be characterized by the highest rates of turnover. At the ecosystem level the gradient between S- and R-selection (resource economics) controls a continuum of temporal effects such as nutrient turnover rates.

C-selected organisms are characterized by mobility and foraging ability, the transfer of matter between sites, and thus the spatial consolidation of resources (they actively seek and gather rather than gradually stockpiling). They attain relatively large body sizes comparatively rapidly, supported by high instantaneous resource availabilities, and thus represent a consolidation of resources in space, with a relatively medium-term temporal impact – C-selected organisms form the apex of a gradient of spatial, rather than temporal, effects on ecosystem processes.

Thus each particular adaptive strategy has characteristic **gathering** (C-selected), **stockpiling** (S-selected) and **spending** (R-selected) effects that delimit where and when matter and energy can move within the overall biotic compartment of the ecosystem. S-selected organisms provide a slow, limited-range route, R-selected organisms have fast, limited-range effects, and C-selected organisms provide a wide-ranging but only moderately rapid pathway. When organisms of contrasting adaptive strategies coexist, biotic ecosystem processes are the net result of the flux of matter and energy coursing through these multiple spatio-temporal pathways, the physical conduits of which are the bodies of the organisms themselves.

These effects are particularly important for plants. Not only are plants usually the primary producers, and thus the main source of carbon and energy for terrestrial ecosystems; dominant plant species usually have the highest mass ratio within the ecosystem (which we have just seen means that they also dominate ecosystem processes). Plants may also drastically alter the structure and chemistry of soil, and the carbon they release is one of the main limitations to soil microbial activity (Chen & Stark, 2000). Indeed, altered management regimes in semi-natural ecosystems induce changes in the plant community that then trickle down to alter microbial community structure (Donnison *et al.*, 2000).

How do plants control the cycling of matter and energy between different communities in terrestrial ecosystems?

First, leaves eventually fall to the ground as litter, and the leaves of different plant species do not all have the same contents of the organic compounds that soil microorganisms require for growth. We have seen (Box 3.2) that a spectrum of leaf economics exists, whereby some leaves are denser, relatively nitrogen-poor and rich in recalcitrant carbon compounds; this leaf economics spectrum is correlated with rates of decomposition (Cornelissen & Thompson, 1997; Santiago, 2007). Indeed, plant species with low relative growth rates and low-quality leaf litter may be associated with particular mycorrhizas and low rates of ecosystem carbon turnover (Cornelissen *et al.*, 2001). Quested *et al.* (2002, 2003) determined that relatively S-selected dwarf shrubs, grasses and sedges had lower nitrogen and phosphorus contents in both leaves and leaf litter with respect to R-selected parasitic plants. Parasite litter decomposed more rapidly (i.e. lost mass more rapidly and led to increased rates of soil respiration) and the stimulation of soil microbial activity released these nutrients more readily into the soil. This augmented the growth of neighbouring plants to a far greater extent than litter of S-selected species. Furthermore, the neighbouring plant species ultimately benefiting from this injection of resources exhibited different growth responses and thus had different capacities to capitalize on this opportunity.

This is a global phenomenon: Cornwell *et al.* (2008) pooled the results of 66 similar experiments for 818 plant species growing from polar to tropical zones over six continents, and found that variability in local leaf litter decomposition rates depends more on the leaf economics traits evident within the community than it does on climate. As the worldwide leaf economics spectrum represents one of the principal spectra of variation in plant adaptive strategies (see Chapter 3) this represents strong evidence that plant CSR-strategies may have controlling effects in terrestrial ecosystems by determining how readily plants decompose and release substrates exploited by soil organisms.

As an example, Eskelinen *et al.* (2009) studied a range of tundra ecosystems and found that:

> . . . above- and below-ground systems are strongly interconnected with productive, fast-growing plant species (forbs) producing high-quality, N-rich organic matter that supports bacteria-based microbial communities and high nutrient availability, and less productive, slow-growing plant species (ericoid dwarf shrubs) producing phenol-rich and N-poor organic matter that favours fungi-based microbial communities and slow nutrient cycling.

This association of fungi with nitrogen-poor litter and low fertilities, with bacteria more prevalent at higher fertilities/richer leaf litter, is a recurring theme in soil ecology (Bardgett *et al.,* 1999; Bardgett & Walker, 2004; Innes *et al.,* 2004; Wardle *et al.,* 2004). Fungi, as a group, have two particular adaptations that can explain their prevalence at low fertilities. First, the ability to produce enzymes capable of degrading recalcitrant carbon forms such as lignin and phenolic compounds such as tannins that are more common in the litter of S-selected plants. Secondly, a lower respiratory requirement for carbon that allows fungi to survive in carbon-poor niches (De Deyn *et al.* (2008) and references therein). Thus leaf litter acts as a vector transporting matter between the two compartments of the ecosystem, ensuring that the fast-slow continuum of plant growth is translated into the overall character of the microbial community and the vigour with which microbes process organic compounds[1].

---

[1] There exist strong parallels between leaf litter and coral 'death assemblages', which are essentially the pieces of dead coral littering the reef. R-selected corals are under-represented in death assemblages compared to the live coral assemblage (Pandolfi & Greenstein, 1997), possibly because they are less likely to fragment during storms than C-selected corals such as the brittle, branched *Acropora palmata* (Lirman, 2000). Different coral species are associated with characteristic microbial communities when alive, and death assemblages have characteristic microbial communities that are distinct from those of living holobionts of the same species (Yakimov *et al.,* 2006). This suggests a route by which coral adaptive strategies may affect the resources available to the marine decomposer community.

Aside from the effects of leaf litter, plant roots have profound effects on the physical and chemical environment in the soil in the immediate vicinity (the rhizosphere). Physically, they affect the degree of aggregation of soil particles and thus soil structure – the finer, less aggregated particles associated with roots provide a greater surface area for colonization and more niches. These physical effects may become more pronounced along succession (Gros *et al.*, 2006), and thus presumably change in response to changes in plant function. Plant roots also have a range of effects on rhizosphere chemistry due to the preferential uptake of water and ions, the release of H^+, HCO_3^- and CO_2 (which alter pH), and the uptake or release of O_2 (Marschner, 1995).

However, one of the main ways in which plant roots affect the resources available to microorganisms is via **rhizodeposition** processes, which essentially involve the release of organic matter into the soil surrounding the root. This can take the form of sloughed-off cells, enzymes, mucilage and root exudates such as organic acids, many of which are used by the plant to stimulate or suppress microorganisms and mobilize mineral nutrients (reviewed by Dakora & Phillips, 2002) and all of which influence microbial activity in some way.

Do different plant adaptive strategies stimulate characteristic soil activities via rhizodeposition processes? Patra *et al.* (2006) found that root exudates of C-selected *Arrhenatherum elatius* were associated with greater rates of soil nitrogen fixation, but lower nitrate concentrations and rates of denitrification with respect to two CSR-selected grass species. Indeed, De Deyn *et al.* (2008) point out that differences in the relative growth rates of contrasting plant adaptive strategies are perhaps the strongest regulator of the rate at which carbon enters and leaves soil, because faster-growing species release carbon into the soil at greater rates but this also stimulates greater microbial respiratory rates and carbon loss. Thus, just as we have seen that the overall effect of richer leaf litter is to increase the rate of nutrient cycling between communities, plants with rapid growth rates may also accelerate ecosystem processes via greater root exudation. Studies such as these provide strong evidence that the high mass ratios of dominant plant strategies regulate soil communities in terrestrial ecosystems.

This is not to say that the microbial community has no influence over the character of the ecosystem. Microorganisms provide specific enzymes required to make different mineral nutrients available to plants and potentially influence the character of the plant community (Reynolds *et al.*, 2003; van der Heijden *et al.*, 2008). This hypothesis of **microbially mediated resource partitioning** suggests that the microbial community expands the range and character of the niches available to plants. Indeed, Wardle *et al.* (2004) suggest different ways, involving both positive and negative feedbacks, in which various players within the soil community may alter the plant humped-back diversity/productivity relationship, summarized in Fig. 6.4.

The relationship between plants and soil organisms is often mutualistic, with resources exchanged between partners. This prompted Broeckling *et al.* (2008) to investigate the hypothesis that root exudates of a particular plant species will support the microbial community typically associated with soil in which that plant grows, but not the communities of other soils. They found that plants

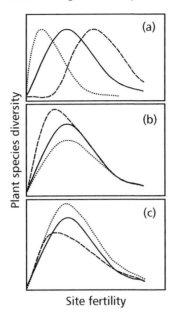

Fig. 6.4 Different regulatory effects of the soil community on the plant diversity/ productivity relationship. **(a)** Soil decomposers can diminish or enhance soil fertility, shifting the humped-back curve to the left (dotted line) or right (dashed line), respectively. **(b)** Arbuscular mycorrhizal fungi increase diversity of subordinate plant species (dashed) whereas pathogens and herbivores reduce diversity (dotted line). **(c)** Arbuscular mycorrhizal fungi reduce the diversity of dominant species (dashed) but pathogens and herbivores increase diversity of dominants (dotted). (From Wardle *et al.*, 2004. Reproduced with permission from AAAS.)

– and even root exudates in isolation – were indeed able to support their typical soil fungal community, but not the fungal community of soils to which each plant species was not native. Thus the exchange of matter between plant and soil communities probably represents a stricter, species-specific mutualism than has previously been assumed, raising the intriguing possibility that the truth is complex chicken-and-egg situation whereby both communities have been subject to a high degree of co-evolution.

The exchange of resources between communities forms the mechanism by which communities feed back on one another to control the rate of ecosystem processes. It has been predicted (Reynolds *et al.*, 2003) that positive, mutualistic feedbacks between plant and soil communities may be important during the early stages of succession and at high latitudes and altitudes, whilst negative feedbacks may be important for the diversity and stability of late-successional ecosystems. Indeed, positive feedbacks may be particularly important to the success of plants invading ecosystems, which may fail if not accompanied by the appropriate association of mutualistic microorganisms (Nuñez *et al.*, 2009). In contrast, feedbacks in late-successional mixed grassland communities are consistently negative, which has been ascribed to the build up of species-specific pathogens

(Harrison & Bardgett, 2010). In a twist on the Janzen-Connell hypothesis (that the mortal attentions of specialist herbivores select against high-density monospecific stands of tropical trees, resulting in high tree diversities) it appears that low densities of single species in a species rich community may be advantageous by avoiding soil pathogen accumulation and seedling mortality (Reynolds *et al.*, 2003). Thus late successional communities could, in part, owe their high diversities to a kind of natural 'crop rotation' that effectively allows species to avoid both soil pathogens and specialist herbivores – a negative feedback that may increase biodiversity.

However, feedbacks are not always evident. Tscherko *et al.* (2005) found that the soil microbial community in pioneer stages of a primary succession was essentially that of bulk soil (i.e. not influenced by R-selected pioneer plants) and that as vegetation cover developed, so did the microbial community: 'based on the CSR framework (Grime, 2001), the competitive plant communities of the later stages are characterized by high productivity, which in turn favoured the rhizosphere microbiota'. This was attributed to consistently high rates of carbon release by relatively C-selected perennial plants in late succession compared to R-selected pioneers. Thus feedbacks between plant and soil communities may differ not only in character (positive/negative) but also in strength, depending in part on the adaptive strategies in the plant community.

Based on this evidence, we suggest that plant-soil microbial community feedbacks become important to community structure and ecosystem functioning after plant species have successfully negotiated the CSD filter, and represent the fine-tuning effects of the proximal filter. Stress, competition and disturbance and resulting plant adaptive strategies have controlling effects over terrestrial ecosystem processes: plants act as sites where nutrients are concentrated, retained and transported over huge distances (huge in comparison to individual microorganisms). Microorganisms are capable of affecting large amounts of matter but in a less coordinated manner and with each individual affecting smaller distances and timescales. Thus the adaptive strategies of single plant species are more likely than single microbial species to determine the gross fluxes of matter and energy within the ecosystem as a whole – the exceptions being ecosystems in which microbes are the dominant life forms or the primary producers, such as desert soil crusts.

So far we have introduced the concept of feedbacks between different ecosystem components by reviewing the principal interactions between plant and soil communities. However, soil communities and animal communities also feed back to affect one another, and although this may occasionally occur directly plants usually act as middlemen.

Wardle *et al.* (2004) suggest that interactions between above-ground herbivores and plants occur at the slowest rates in unproductive habitats (both in terms of the consumption of plant biomass and the return of fecal matter), and that this is attributable to plant traits such as low forage quality, low specific leaf area, long leaf life-span, and an overall slow-growing, long-lived plant strategy. There is an obvious parallel, also noted by Wardle *et al.*, between the pace of these above-ground processes and the way in which we have seen plant traits regulate the rate of below-ground processes. These processes are, inevitably, all linked, with the high lability of fecal matter in productive habitats ultimately feeding back down to the soil community to favour decomposers such as bacteria and earthworms. Grazing also stimulates plant regrowth, which can introduce additional carbon into the soil. In unproductive habitats nutrient returns mainly take the form of recalcitrant leaf litter, encouraging not bacteria and earthworms but fungi and arthropods, and thus a qualitatively and functionally different soil community. We have already seen the ways in which microbially mediated resource partitioning can affect plant diversity and productivity (see Fig. 6.4 above) and, by affecting the vegetation, soil communities can thus feed back to affect the herbivore community. Thus the plant community forms a vital conduit by which matter and energy pass between above-ground and below-ground compartments, and plant adaptive strategies are crucial to the rate at which these interactions occur and therefore the functioning of the entire ecosystem.

Top-down control by herbivores

It is especially difficult to test hypotheses relating to the role of herbivores in the assembling and maintenance of ecosystems associated with particular environmental conditions. One approach to this problem, advocated by several ecologists (Heal & Grime, 1991; Beyers & Odum, 1993; Lawton, 1995) is to create simplified ecosystems in microcosms that permit controlled input of organisms and manipulation of climate and resource supply. In several of these experiments (Fraser & Grime, 1998b, 1999; Buckland & Grime, 2000) interesting insights have been gained into the critical role of herbivores in determining the functional composition of the vegetation developed under different levels of soil fertility. Figure 6.5 compares the ten highest contributors to the above-ground plant biomass in an outdoor microcosm experiment in which the ecosystem was allowed to assemble over a period of two years at high and low fertility and in the presence and absence of herbivorous slugs and aphids.

The results show that at high fertility the herbivores caused some minor changes in the plant species hierarchy, but the assemblage in the presence and absence of herbivores closely resembles those commonly observed on productive farmland in Western Europe. When we turn to the low-fertility microcosms however we find that, in the absence of herbivores, the vegetation is again dominated by potentially fast-growing broad-leaved grasses (*Poa annua* and *Lolium perenne*) and there is a complete absence from the top ten of the stress-tolerant species included in the seed mix. Only in the presence of the slugs and aphids do we see a demotion of *Poa annua* and *Lolium perenne* and the appearance of the stress-tolerator, *Festuca ovina* and the S/CSR-strategist *Pilosella officinarum*.

These results provide no support for the idea (Tilman, 1988) that the vegetation of unproductive ecosystems is determined by the superior ability of stress-tolerators to compete for mineral nutrients on poor soils. We see that even when severely stunted by low nutrient supply, potentially fast-growing R- and C-strategists prevail. It is only in the presence of the herbivores that the fitness of the well-defended stress-tolerators begins to become apparent.

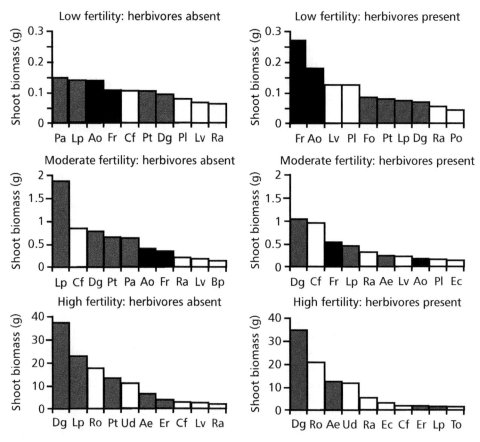

Fig. 6.5 The ten highest contributors to the vegetation developed over two growing seasons in closed outdoor microcosms sown with a standardized seed mixture of 48 species of a wide range of plant functional types. Microcosms providing high and low soil fertility were compared and, at each fertility level, vegetation development was examined in the presence and in the absence of generalist herbivores (slugs and grass aphids). Black bars = slow-growing grasses, Grey bars = fast-growing grasses, White bars = forbs. Ae, False oat-grass (*Arrhenatherum elatius*), Ao, Sweet vernal-grass (*Anthoxanthum odoratum*), Cf, Mouse-ear chickweed, (*Cerastium fontanum*), Cock's foot (*Dactylis glomerata*), Ec, American willow-herb (*Epilobium ciliatum*), Er, Couch-grass (*Elytrigia repens*), Fo, Sheep's fescue (*Festuca ovina*), Fr, Red fescue (*Festuca rubra*), Lp, Perennial rye-grass (*Lolium perenne*), Lv, Ox-eye daisy (*Leucanthemum vulgare*), Pa, Annual meadow-grass (*Poa annua*), Po, Mouse-ear hawkweed (*Pilosella officinalis*), Pt, Rough meadow-grass (*Poa trivialis*), Pl, Ribwort plantain (*Plantago lanceolata*), Ra, Common Sorrel (*Rumex acetosa*), Ro, Broad-leaved dock (*Rumex obtusifolius*), To, Dandelion (*Taraxacum officinale*), Ud, Stinging nettle (*Urtica dioica*). (Redrawn with permission after Buckland & Grime, 2000. Copyright © Blackwell Publishing Ltd.)

Top-down control by carnivores

The trophic relationships involved in ecosystem processes are often complicated by a variety of secondary downstream effects; most notably, both plants and herbivores may also experience top-down control by carnivores. Carnivores may influence plant community structure because they alter both herbivore numbers (diversity and population sizes) and behaviour. Herbivore behavioural response to the threat of ambush predation may essentially be a paranoid state of high alert, resulting in densely concentrated herbivore groups. This has an intense local impact on the vegetation. Active hunters are more easily avoided, resulting in more widely distributed herbivores and a relatively diffuse impact on vegetation (Schmitz, 2009). Thus one of the most important aspects of predation is **hunting mode**; whether the predator is an energy-conserving ambush hunter (i.e. more likely to be S-selected) or a cursorial, running hunter (C-selected).

Schmitz (2009) investigated the effect of these contrasting hunting modes on the behaviour of the herbivorous Red-Legged Grasshopper (*Melanoplus femur-rubrum*) in a grassland ecosystem in Connecticut, USA. The Nursery Web Spider (*Pisaurina mira*) is an ambush hunter that waits in the prairie canopy and forces the grasshopper to move from its preferred food, the dominant plant Smooth Meadow-grass (*Poa pratensis*), downwards to hide among the broader leaves of the Goldenrod (*Solidago rugosa*), which is eaten instead. Hunting by this spider thus favours the growth of the dominant grass, negatively impacts on the Goldenrod, and favours a range of smaller herbs that are usually suppressed by the Goldenrod. In contrast, the more active Jumping Spider (*Phidippus rimator*) hunts throughout most of the vegetation but not the canopy, and does not cause spatial shifts in grasshopper foraging. Predation by the Jumping Spider thus indirectly favours the growth of both the dominant grass and the Goldenrod, which in turn suppresses the smaller herbs.

Schmitz (2009) manipulated the relative abundance of active- and ambush-hunters in experimental microcosms, allowing the extent of each hunting mode to be correlated statistically with shifts in plant community structure. He found that ambush hunting, by favouring both the small herbs and the dominant grass, increased plant species evenness after two years, resulting in declines in leaf litter quality (i.e. greater C:N), soil nitrogen mineralisation rates and ultimately above-ground net primary productivity. Schmitz concluded that the adaptive strategies of different predators impose differing degrees of top-down control, which is also diluted as the effects cascade down through trophic levels – a subtle feedback mechanism that alters the basic underlying diversity/productivity relationship.

However, there are strong grounds for the idea that top-down control by carnivores can have more blatant effects that interact directly with plant CSR strategies to determine gross community structure, and operate by controlling herbivore species diversity rather than the behaviour of single species. Furthermore, these relationships may change predictably along the productivity gradient.

The Fretwell/Oksanen hypothesis (Fretwell, 1977, 1987; Oksanen, 1981) seeks to explain how top-down control by carnivores varies according to habitat productivity. It suggests that top-down control has the greatest impact in highly productive habitats because palatable, fast-growing (C-selected) plants that do not invest in defences can persist only where biomass losses to herbivores are reduced by the activities of carnivores, which are more easily supported at high productivities. Plant species persisting in unproductive habitats are less likely to rely on protection by other organisms because they exhibit inherent robustness and defence against herbivory; plant community structure is under less influence from top-down control. At intermediate productivities a range of plant CSR strategies is present, including palatable and non-palatable species that can together support a discrete community of generalist herbivores but are insufficient to support large populations of carnivores, releasing herbivores somewhat from predation pressure. Thus carnivores impose particularly strong top-down control of the plant community in productive habitats, favouring C-selected plants, whereas herbivores tend to impose top-down control at intermediate productivities, favouring plant adaptive strategies that include some inherent defence at the expense of growth rates and competitive ability.

In order to test this idea, Fraser & Grime (1998a) found that when they placed maggots and pieces of lettuce in natural plant communities of differing productivities, maggots were consumed to a greater extent in high productivity communities (confirming the prominence of carnivores), whereas mainly lettuce was consumed at intermediate productivities (confirming the prominence of herbivores). When these communities were repeatedly doused in pesticides the plant communities at intermediate productivities exhibited the greatest increases in biomass accumulation, confirming that herbivores had the greatest suppressive effect at intermediate productivities, as predicted by the Fretwell/Oksanen hypothesis. In a further test of the hypothesis Fraser & Grime (1998b) constructed invertebrate-proof microcosms consisting of three grass species, the Grass Aphid (*Sitobion avenae*) and its predator, the Seven Spot Ladybird (*Coccinella septempunctata*). Soil fertility was manipulated to simulate conditions of differing productivity. When the predatory ladybird was introduced into the microcosms the yield of the most palatable grass increased at high productivity – an effect not seen at lower soil fertilities/productivities, at which predator activity was minimal.

These experiments suggest that top-down control by carnivores may affect community structure and ecosystem processes directly by favouring particular plant CSR strategies, and the nature of carnivore adaptive strategies is likely to subtly fine-tune this process via differences in behavioural ecology.

The key role of eco-evolutionary dynamics

Predators provide the final key insight in our search for links between evolution and the ecosystem. In the previous chapter we saw the example of adaptive radiation in the Threespine Stickleback, which can exhibit a generalized form and two specialized forms (surface and bottom feeders). Harmon *et al.* (2009) devised an experiment to test the idea that, via differences in the extent of top-down control on either herbivorous or omnivorous zooplankton prey, these contrasting stickleback lifestyles elicit **trophic cascades** throughout the foodweb that ultimately affect algal biomass and primary productivity. What they actually found was slightly more complicated, as changes to the zooplankton community and the cycling of nutrients in the ecosystem also resulted in changes in the type of organic carbon dissolved in the water and thus light transmission, algal photosynthesis and primary productivity. Crucially, it was evident that the evolution of different stickleback forms had contrasting modifying effects on ecosystems, thereby altering the environments in which the sticklebacks evolved.

Evolutionary changes occur in small steps defined by the life-spans of individuals and, particularly where individuals have short life-spans, evolutionary effects may become evident over the same kind of relatively brisk timescales as ecological processes. Over much longer timescales macroevolution averages out these rapid microevolutionary resonances with immediate ecological circumstances (or 'oscillating directional selection'; Grant & Grant, 2008). Such feedbacks between microevolution and ecology are known as '**eco-evolutionary dynamics**'.

Numerous examples of contemporary evolution, or adaptation over a small number of generations, now exist (Kettlewell, 1961; Hendry & Kinnison, 1999; Kinnison & Hendry, 2001; Reznick & Ghalambor, 2001; Elmer *et al.*, 2010) and although examples of feedbacks between contemporary evolution and the ecosystem are currently extremely rare they are starting to be discovered (see reviews by Fussman *et al.*, 2007; Bailey *et al.*, 2009; Pelletier *et al.*, 2009; Post & Palkovacs 2009; Schoener, 2011). Eco-evolutionary dynamics are thought to affect biological invasions (Kinnison *et al.*, 2008), predator-prey community dynamics (Becks *et al.*, 2010) and the persistence of populations and thus species extinction (Kinnison & Hairston, 2007; Weese *et al.*, 2011).

In short, adaptive radiation and extinction shape the species pool to govern the material that is then subject to ecological filters, assembled into the community and that affects the ecosystem, but conditions within the ecosystem may affect the opportunities for adaptive radiation and thus feed back to regulate evolution. Both species and their niches may be dynamic:

> The traditional view of adaptive radiation is one of ecological opportunity, whereby lineages diversify until all available niches are filled. However, this view largely ignores the role that organisms play in shaping their environments. . . . Rather than lineages simply diversifying to fill available niches, ecological niches themselves may be diversifying. (Palkovacs & Hendry, 2010)

Harmon *et al.*'s (2009) sticklebacks demonstrate that the degree of specialization and the type of specialization are important factors influencing eco-evolutionary dynamics (i.e. the generalist form had more drastic ecosystem effects than either of the two highly specialized forms). Feedbacks are also more likely for species with controlling effects on ecosystem properties, such as dominant tree species with high mass ratios (Post & Palkovacs, 2009). These studies suggest that the character of adaptive strategies – particularly of dominant species – plays a key part in the eco-evolutionary feedbacks that regulate ecosystems.

In the next, and final, chapter we shall see that the concept of eco-evolutionary dynamics allows us to fully integrate evolutionary and ecological processes into a single model.

Summary

1   Recognition of universal patterns of variation in plant life-history and physiology (Ramenskii, 1938; Grime, 1977), allowed Chapin (1980) to propose that the productivity of an ecosystem, the tempo of ecosystem functioning and responses to disturbance vary in concert with variation in CSR-related plant traits such as potential rates of growth, tissue turnover, mineral nutrient concentrations, anti-herbivore defence and rates of decomposition. On this basis it could be proposed that these core traits are simultaneously integral to the functioning and integrity of plants and ecosystems. Moreover, since screening experiments showed that values for these particular traits exhibit strong convergence within plant communities, it follows that the dominant controls on ecosystem functioning would be exerted by those species that are most abundant in the community: this principle is now recognized as the **mass ratio effect**. This can explain why the strength and direction of ecosystem processes is not correlated in a simple, unidirectional manner with species diversity.

2   Many features of ecosystems and of their responses to changing conditions are predictable from the CSR strategies of the dominant components of their plant communities. These features include primary production, losses to herbivores, litter decomposition, carbon sequestration, retention of mineral nutrients and pollutants and resistance and resilience following perturbation. Plant adaptive strategies, via differences in leaf litter quality and rhizodeposition, have controlling effects on soil microbial community structure. Recent experiments suggest that in future it will be possible to use the traits of dominant plant species to devise predictions of the resistance of ecosystems to specified changes in climate.

3   It is also increasingly evident that sets of traits prominent in the plant community are capable of selecting predictable sets of traits of life-history and physiology in associated herbivores and decomposers. Butterflies provide a classic example of this phenomenon.

4   In terrestrial habitats plant CSR strategies affect the rates of resource supply, particularly carbon, to soil microbial communities and they also govern the character of the soil community by altering soil structure and chemistry. The activity of the soil community, by dint of the specialized biochemical pathways available to different microbial species, can alter the soil resources available to plants, thereby affecting the plant community.

5   There is strong evidence that animal strategies can control vegetation not only through the direct effects of herbivores on plants but also by the modifying effects of carnivores and parasitoids on the feeding behaviour of herbivores.

6   The greatest difficulties in understanding and predicting the functioning of ecosystems occur in circumstances where several neighbouring ecosystems occupy small areas and exchange resources and organisms. Problems also arise where ecosystems experience complex or intermittent effects of human interference, insect outbreaks, visits by migrating herds of mammals or extreme climatic events.

7   As organisms evolve the ecology of species changes, with changing ecology in turn feeding back to affect evolution. Such eco-evolutionary dynamics are key to integrating evolutionary and ecological processes.

7

The Path from Evolution to Ecology

During the course of this book we have attempted to lay a path by which we can negotiate from the evolution of adaptive strategies to the ecology of ecosystems. At times the temptation to step briefly from the path into the details of crab physiology, the latest techniques for studying microbial diversity or even the rival attempts to unify ecology has been irresistible. Now, and for the last time, we will squint to apply MacArthur's blurred vision and ask three questions:

- What has been learned by this particular pursuit of the ecosystem?
- What are the implications for conservation and management of ecosystems and even larger areas of terrestrial and aquatic habitat?
- What will be the most rewarding goals in the next decade of ecological research?

What has been learned?

We submit that the limits to variation in the physiologies and life-styles evident for plants, animals and microorganisms reviewed in Chapters 2–4 are consistent with the idea that all organisms face a three-way trade-off with respect to the investment of resources. Inescapable dilemmas lie between enhancement of resource acquisition (C-selection), maintenance of long-lived individuals and tissues (S-selection), and commitment to early, often lethal, reproduction (R-selection). These three alternative paths of evolution and ecology recur widely

The Evolutionary Strategies that Shape Ecosystems, First Edition. J. Philip Grime, Simon Pierce.
© 2012 John Wiley & Sons, Ltd. Published 2012 by John Wiley & Sons, Ltd.

within the tree of life and indicate that similar basic constraints of competition, stress and disturbance act across the world on a set of core organismal traits justifying the designation universal adaptive strategy theory (UAST). This confirms our initial suggestion (Chapter 2) that, when properly defined and understood, stress, disturbance and competition have extraordinary capacity to predict constraints on selection and paths of adaptive response.

A relationship between UAST and the species diversity of communities can be defined in the form of the humped-back model in which, at scales that vary enormously according to habitat and ecosystem, assemblages reach maximum potential species-richness and evenness in a corridor that lies between high biomass (a high degree of C-specialization) and low biomass (highly S- and/or R-specialized). Actual levels of species-richness in specific circumstances within the corridor are further affected by niche differentiation and the size and nature of the local species pool and are compatible with the centrifugal model (Keddy, 1990). Whilst we have mainly concerned ourselves with species identity the second face of biodiversity, species relative abundance, has recently come under investigation using a method that appears to be compatible with UAST (Shipley *et al.*, 2006; Shipley, 2010).

Evidence connecting adaptive strategies to variation in communities and ecosystems is being collated for plants, is becoming available for particular groups of animals and is 'a gleam in the eye' of many microbiologists. Large-scale reviews of plant traits such as the Integrated Screening Programme and the worldwide leaf economics spectrum have revealed the fact that identifiable sets of inter-correlated traits that control ecosystem functioning such as element cycling and storage, gas fluxes, and ecosystem resistance and resilience, are subject to a filter admitting or excluding plants from each community. This coarse filter is determined by the balance between competition, stress and disturbance (the CSD-equilibrium) at each site. Entry to the community is also controlled by a proximal filter that operates on a variety of traits affecting specific aspects of regeneration, dispersal, phenology, specialized defence and physiological tolerance that are not entrained in the CSD-equilibrium but determine which of many candidate species in the local pool, and with similar potential effects on the ecosystem, are actually recruited into each community and are capable of persisting. This evidence that admission to the community requires penetration through two types of filter only one of which seriously impacts on ecosystem functioning promises resolution of an old conundrum:

> How similar must two organisms be to exploit the same environment and how different to coexist? Anon.

Whilst the twin-filter model describes the entry of species into the community we must reconcile this with natural selection, which does not select against species but against individuals within populations of species. Indeed, any general model of community assembly must be able to take into account both inter- and intra-specific variability in phenotypic traits. Figure 7.1 shows a more detailed version of the twin-filter model demonstrating the entry of species into the community based on the capability of different individuals to establish in suitable

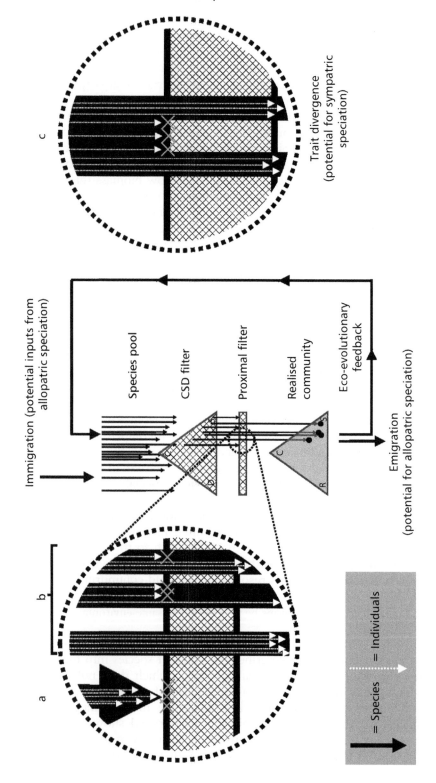

Immigration (potential inputs from allopatric speciation)

Species pool

CSD filter

Proximal filter

Realised community

Eco-evolutionary feedback

Emigration (potential for allopatric speciation)

Trait divergence (potential for sympatric speciation)

= Species

= Individuals

Fig. 7.1 Detail of the twin-filter model showing how ecological filters (which select against species entry into communities) can also act as evolutionary selection pressures (selecting against individuals): **(a)** a species (black arrow) cannot pass the filter and enter the community because none of the individuals (white arrows) are suitably adapted, in contrast to **(b)** species entering the community because all individuals or just some (due to intraspecific trait variability) are suitably adapted. For the species in **(c)** two sub-populations with extreme trait values find suitable niches, but individuals with average trait values do not. This micro-evolutionary trait divergence in response to the proximal filter potentially initiates sympatric speciation. Eco-evolutionary feedback then refreshes the species pool, and as species diverge more widely from one another the CSD filter begins to select against entire suites of traits. Emigration and immigration are important shorter-term processes, but also provide the potential for allopatric speciation over the longer-term. When the species pool is updated by the emergence or arrival of new species biotic interactions change, or species modify abiotic conditions, altering the range of suitable phenotypes that can pass the CSD filter. Local extinctions occur when a species no longer includes individuals that are suitably adapted to pass local filters, and total extinction occurs when a local extinction represents the loss of the last remaining population of a species. In this way the processes of natural selection, speciation via adaptive radiation, and extinction can be reconciled with the concepts of the ecological niche and the adaptive strategies predicted by universal adaptive strategy theory.

niches due to intraspecific trait variability, reflecting the effects of genetic diversity (Booth & Grime, 2003; Fridley *et al.*, 2007; Whitlock *et al.*, 2007). At the heart of this model lie natural selection and universal adaptive strategy theory, because adaptive strategies represent the general directions of life-history specialization that initially allow entry into the community. At fine scales, these relationships are then modified by differences in small sets of functional traits between species of equivalent overall adaptive strategy. The inputs, outputs and processes of the twin-filter model can then be seen as dynamic processes involving adaptive radiation (including both sympatric and allopatric speciation) and eco-evolutionary feedbacks.

We do not pretend that this model can explain every impact of species evolution on community assembly and ecosystem functioning. There will always be a point at which ecological theory encounters difficulty in encompassing all of the variation evident in nature, either between individuals or even between species, and which confounds attempts to produce a straightforward, one-size-fits-all set of models. Nonetheless, by broadly defining the mechanisms by which adaptive strategies determine the coarse structure of communities and the functioning of ecosystems we have sketched a framework consistent with empirical studies within which we can start to understand the wider implications for ecology of the theories of evolution by means of natural selection and adaptive radiation.

Has this ecological framework any implications for the conduct of research in evolutionary biology? Some insights into evolutionary processes may arise from the concepts encapsulated in the twin-filter model. The CSD filter selects between variants associated with the mundane, mainstream gross features of all organisms that 'appear to involve relatively intractable, multigenic linkages and trade-offs between different aspects of the core physiology' (Grime, 2006). This leads to the suspicion that the CSD filter is more likely to produce evolutionary developments and ecological solutions that are more substantive and persistent in the long term. The corollary of this hypothesis applies in the many forms of microevolutionary selection predicted to arise from the operation of the proximal filter. For plants it is well documented that there is enormous scope for interspecific and intraspecific variation in specific traits and small sets of inter-linked traits that effect minor evolutionary changes such as those affecting germination biology (Baskin & Baskin, 1998), shoot architecture and phenology (Bakker, 1989) and tolerances of heat, cold or desiccation (Pierce *et al.*, 2005). We suggest that differences in these non-CSR-entrained traits are relatively pliable and often reflect evolutionary events during which species' ecologies start to diverge from one another (i.e. speciation is most likely to be initiated by the kind of subtle differences in traits that are acted upon by the proximal filter).

This is in agreement with a number of studies of both plants and animals (reviewed by Ackerly *et al.*, 2006) suggesting that the sympatric divergence of traits involved in fine-scale niche differentiation usually precedes evolutionary changes in broader ecological requirements. Thus incremental micro-evolutionary responses to proximal filter selection pressures could continually feed back to ultimately influence macroevolution of the overall adaptive strategy. This process can be represented by a split in a 'species arrow' (see Fig. 7.1c), followed by a series of divergences in the trajectories of daughter sub-taxa, occurring with each eco-evolutionary feedback to eventually produce strategically distinct taxa in the species pool.

What are the implications for conservation and management?

Our results support the hypothesis that the approach of plant functional groups based on Grime's CSR scheme is suitable to gain a better understanding of the effects of landscape fragmentation on plant populations. (Körner & Jeltsch, 2008)

The combination of Ellenberg indicator values with Grime's strategies can usefully predict habitat invasibility. (Simonová & Lososová, 2008)

Recognition of widespread decline and loss of familiar animals and plants in many parts of the world has become a powerful motivating force for the conservation and protection of biodiversity. The current challenge for ecologists is to advise the general public and those who manage natural resources about how to translate such rather general concerns into specific actions that 'make a difference'. We suggest that the development of general theories of adaptive strategies capable of classifying and understanding variation in populations, communities and ecosystems constitutes the crucial next step if ecology is to fulfil this role. General theory, providing that it is a sufficiently accurate reflection of how patterns arise and change in nature, can help conservation and management in several respects.

First it can explain why some organisms are declining and others are expanding. In a world experiencing increasing impacts of disturbance and eutrophication it has been no great surprise to witness the expansion of R- and C-selected plant species at the expense of stress-tolerators. As human population densities rise and the drive toward more intensive methods of food production continues this functional shift seems destined to affect an increasing proportion of the landscape. It is also predictable (Chapters 5 and 6) that an additional consequence of this shift will be to produce a vegetation cover that is labile, responsive to climate change and open to colonisation by invasive species (Davis, 2009). It is unlikely that these effects will be confined to plants. Myers (1996), for example has reported rapid expansions in animals that 'exploit newly vacant niches by making widespread use of food resources; they are generally short-lived, with brief gaps between generations; they feature high rates of population increase; and they are adaptable to a wide range of environments.'

The second benefit to conservation and environmental management arising from the development of unifying general theory in ecology is the potential to widen the focus of attention from individual organisms to the ecosystem itself. As the central thesis of this book, we have emphasized that organisms conform in their life-histories, resource dynamics and core physiologies to recurring patterns of specialization that are valuable clues to the functional characteristics of the ecosystems they occupy. Indeed, a mainstream view is now developing that key traits of organisms, and in particular the functional diversity of dominant plants, have the potential to characterize ecosystem properties that predict sensitivity to pollution, climate change, effects of management and maintenance of ecosystem services (e.g. Díaz & Cabido, 2001; de Bello *et al.*, 2010; Vandewalle *et al.*, 2010). However, as we pointed out in Chapter 3, investigation of traits alone does not explain why consistent relationships have evolved between traits and ecosystem characteristics. Indeed, because adaptive strategies regulate ecosystem processes a range of ecosystem services can be seen as dependent not on traits *per se* but on the evolutionary histories and strategies of component organisms; the true mediators of regulating services (including carbon sequestration and climate regulation, waste detoxification, air and water purification, pest and disease control), supporting services (nutrient dispersal and cycling, primary production) and also provisioning services such as energy (biomass fuels) and food.

Perhaps the chief benefit of general theory is the lateral transfer of knowledge from ecosystems that are familiar and well researched to others that are only scarcely investigated, embuing ecology with the power to predict.

Research priorities for the next decade

In the previous section we acknowledged a trend in modern research towards the investigation of traits rather than the use of general life-history theory to understand how suites of traits affect survival. It is only rarely (e.g. Shipley, 2010) that traits are explicitly considered as part of adaptive strategies and these are in turn viewed in the context of evolutionary processes. We suggest that lack of context is symptomatic of a potentially insidious phenomenon characterized by impatience with the often slow and methodical testing of theory against empirical data.

This is exemplified by the leaf-height-seed (LHS) scheme, which aims to classify seed plants using just three simple traits and was devised by Westoby (1998) as a specific reaction against theory, and in particular against CSR theory: 'impediments to wider implementation of the CSR scheme arise from its defining axes by reference to concepts'. This censure of concepts has, we would argue, been influential in encouraging the recent retreat from general ecological theory and reliance on observations of trait variability without reference to evolutionary context (e.g. Reich *et al.*, 1999; Lavorel & Garnier, 2002; Garnier *et al.*, 2006). Although we by no means wish to deride these studies – many have proven invaluable in confirming that the axes of trait variability seen by the ISP are a global phenomenon – they unfortunately cannot predict how such trait syndromes could evolve, why certain suites of traits consistently occur in particular situations, nor why trade-offs in leaf economics may be associated with other tradeoffs, such as trade-offs in reproductive development (e.g. Cerabolini *et al.*, 2010a). Traits alone have little broader explanatory power. In the absence of theory, links between traits and the habitat, such as between canopy height, competition and productivity, become vague inferences rather than explicit evidence of adaptation and ecological specialization.

To draw an analogy the observation, via the Hubble space telescope, that stars may sometimes appear to be surrounded by rings of light is an unusual fact. It is only within the wider framework of the theory of general relativity that this gravitational lensing (the deflection of light by a large mass positioned between the telescope and the star) can be interpreted as evidence of a fundamental property of nature, in this case that gravity distorts space-time. Similarly, measurements of traits, of trait variability (such as that encompassed by the worldwide leaf economics spectrum or the organisation of trait combinations according to the LHS scheme) all represent interesting observations but these can only be defined by reference to concepts such as adaptive strategies or natural selection. Only theory can explain the relevance of traits to an organism's ecology, and suggest how analogous suites of traits may have evolved.

Now, as we move forward to discuss research priorities for the next decade, we consider it necessary to highlight the need for future research to return to a healthy balance between empirical and theoretical enquiry. What, then, does the future hold? First, we forecast a true consolidation of plant, microbial and animal ecological theory within the framework we have presented here. Although by far the longest chapter in this book concerns animals we are conscious of the fact that most of the figures and tables of data refer to plants. This is explicable in part because both of the authors are plant ecologists. It is also because, as we have noted several times before, plants stand still and wait to be counted (Harper, 1977) and it has been possible in the present era to begin to assemble large databases and to conduct objective searches for patterns of adaptive and ecological specialization. We are not the first authors to be frustrated by the patchy nature of ecological information and we note the observation of Charles Krebs upon completing the fifth edition of his textbook, *Ecology*:

It is not always easy to see where the pieces fit in ecology, and we shall encounter many isolated parts . . . not clearly connected to anything else. This is typical of a young science. (Krebs, 2001)

How should we address the problems arising from the fragmentary nature of ecology? Do we need for animals, microbes and a wider range of plants, the kinds of extensive screening of key functional traits that were involved in the Integrated Screening Programme and in detecting the worldwide leaf economics spectrum? Our response to this question is a resounding 'Yes!'. Only through such coordinated action can we discover the compass and the common vocabulary to speed our progress. Preparation of this book has revealed some cause for optimism. Growing links and parallels across very different circumstances, particularly between the character of adaptive strategies, communities and ecosystems, are becoming increasingly apparent. We have also, among others, predicted that increasing computing power will provide novel tools with which we can follow the precise details of actual or simulated community assembly for communities of microorganisms, plants and animals, without losing sight of general principles. We suggest that this will finally allow formulation of the assembly rules and functional characteristics of communities and ecosystems. This is an ambitious but achievable goal. More than 150 years after the publication of biology's grand unifying theory, ecology is at last starting to scale up.

References

Aarssen, L.W. & Turkington, R. (1983) What is community evolution? *Evolutionary Theory* **6**, 211–217.

Ackerly, D.D., Schwilk, D.W. & Webb, C.O. (2006) Niche evolution and adaptive radiation: testing the order of trait divergence. *Ecology* **87**, S50–S61.

Adler, P.B., Seabloom, E.W., Borer, E.T., *et. al.* 2011. Productivity is a poor predictor of plant species richness. *Science* **333**, 1750–1753.

Affonso, E.G., Polez, V.P., Correa, C.F., Mazon, A.F., Araujo, M.R.R., Moraes, G. & Rantin, F.T. (2002) Blood parameters and metabolites in the teleost fish *Colossoma macropomum* exposed to sulfide or hypoxia. *Comparative Biochemistry and Physiology C – Toxicology & Pharmacology* **133**, 375–382.

Agawin, N.S.R. & Agusti, S. (2005) *Prochlorococcus* and *Synechococcus* cells in the central Atlantic Ocean: distribution, growth and mortality (grazing) rates. *Life and Environment* **55**, 165–175.

Allcock, K.G. & Hik, D.S. (2003) What determines disturbance-productivity-diversity relationships? The effect of scale, species and environment on richness patterns in an Australian woodland. *Oikos* **102**, 173–185.

Al-Mufti, M.M., Sydes, C.L., Furness, S.B., Grime, J.P. & Band, S.R. (1977) A quantitative analysis of shoot phenology and dominance in herbaceous vegetation. *Journal of Ecology* **65**, 759–791.

Andersen, A.N. (1995) A classification of Australian ant communities, based on functional groups which parallel plant life-forms in relation to stress and disturbance. *Journal of Biogeography* **22**, 15–29.

Antonovics, J. (1976) The population genetics of mixtures. In: Wilson, J.R. (ed.) *Plant Relations in Pastures*, pp. 233–252. CSIRO, Melbourne, Australia.

Arezo, M.J., Pereiro, L. & Berois, N. (2005) Early development in the annual fish *Cynolebias viarius*. *Journal of Fish Biology* **66**, 1357–1370.

Aride, R.H.P, Roubach, R. & Val, A.L. (2007) Tolerance response of tambaqui *Colossoma macropomum* (Cuvier) to water pH. *Aquaculture Research* **38**, 588–594.

Arnold, W. & Dittami, J. (1997) Reproductive suppression in male alpine marmots. *Animal Behaviour* **53**, 53–66.

The Evolutionary Strategies that Shape Ecosystems, First Edition. J. Philip Grime, Simon Pierce.
© 2012 John Wiley & Sons, Ltd. Published 2012 by John Wiley & Sons, Ltd.

Aronson, R.B. & Precht, W.F. (1995) Landscape patterns of reef coral diversity: a test of the intermediate disturbance hypothesis. *Journal of Experimental Marine Biology and Ecology* **192**, 1–14.

Atkinson, C.J. & Farrar, J.F. (1983) Allocation of photosynthetically-fixed carbon in *Festuca ovina* L. and *Nardus stricta* L. *New Phytologist* **95**, 519–531.

Avilés, L. (1997) Causes and consequences of cooperation and permanent-sociality in spiders. In: Choe, J. & Crespi, B. (eds.) *The Evolution of Social Behaviour in Insects and Arachnids*, pp. 476–498. Cambridge University Press, Cambridge, UK.

Avilés, L., Maddison, W.P. & Agnarsson, I. (2006) A new independently derived social spider with explosive colony proliferation and a female size dimorphism. *Biotropica* **38**, 743–753.

Ayasse, M., Schiestl, F.P., Paulus, H.F., Löfstedt, C., Hansson, B., Ibarra, F. & Francke, W. (2000) Evolution of reproductive strategies in the sexually deceptive orchid *Ophrys sphegodes*: how does flower-specific variation of odor signals influence reproductive success? *Evolution* **54**, 1995–2006.

Bailey, J.K., Hendry, A.P., Kinnison, M.T., Post, D.M., Palkovacs, E.P., Pelletier, F., Harmon, L.J. & Schweitzer, J.A. (2009) From genes to ecosystems: an emerging synthesis of eco-evolutionary dynamics: *New Phytologist* **184**, 746–749.

Baker, H.G. (1972) Seed weight in relation to environmental conditions in California. *Ecology* **53**, 997–1010.

Bakker, J.P. (1989) *Nature Management by Grazing and Cutting*. Kluwer, Dordrecht, The Netherlands.

Baltz, D.M. (1984) Life-history variation among female surfperches (Perciformes: Embiotocidae). *Environmental Biology of Fishes* **10**, 159–171.

Bardgett, R.D. (2002) Causes and consequences of biological diversity in soil. *Zoology* **105**, 367–374.

Bardgett, R.D., Lovell, R.D., Hobbs, P.J. & Jarvis, S.C. (1999) Seasonal changes in soil microbial communities along a fertility gradient of temperate grasslands. *Soil Biology & Biochemistry* **31**, 1021–1030.

Bardgett, R.D. & Walker, L.R. (2004) Impact of colonizer plant species on the development of decomposer microbial communities following deglaciation. *Soil Biology & Biochemistry* **36**, 555–559.

Bar Zeev, E., Yogev, T., Man-Aharonovich, D., Kress, N., Herut, B., Beja, O. & Berman-Frank, I. (2008) Seasonal dynamics of the endosymbiotic, nitrogen-fixing cyanobacterium *Richelia intracellularis* in the eastern Mediterranean Sea. *ISME Journal* **2**, 911–923.

Baskin, C.C. & Baskin, J.M. (1998) *Seeds: Ecology, Biogeography and Evolution of Dormancy and Germination*. Academic Press, San Diego, USA.

Beck, J. & Chey, V.K. (2008) Explaining the elevational diversity pattern of geometrid moths from Borneo: a test of five hypotheses. *Journal of Beogeography* **35**, 1452–1464.

Becks, L., Ellner, S.P., Jones, L.E. & Hairston, N.G. Jr (2010) Reduction of adaptive genetic diversity radically alters eco-evolutionary community dynamics. *Ecology Letters* **13**, 989–997.

Bennett, M.D. (1972) Nuclear DNA content and minimum generation time. *Proceedings of the Royal Society of London B* **181**, 109–135.

Bentley, P.J. & Blumer, F.C. (1962) Uptake of water by the lizard, *Moloch horridus*. *Nature* **194**, 699–700.

Benzing, D.H. (2000) *Bromeliaceae: Profile of an Adaptive Radiation*. Cambridge University Press, Cambridge, UK.

Berg, G. & Smalla, K. (2009) Plant species and soil type cooperatively shape the structure and function of microbial communities in the rhizosphere. *FEMS Microbiology Ecology* **68**, 1–13.

Bernalier, A., Fonty, G., Bonnemoy, F. & Gouet, P. (1992) Degradation and fermentation of cellulose by the rumen anaerobic fungi in cultures or in association with cellulolytic bacteria. *Current Microbiology* **25**, 145–148.

Bernalier A, Fonty G, Bonnemoy F & Gouet P. (1993) Inhibition of the cellulolytic activity of *Neocallimastix frontalis* by *Ruminococcus flavifaciens*. *Journal of General Microbiology* **139**, 873–880.

Beyers, R.J. & Odum, H.T. (1993) *Ecological Microcosms*. Springer-Verlag, New York, USA.

Bezemer, T.M., Lawson, C.S., Hedlund, K., Edwards, A.R., Brook, A.J., Igual, J.M., Mortimer, S.R. & van der Putten. W.H. (2006) Plant species and functional group effects on abiotic and microbial soil properties and plant-soil feedback responses in two grasslands. *Journal of Ecology* **94**, 893–904.

Bhattarai, K.R., Vetaas, O.R. & Grytnes, J.A. (2004) Relationship between plant species richness and biomass in an arid sub-alpine grassland of the Central Himalayas, Nepal. *Folia Geobotanica* **39**, 57–71.

Bierbaum, R.M. & Ferson, S. (1986) Do symbiotic pea crabs decrease growth rate in mussels? *Biological Bulletin* **170**, 51–61.

Bigelow, H.B. & Schroeder, W.C. (1948) Sharks. *Memoir Sears Foundation for Marine Research* **1**, 59–546.

Bilton, M.C., Whitlock, R., Grime, J.P., Marion, G. & Pakeman, R.J. (2010) Intraspecific trait variation in grassland plant species reveals fine-scale strategy trade-offs and size differentiation that underpins performance in ecological communities. *Botany* **88**, 939–952.

Blair Hedges, S. (2008) At the lower size limit in snakes: two new species of threadsnakes (Squamata: Leptotyphlopidae: *Leptotyphlops*) from the Lesser Antilles. *Zootaxa* **1841**, 1–30.

Blakemore, R.J., Csuzdi, C.S., Ito, M.T., Kaneko, N., Paoletti, M.G., Spiridonov, S.E., Uchida, T. & Van Praagh, B.D. (2007) *Megascolex* (*Promegascolex*) *mekongianus* Cognetti, 1922 – its extent, ecology and allocation to *Amynthas* (Clitellata/Oligochaeta: Megascolecidae). *Opuscula Zoologica (Budapest)* **36**, 19–30.

Blanchard, F.N. (1937) Data on the natural history of the red-bellied snake *Storeria occipitomaculata* (Storer), in northern Michigan. *Copeia* **1937**, 151–162.

Blaum, N., Rossmanith, E., Schwager, M. & Jeltsch, F. (2007) Responses of mammilian carnivores to land use in arid savanna rangelands. *Basic and Applied Ecology* **8**, 552–564.

Bolker, B.M. & Pacala, S.W. (1999) Spatial moment equations for plant competition: understanding spatial strategies and the advantages of short dispersal. *American Naturalist* **153**, 575–602.

Bonilla, S., Conde, D., Aubriot, L. & del Carmen Pérez, M. (2005) Influence of hydrology on phytoplankton species composition and life strategies in a subtropical coastal lagoon periodically connected with the Atlantic ocean. *Estuaries* **28**, 884–895.

Booth, R.E. & Grime, J.P. (2003) Effects of genetic impoverishment on plant community diversity. *Journal of Ecology* **91**, 721–730.

Booth, T., Gorrie, S. & Muhsin, T.M. (1998) Life strategies among fungal assemblages on *Salicornia europaea* aggregate. *Mycologia* **80**, 176–191.

Bornhofen, S., Barot, S. & Lattaud, C. (2011) The evolution of CSR life-history strategies in a plant model with explicit physiology and architecture. *Ecological Modelling* **222**, 1–10.

Bornhofen, S. & Lattaud, C. (2008) Evolving CSR Strategies in virtual plant communities. In: Bullock, S., Noble, J., Watson, R. & Bedaum M.A. (eds) *Artificial Life XI*, pp. 72–79. MIT Press, Cambridge, Massachusetts, USA.

Box, E.O. (1981) *Macroclimate and Plant Forms: An Introduction to Predictive Modelling in Phytogeography*. Junk, The Hague, The Netherlands.

Boyle-Yarwood, S.A., Bottomley, P.J. & Myrold, D.D. (2008) Community composition of ammonia-oxidising bacteria and archaea in soils under stands of red alder and Douglas fir in Oregon. *Environmental Microbiology* **10**, 2956–2965.

Braby, M.F. (2002) Life-history strategies and habitat templets of tropical butterflies in north-eastern Australia. *Evolutionary Ecology* **16**, 399–413.

Braun-Blanquet, J. (1928) *Pflanzensoziologie. Grundzuge der Vegetationskunde.* Springer Verlag, Berlin, Germany.

Bremer, C., Braker, G., Matthies, D., Reuter, A., Engels, C. & Conrad, R. (2007) Impact of plant functional group, plant species, and sampling time on the composition of *nirK*-type denitrifier communities in soil. *Applied and Environmental Microbiology* 73, 6876–6884.

Briggs, D. & Walters, S.M. (1969) *Plant Variation and Evolution.* Weidenfield & Nicholson, London, UK.

Brischoux, F. & Bonnet, X. (2008) Estimating the impact of sea kraits on the anguilliform fish community (congridae, Muraenidae, Ophichthidae) of New Caledonia. *Aquatic Living Resources* 21, 395–399.

Brock, T.D. (1978) *Thermophilic Microorganisms and Life at High Temperatures.* Springer-Verlag. New York, USA.

Brockhurst, M.A., Colegrave, N., Hodgson, D.J. & Buckling, A. (2007) Niche occupation limits adaptive radiation in experimental microcosms. *PLoS ONE* 2, e193.

Broeckling, C.D., Broz, A.K., Bergelson, J., Manter, D.K. & Vivanco, J.M. (2008) Root exudates regulate soil fungal community composition and diversity. *Applied and Environmental Microbiology* 74, 738–744.

Brown, G.P. & Shine, R. (2002) Reproductive ecology of a tropical natricine snake, *Tropidonophis mairii* (Colubridae). *Journal of Zoology* 258, 63–72.

Buckland, S.M. & Grime, J.P. (2000) The effects of trophic structure and soil fertility on the assembly of plant communities: a microcosm experiment. *Oikos* 91, 336–352.

Budel, B., Bendix, J., Bicker, F.R. & Green, T.G.A. (2008) Dewfall as a water source frequently activates the endolithic cyanobacterial communities in the granites of Taylor Valley, Antarctica. *Journal of Phycology* 44, 1415–1424.

Bulleri, F. (2009) Facilitation research in marine systems: state of the art, emerging patterns and insights for future developments. *Journal of Ecology* 97, 1121–1130.

Bunce, R.G.H., Barr, C.J., Gillespie, M.K., Howard, D.C., Scott, W.A., Smart, S.M., Van de Poll, H.M. & Watkins, J.W. (1999) Vegetation of the British Countryside; The Countryside Vegetation System. *ECOFACT, Volume 1, DETR, London.*

Burch, T.L. (1979) The importance of communal experience to survival for spiderlings of *Araneus diadematus* (Araneae: Araneidae). *Journal of Arachnology* 7, 1–18.

Burness, G.P., Diamond, J. & Flannery, T. (2001) Dinosaurs, dragons, and dwarfs: the evolution of maximal body size. *Proceedings of the National Academy of Sciences of the USA* 98, 14518–14523.

Butt, K. & Nuutinen, V. (1998) Reproduction of the earthworm *Lumbricus terrestris* Linné after the first mating. *Canadian Journal of Zoology* 76, 104–109.

Bybee, P.J., Lee, A.H. & Lamm, E-T. (2006) Sizing the Jurassic theropod dinosaur *Allosaurus*: assessing growth strategy and evolution of ontogenetic scaling of limbs. *Journal of Morphology* 267, 347–359.

Caccianiga, M., Luzzaro, A., Pierce, S., Ceriani, R.M. & Cerabolini, B. (2006) The functional basis of a primary succession resolved by CSR classification. *Oikos* 112, 10–20.

Cailliet, G.M., Andrews, A.H., Burton, E.J., Watters, D.L., Kline, D.E. & Ferry-Graham, L.A. (2001) Age determination and validation studies of marine fishes: do deep-dwellers live longer? *Experimental Gerontology* 36, 739–764.

Calder, N. (2003) *Magic Universe: The Oxford Guide to Modern Science.* Oxford University Press, Oxford, UK.

Callaway, R.M., Brooker, R.W., Choler, P., Kikvidze, Z., Lortie, C.J., Michalet, R., Paolini, L., Pugnaire, F.I., Newingham, B., Aschehoug, E.T., Armas, C., Kikodze, D. & Cook, B.J. (2002) Positive interactions among alpine plants increase with stress. *Nature* 417, 844–848.

Campbell, B.D., Grime, J.P. and Mackey, J.M.L. (1991) A trade-off between scale and precision in resource foraging. *Oecologia* 87, 532–538.

Campbell. L, & Vaulot, D. (1993) Photosynthetic picoplankton community structure in the subtropical North Pacific Ocean near Hawaii (station ALOHA). *Deep Sea Research Part 1: Oceanographic Research Papers* **40**, 2043–2060.

Canals, M., Salazar, M.J., Durán, C., Figueroa, D. & Veloso, C. (2008) Respiratory refinements in the mygalomorph spider *Grammostola rosea* Walckenaer 1837 (Araneae, Theraphosidae). *The Journal of Arachnology* **35**, 481–486.

Cardinale, B.J., Matulich, K.L., Hooper, D.U., Byrnes, J.E., Duffy, E., Gamfeldt, L., Balvanera, P., O'Connor, M.I. & Gonzalez, A. (2011) The functional role of producer diversity in ecosystems. *American Journal of Botany* **98**, 572–592.

Carey, F.G., Kanwisher, J.W., Brazier, O., Gabrielson, G., Casey, J.G. & Pratt, H.L. Jr (1982) Temperature and activities of a white shark, *Carcharodon carcharias*. *Copeia* **1982**, 254–260.

Carpenter, J. (2005) Competition for food between an introduced crayfish and two fishes endemic to the Colorado River basin. *Environmental Biology of Fishes* **72**, 335–342.

Carvalho, C.D. & Bessa, E.C.D. (2008) Life-history strategy of *Bradybaena similaris* (Ferussac, 1821) (Mollusca, Pulmonata, Bradybaenidae). *Molluscan Research* **28(3)**, 171–174.

Cary, S.C., Shank, T. & Stein, J. (1998) Worms bask in extreme temperatures. *Nature* **391**, 545–546.

Caswell, H. (1989) *Matrix Population Models*. Sinauer, Sunderland, UK.

Cavedon, K. & Canale-Parola, E. (1992) Physiological interactions between a mesiphilic cellulolytic *Clostridium* and a non-cellulolytic bacterium. *FEMS Microbial Ecology* **86**, 237–245.

Cerabolini, B., Brusa, G., Ceriani, R.M., De Andreis, R., Luzzaro, A. & Pierce, S. (2010a) Can CSR classification be generally applied outside Britain? *Plant Ecology* **210**, 253–261.

Cerabolini, B., Pierce, S., Luzzaro, A. & Ossola, A. (2010b) Species evenness affects ecosystem processes *in situ* via diversity in the adaptive strategies of dominant species. *Plant Ecology* **207**, 333–345.

Chalcraft, D.R., Wilsey, B.J., Bowles, C. & Willig, M.R. (2009) The relationship between productivity and multiple aspects of biodiversity in six grassland communities. *Biodiversity and Conservation* **18**, 91–104.

Champion de Crespigny, F.E., Herberstein, M.E. & Elgar, M.A. (2001) Food caching in orb-web spiders (Araneae: Araneoidea). *Naturwissenschaften* **88**, 42–45.

Chapin, F.S. III. (1980) The mineral nutrition of wild plants. *Annual Review of Ecology and Systematics* **11**, 233–260.

Charmantier., A, McCleery, R.H., Cole, L.R., Perrins, C., Kruuk, E.B. & Sheldon, B.C. (2008) Adaptive phenotypic plasticity in response to climate change in a wild bird population. *Science* **320**, 800–803.

Chase, T.E., Nelson, J.A., Burbank, D.E. & Van Etten, J.L. (1989) Mutual exclusion occurs in a *Chlorella*-like green alga inoculated with two viruses. *Journal of General Virology* **70**, 1829–1836.

Chausson, F., Bridges, C.R., Sarradin, P-M., Green, B.N., Riso, R., Caprais, J-C. & Lallier, F.H. (2001) Structural and functional properties of hemocyanin from *Cyanagraea praedator*, a deep-sea hydrothermal vent crab. *PROTEINS: Structure, Function and Genetics* **45**, 351–359.

Chawla, A., Rajkumar, S., Singh, K.N., Lal, B., Singh, R.D. & Thukral, A.K. (2008) Plant species diversity along an altitudinal gradient of Bhabha Valley in Western Himalaya. *Journal of Mountain Science* **5**, 157–177.

Chen, J. & Stark, J.M. (2000) Plant species effects and carbon and nitrogen cycling in a sagebrush-crested wheatgrass soil. *Soil Biology & Biochemistry* **32**, 47–57.

Chen, L.Z., Wang, G.H., Hong, S., Liu, A., Li, C. & Liu, Y.D. (2009) UV-B-induced oxidative damage and protective role of exopolysaccharides in desert cyanobacterium *Microcoleus vaginatus*. *Journal of Integrative Plant Biology* **51**, 194–200.

Chiappe, L.M., Coria, R.M., Dingus, L., Jackson, F., Chinsamy, A. & Fox, M. (1998) Sauropod dinosaur embryos from the Late Cretaceous of Patagonia. *Nature* **396**, 258–261.

Choler, P., Michalet, R. & Callaway, R.M. (2001) Facilitation and competition on gradients in alpine plant communities. *Ecology* **82**, 3295–3308.

Chornesky, E.A. (1991) The ties that bind: inter-clonal cooperation may help a fragile coral dominate shallow high-energy reefs. *Marine Biology* **109**, 41–51.

Chow, S.S., Wilke, C.O., Ofria, C., Lenski, R.E. & Adami, C. (2004) Adaptive radiation from resource competition in digital organisms. *Science* **305**, 84–86.

Chown, S.L. & Gaston, K.J. (1999) Patterns in procellariiform diversity as a test of species-energy theory in marine systems. *Evolutionary Ecology Research* **1**, 365–373.

Chown, S.L. & Nicolson, S.W. (2004) *Insect Physiological Ecology: Mechanisms and Patterns*. Oxford University Press, Oxford, UK.

Claessens, L.P.A.M., O'Connor, P.M. & Unwin, D.M. (2009) Respiratory evolution facilitated the origin of pterosaur flight and aerial gigantism. *PLoS ONE* **4**, e4497.

Claverie, J.M., Ogata, H., Audic, S., Abergel, C., Suhre, K. & Fournier, P.E. (2006) *Mimivirus* and the emerging concept of 'giant' virus. *Virus Research* **117**, 133–144.

Colasanti, R.L., Hunt, R. & Askew, A.P. (2001) A self-assembling model of resource dynamics and plant growth incorporating plant functional types. *Functional Ecology* **15**, 676–687.

Cole, L.C. (1954) The population consequences of life-history phenomena. *Quarterly Review of Biology* **29**, 103–137.

Cole, S.K. & Gilbert, D.L. (1970) Jet propulsion of squid. *Biol.Bull.Mar.Biol.Lab.Woods Hole* **138**, 245–246.

Compagno, L.J.V. (1984) FAO species catalogue. Vol. 4. Sharks of the world. An annotated and illustrated catalogue of shark species known to date. Part 1: Hexanchiformes to Lamniformes. *FAO Fish. Synop.* **125**, 1–249.

Comte, J. & del Giorgio, P.A. (2009) Links between resources, C metabolism and the major components of bacterioplankton community structure across a range of freshwater ecosystems. *Environmental Microbiology* **11**, 1704–1716.

Conway-Morris, S. (2003) *Life's Solution: Inevitable Humans in a Lonely Universe.* Cambridge University Press, Cambridge, UK.

Cook, R.C. & Rayner, A.D.M. (1984) *Ecology of Saprophytic Fungi.* Longman, London, UK.

Cornelissen, J.H.C., Aerts, R., Cerabolini, B., Werger, M.J.A. & van der Heijden, M.G.A. (2001) Carbon cycling traits of plant species are linked with mycorrhizal strategy. *Oecologia* **129**, 611–619.

Cornelissen, J.H.C. & Thompson, K. (1997) Functional leaf attributes predict litter decomposition rate in herbaceous plants. *New Phytologist* **135**, 109–114.

Cornell, H.V. & Karlson, R.H. (2000) Coral species richness: ecological versus biogeographical influences. *Coral Reefs* **19**, 37–49.

Cornwell, W.K. & Grubb, P.J. (2003) Regional and local patterns in plant species richness with respect to resource availability. *Oikos* **100**, 417–428.

Cornwell.W.K., Cornelissen, J.H.C., Amatangelo, K., Dorrepaal, E., Eviner, V.T., Godoy. O., Hobbie, S.E., Hoorens, B., Kurokawa, H., Pérez-Harguindeguy, N., Quested, H.M., Santiago, L.S., Wardle, D.A., Wright, I.J., Aerts, R., Allison, S.D., van Bodegom, P., Brovkin, V., Chatain, A., Callaghan, T.V., Díaz, S., Garnier, E., Gurvich, D.E., Kazakou, E., Klein, J.A., Read, J., Reich, P.B., Soudzilovskaia, N.A., Vaieretti, M.V. & Westoby, M. (2008) Plant species traits are the predominant control on litter decomposition rates within biomes worldwide. *Ecology Letters* **11**, 1065–1071.

Coyne, J.A. (2009) *Why Evolution is True.* Oxford University Press, Oxford, UK.

Crawford, R.M.M. (1989) *Studies in Plant Survival.* Blackwell, Oxford, UK.

Crawley, M.J. (ed.) (1986) *Plant Ecology.* Blackwell, Oxford, UK.

Curtis, J.T. (1959) *The Vegetation of Wisconsin: An Ordination of Plant Communities.* University of Wisconsin Press, Wisconsin, USA.

Curtis, T. (2007) Theory and the microbial world. *Environmental Microbiology* **9**, 1.

Cushman, J.C. (2001) Crassulacean acid metabolism: a plastic photosynthetic adaptation to arid environments. *Plant Physiology* **127**, 1439–1448.

Cushman, J.C. & Borland, A.M. (2002) Induction of Crassulacean acid metabolism by water limitation. *Plant, Cell & Environment* **25**, 295–310.

Dafni, A., Ivri, Y. & Brantjes, N.B.M. (1981) Pollination of *Serapias vomeracea* Briq. (Orchidaceae) by imitation of holes for sleeping solitary male bees (Hymenoptera). *Acta Botanica Neerlandica* **30**, 69–73.

Dakora, F.D. & Phillips, D.A. (2002) Root exudates as mediators of mineral acquisition in low-nutrient environments. *Plant and Soil* **245**, 35–47.

Danchin, E., Giraldeau, L-A. & Wagner, R.H. (2008) Animal aggregations: hypotheses and controversies. In: Danchin, E., Giraldeau, L-A. & Cézilly, F. (eds) *Behavioural Ecology*, pp. 503–545. Oxford University Press, Oxford, UK.

Dansereau, P. (1951) Description and recording of vegetation upon a structural basis. *Ecology* **32**, 172–229.

Darwin, C. (1845, 2nd edn) *Journal of researches into the natural history and geology of the countries visited during the voyage of H.M.S. Beagle round the world, under the command of Capt. Fitz Roy, R.N.* John Murray, London, UK.

Darwin, C. (1859) *The Origin of Species by means of Natural Selection or the Preservation of Favoured Races in the Struggle for Life.* Penguin Classics Reprint, 1985. Penguin Books, London, UK.

Darwin, C. & Wallace, A.R. (1858) On the tendency of species to form varieties; and on the perpetuation of varieties and species by natural means of selection. *Journal of the Linnean Society of London (Zoology)* **3**, 45–62.

Da Silva, J.A.M., Pereira, M. & de Oliveira-Pereira, M.I. (2003) Fruits and seeds consumed by tambaqui (*Colossoma macropomum*, Cuvier 1818) incorporated in the diet: gastrointestinal tract digestibility and transit velocity. *Brazilian Journal of Animal Science* **32**, 1815–1824 Suppl. 2.

Davis, M.A. (2009) *Invasion Biology.* Oxford University Press, Oxford, UK.

Davis, M.R.H., Zhao, F.J. & McGrath, S.P. (2004) Pollution-induced community tolerance of soil microbes in response to a zinc gradient. *Environmental Toxicology and Chemistry* **23**, 2665–2672.

Dawkins, R. (1982) *The Extended Phenotype.* Oxford University Press, Oxford, UK.

Dawkins, R. (2004) *The Ancestor's Tale: A Pilgrimage to the Dawn of Life.* Weidenfield & Nicholson, London, UK.

Dawkins, R. (2009) *The Greatest Show on Earth: The Evidence for Evolution.* Bantam Press, London, UK.

De Almeida, L.C. & Franco, M.R.B. (2006) Determination of essential fatty acids in captured and farmed tambaqui (*Colossoma macropomum*) from the Brazilian Amazonian Area. *Journal of the American Oil Chemists Society* **83**, 707–711.

De Almeida, L.C., Lundstedt, L.M. & Moraes, G. (2006) Digestive enzyme responses of tambaqui (*Colossoma macropomum*) fed on different levels of protein and lipid. *Aquaculture Nutrition* **12**, 443–450.

De Bello, F., Lavorel, S., Díaz, S., Harrington, R., Cornelissen, J.H.C., Bardgett, R.D., Berg, M.P., Cipriotti, P., Feld, C.K., Hering, D., da Silva, P.M., Potts, S.G., Sandin, L., Sousa, J.P., Storkey, J., Wardle, D.A. & Harrison, P.A. (2010) Towards an assessment of multiple ecosystem processes and services via functional traits. *Biodiversity Conservation* **19**, 2873–2893.

De Deyn, G.B., Cornelissen, H.C. & Bardgett, R.D. (2008) Plant functional traits and soil carbon sequestration in contrasting biomes. *Ecology Letters* **11**, 516–531.

DeLong, E.F., Preston, C.M., Mincer, T., Rich, V., Hallam, S.J., Frigaard, N-U., Martinez, A., Sullivan, M.B., Edwards, R., Brito, B., Chisholm, S.W. & Karl, D.M. (2006) Community genomics among stratified microbial assemblages in the ocean's interior. *Science* **311**, 496–503.

Delbrück, M. (1945) Interference between bacterial viruses. III. The mutual exclusion effect and the depressor effect. *Journal of Bacteriology* **50**, 151–170.

de Magalhães, J.P. & Costa, J. (2009) A database of vertebrate longevity records and their relation to other life-history traits. *Journal of Evolutionary Biology* **22**, 1770–1774.

De Miguel, J.M., Casado, M.A., Del Pozo, A., Ovalle, C., Moreno-Casasola, P., Travieso-Bello, A.C., Barrera, M., Ricardo, N., Tecco, P.A. & Acosta, B. (2010) How reproductive, vegetative and defensive strategies of Mediterranean grassland species respond to a grazing intensity gradient. *Plant Ecology* **210**, 97–110.

Dennis, R.L.H., Hodgson, J.G., Grenyer, R., Shreeve, T.G. & Roy, D.B. (2004) Host plants and butterfly biology. Do host–plant strategies drive butterfly status? *Ecological Entomology* **29**, 12–26.

Desbruyères, D., Chevaldonnè, P., Alayse, A-M., Jollivet, D., Lallier, F.H., Jouin-Toulmond, C., Zal, F., Sarradin, P-M., Cosson, R., Caprais, J-C., Arndt, C., O'Brien, J., Guezennec, J., Hourdez, S., Riso, R., Gaill, F., Laubier, L. & Toulmond, A. (1998) Biology and ecology of the 'Pompeii worm' (*Alvinella pompejana* Desbruyères and Laubier), a normal dweller of an extreme deep-sea environment: a synthesis of current knowledge and recent developments. *Deep-Sea Research Part II* **45**, 383–422.

Díaz, S. & Cabido, M. (1997) Plant functional types and ecosystem function in relation to global change. *Journal of Vegetation Science* **8**, 463–474

Díaz, S. & Cabido, M. (2001) Vive la différence: plant functional diversity matters to ecosystem processes. *TRENDS in Ecology & Evolution* **16**, 646–655.

Díaz, S., Hodgson, J.G., Thompson, K., Cabido, M., Cornelissen, J.H.C., Jalili, A, Montserrat-Martí, G., Grime, J.P., Zarrinkamar, F., Asri, Y., Band, S.R., Basconcelo, S., Castro-Díez, P., Funes, G., Hamzehee, B., Khoshnevi, M., Pérez-Harguindeguy, N., Pérez-Rontomé, M.C., Shirvany, F.A., Vendramini, F., Yazdani, S., Abbas-Azimi, R., Bogaard, A., Boustani, S., Charles, M., Dehghan, M., de Torres-Espuny, L., Falczu, V., Guerrero-Campo, J., Hynd, A., Jones, G., Kowsary, E., Kazemi-Saeed, F., Maestro-Martínez, M., Romo-Díez, A., Shaw, S., Siavash, B., Villar-Salvador, P. & Zak, M.R. (2004) The plant traits that drive ecosystems: evidence from three continents. *Journal of Vegetation Science* **15**, 295–304.

Dickinson, G. & Murphy, K. 2007. *Ecosystems*. *2ⁿᵈ Edition*. Routledge, Abingdon, UK.

Dixon, D.R., Simpson-White, R. & Dixon L.R.J. (1992) Evidence for thermal stability of ribosomal DNA sequences in hydrothermal-vent organisms. *Journal of the Marine Biological Association of the United Kingdom* **72**, 519–527.

Dobzhansky, T. (1973) Nothing in biology makes sense except in the light of evolution. *American Biology Teacher* **35**, 125–129.

Dodd, A.N., Borland, A.M., Haslam, R.P., Griffiths, H. & Maxwell, K. (2002) Crassulacean acid metabolism: plastic fantastic. *Journal of Experimental Botany* **53**, 569–580.

Dodson, S.I., Arnott, S.E. & Cottingham, K.L. (2000) The relationship in lake communities between primary productivity and species richness. *Ecology* **81**, 2662–2679.

Dong, Z-M. & Currie, P.J. (1996) On the discovery of an oviraptorid skeleton on a nest of eggs at Bayan Mandahu, Inner Mongolia, People's Republic of China. *Canadian Journal of Earth Science* **33**, 631–636.

Donnison, L.M, Griffith, G.S., Hedger, J., Hobbs, P.J. & Bardgett, R.D. (2000) Management influences on soil microbial communities and their function in botanically diverse haymeadows of northern England and Wales. *Soil Biology & Biochemistry* **32**, 253–263.

Duckworth, J.C., Kent, M. & Ramsay, P.M. (2000) Plant functional types – an alternative to taxonomic plant community description in biogeography? *Progress in Physical Geography* **24**, 515–542.

Dufresne, A., Salanoubat, M., Partensky, F., Artiguenave, F., Axmann, I.M., Barbe, V., Duprat, S., Galperin, M.Y., Koonin, E.V., Le Gall, F., Makarova, K.S., Ostrowski, M., Oztas, S., Robert, C., Rogozin, I.B., Scanlan, D.J., Tandeau de Marsac, N., Weissenbach, J., Wincker, P., Wolf, Y.I. & Hess, W.R. (2003) Genome sequence of the cyanobacteriam *Prochlorococcus marinus* SS120, a nearly minimal oxyphototrophic genome. *PNAS* **100**, 10020–10025.

Eberth, D.A., Xing, X. & Clark, J.M. (2010) Dinosaur death pits from the Jurassic of China. *PALAIOS* **25**, 112–125.

Ehrlich, P.R. (1993) Biodiversity and ecosystem function: need we know more? In: Schulze, E-D. & Mooney, HA. (eds) *Biodiversity and Ecosystem Function*. Springer-Verlag, Berlin, Germany, vii–x.

Ehrlich, P.R. & Ehrlich, A.H. (1981) *Extinction: The Causes and Consequences of the Disappearance of Species*. Random House, New York, USA.

Eigen, M., Biebricher, C.K. & Gebinoga, M. (1991) The hypercycle: coupling of RNA and protein biosynthesis in the infection cycle of an RNA bacteriophage. *Biochemistry* **30**, 11005–11018.

Ellis, D.G., Bizzoco, R.W. & Kelley, S.T. (2008) Halophilic Archaea determined from geothermal steam vent aerosols. *Environmental Microbiology* **10**, 1582–1590.

Elliott, J.A., Reynolds, C.S. & Irish, A.E. (2001) An investigation of dominance in phytoplankton using the PROTECH model. *Freshwater Biology* **46**, 99–108.

Elliott, J.A., Reynolds, C.S., Irish, A.E. & Tett, P. (1999) Exploring the potential of the PROTECH model to investigate phytoplankton community theory. *Hydrobiologia* **414**, 37–43.

Elmer, K.R., Lehtonen, T.K., Kautt, A.F., Harrod, C. & Meyer, A. (2010) Rapid sympatric ecological differentiation of crater lake cichlid fishes within historic times. *BMC Biology* **8**, 60.

El Sharkawi, H., Yamamoto, S. & Honna, T. (2006) Rice yield and nutrient uptake as affected by cyanobacteria and soil amendments – a pot experiment. *Journal of Plant Nutrition and Soil Science* **169**, 809–815.

Enright, N.J., Franco, M. & Silvertown, J. (1995) Comparing life histories using elasticity analysis: the importance of life span and the number of life-cycle stages. *Oecologia* **104**, 79–84.

Eriksen-Hamel, N.S. & Whalen, J.K. (2007) Competitive interactions affect the growth of *Aporrectodea caliginosa* and *Lumbricus terrestris* (Oligochaeta: Lumbricidae) in single- and mixed-species laboratory cultures. *European Journal of Soil Biology* **43**, 142–150.

Erikson, G.M., Makovicky, P.J., Currie, P.J., Norell, M.A., Yerby, S.A. & Brochu, C.A. (2004) Gigantism and comparative life-history parameters of tyrannosaurid dinosaurs. *Nature* **430**, 772–775.

Eskelinen, A., Stark, S. & Männistö, M. (2009) Links between plant community composition, soil organic matter quality and microbial communities in contrasting tundra habitats. *Oecologia* **161**, 113–123.

Espinar, J.L. (2006) Sample size and the detection of a hump-shaped relationship between biomass and species richness in Mediterranean wetlands. *Journal of Vegetation Science* **17**, 227–232.

Farlow, J.O. (1976) A consideration of the trophic dynamics of a late cretaceous large-dinosaur community (Oldham Formation). *Ecology* **57**, 841–857.

Fauquet, C.M., Mayo, M.A., Maniloff, J., Desselberger, U. & Ball, L.A. (2005) *Virus Taxonomy: Classification and Nomenclature of Viruses*. Elsevier Academic Press, Amsterdam, The Netherlands.

Feild, T.S., Arens, N.C., Doyle, J.A., Dawson, T.E. & Donoghue, M.J. (2004) Dark and disturbed: a new image of early angiosperm ecology. *Paleobiology* **30**, 82–107.

Feild, T.S., Chatelet, D.S. & Brodbribb, T.J. (2009) Ancestral xerophobia: a hypothesis on the whole plant ecophysiology of early angiosperms. *Geobiology* **7**, 237–264.

Fenchel, T. & Finlay, B.J. (1995) *Ecology and Evolution in Anoxic Worlds*. Oxford Series in Ecology and Evolution. Oxford University Press, Oxford, UK.

Feuillade, M., Bohatier, J., Bourdier, G., Dufour, Ph., Feuillade, J. & Krupka, H. (1998) Amino acid uptake by a natural population of *Oscillatoria rubescens* in relation to uptake by bacterioplankton. *Archives of Hydrobiology* **113**, 345–358.

Fischer, M.G., Allen, M.J., Wilson, W.H. & Suttle, C.A. (2010) Giant virus with a remarkable complement of genes infects marine zooplankton. *Proceedings of the National Academy of Science* **107**, 19508–19513.

Fisher, C.R., Fitt, W.K. & Trench, R.K. (1985) Photosynthesis and respiration in *Tridacna gigas* as a function of irradiance and size. *Biological Bulletin* **169**, 230–245.

Florindo, L.H., Reid, S.G., Kalinin, A.L., Milsom, W.K. & Rantin, F.T. (2004) Cardiorespiratory reflexes and aquatic surface respiration in the neotropical fish tambaqui (*Colossoma macropomum*): acute responses to hypercarbia. *Journal of Comparative Physiology B – Biochemical, Systematic and Environmental Physiology* **174**, 319–328.

Florindo, L.H., Leite, C.A.C., Kalinin, A.L., Reid, S.G., Milsom, W.K. & Rantin, F.T. (2006) The role of branchial and orobranchial O_2 chemoreceptors in the control of aquatic surface respiration in the neotropical fish tambaqui (*Colossoma macropomum*): progressive responses to prolonged hypoxia. *Journal of Experimental Biology* **209**, 1709–1715.

Fock, H.O. (2009) Deep-sea pelagic ichthyonekton diversity in the Atlantic Ocean and the adjacent sector of the Southern Ocean. *Global Ecology and Biogeography* **18**, 178–191.

Foelix, R.F. (1996) *Reproduction: Biology of Spiders*. Oxford University Press, Oxford, UK.

Forey, E., Touzard, B. & Michalet, R. (2010) Does disturbance drive the collapse of biotic interactions at the severe end of the diversity-biomass gradient? *Plant Ecology* **206**, 287–295.

Forster, R.R. & Platnick, N.I. (1977) A review of the spider family Symphytognathidae (Arachnida, Araneae). *American Museum Novitates* **2619**, 1–29.

Fox, R., Lehmkuhle, D.W. & Westendorf, D.H. (1976) Falcon visual acuity. *Science* **192**, 263–265.

Fraser, L.H. & Grime, J.P. (1998a) Primary productivity and trophic dynamics investigated in a North Derbyshire dale. *Oikos* **80**, 499–508.

Fraser, L.H. & Grime, J.P. (1998b) Top-down control and its effect on the biomass and composition of three grasses at high and low soil fertility in outdoor microcosms. *Oecologia* **113**, 239–246.

Fraser, L.H. & Grime, J.P. (1999) Interacting effects of herbivory and fertility on a synthesized plant community. *Journal of Ecology* **87**, 514–525.

Freschet, G.T., Cornelissen, J.H.C., van Logtestijn, R.S.P. & Aerts, R. (2010) Evidence of the 'plant economics spectum' in a subarctic flora. *Journal of Ecology* **98**, 362–373.

Fretwell, D. (1977) The regulation of plant communities by food chains exploiting them. *Perspectives in Biology and Medicine* **20**, 169–185.

Fretwell, D. (1987) Food chain dynamics: the central theory of ecology? *Oikos* **50**, 291–301.

Fridley, J.D., Grime, J.P. & Bilton, M. (2007) Genetic identity of interspecific neighbours mediates plant responses to competition and environmental variation in a species-rich grassland. *Journal of Ecology* **95**, 908–915.

Friedlander, A.M. & Parrish, J.D. (1998) Habitat characteristics affecting fish assemblages on a Hawaiian coral reef. *Journal of Experimental Marine Biology and Ecology* **224**, 1–30.

Fussmann, G.F., Loreau, M. & Abrams, P.A. (2007) Eco-evolutionary dynamics of communities and ecosystems. *Functional Ecology* **21**, 465–477.

Gadagkar, R. (1991) *Belonogaster, Mischocyttarus, Parapolybia*, and independent-founding *Ropalidia*. In: Ross, K.G. & Matthews, R.W. (eds) *The Social Biology of Wasps*, pp. 149–187. Comstock Publishing Associates, Cornell University Press, Ithaca, USA.

Gaillard, J-M., Pontier, D., Allainé, D., Lebreton, J.D., Trouvilliez, J. & Clobert, J. (1989) An analysis of demographic tactics in birds and mammals. *Oikos* **56**, 59–76.

García, D., Loureiro, M. & Tassino, B. (2008) Reproductive behaviour in the annual fish *Austrolebias reicherti* Loureiro & García 2004 (Cyprinodontiformes: Rivulidae). *Neotropical Ichthyology* **6**, 243–248.

Garcia-Pichel, F., Mechling, M. & Castenholz, R.W. (1994) Diel migrations of microorganisms within a benthic, hypersaline mat community. *Applied and Environmental Microbiology* **60**, 1500–1511.

Garnier, E., Lavorel, S., Ansquer, P., Castro, H., Cruz, P., Dolezal, J., Eriksson, O., Fortunel, C., Freitas, H., Golodets, C., Grigulis, K., Jouany, C., Kazakou, E., Kigel, J., Kleyer, M., Lehsten, V., Lepš, J., Meier, T., Pakeman, R., Papadimitriou, M., Papanastasis, V.P., Quested, H., Quétier, F., Robson, M., Roumet, C., Rusch, G., Skarpe, C., Sternberg, M., Theau, J-P., Thébault, A., Vile, D. & Zarovali, M.P. (2006) Assessing the effects of land-use change on plant traits, communities and ecosystem functioning in grasslands: a standardized methodology and lessons from an application to 11 European sites. *Annals of Botany* **99**, 967–985.

Garpe, K.C. & Öhman, M.C. (2004) Coral and fish distribution patterns in Mafia Island Marine Park, Tanzania: fish-habitat interactions. *Hydrobiologia* **498**, 191–211.

Garvey, J.E., Stein, R.A. & Thomas, H.M. (1994) Assessing how fish predation and interspecific prey competition influence a crayfish assemblage. *Ecology* **75**, 532–547.

Gattinger, A., Höfle, M.G., Scholter, M., Embacher, A., Böhme, F., Munch, J.C. & Labrenz, M. (2007) Traditional cattle manure application determines abundance, diversity and activity of methanogenic Archaea in arable European soil. *Environmental Microbiology* **9**, 612–624.

Gilmour, K.A., Milsom, W.K., Rantin, F.T., Reid, S.G. & Perry, S.F. (2005) Cardiorespiratory responses to hypercarbia in tambaqui *Colossoma macropomum*: chemoreceptor orientation and specificity. *Journal of Experimental Biology* **208**, 1095–1107.

Gimingham, C.H. (1951) The use of life-form and growth-form in the analysis of community structure, as illustrated by a comparison of two dune communities. *Journal of Ecology* **39**, 396–406.

Gimingham, C.H. (1972) *Ecology of Heathlands*. Chapman and Hall, London, UK.

GIROS (2009) *Orchidee d'Italia*. Il Castello, Cornaredo (MI), Italia (in Italian).

Girvan, M.S., Bullimore, J., Ball, A.S., Pretty, J.N. & Osborn, A.M. (2004) Responses of active bacterial and fungal communities in soils under winter wheat to different fertilizer and pesticide regimes. *Applied and Environmental Microbiology* **70**, 2692–2701.

Gobbi, M., Caccianiga, M., Cerabolini, B., De Bernardi, F., Luzzaro, A. & Pierce, S. (2010) Plant adaptive responses during primary succession are associated with functional adaptations in ground beetles on recently deglaciated terrain. *Community Ecology* **11**, 223–231.

Goldbogen, J.A., Calambokidis, J., Oleson, E., Potvin, J., Pyenson, N.D., Schorr, G. & Shadwick, R.E. (2011) Mechanics, hydrodynamics and energetics of blue whale lunge feeding: efficiency dependence on krill density. *The Journal of Experimental Biology* **214**, 131–146.

Golovlev, E.L. (2001) Ecological strategy of bacteria: specific nature of the problem. *Microbiology* **70**, 379–383.

Grace, J.B., Anderson, T.M., Smith, M.D., Seabloom, E., Andelman, S.J., Meche, G., Weiher, E., Allain, L.K., Jutila, H., Sankaran, M., Knops, J., Ritchie, M. & Willig, M.R. (2007) Does species diversity limit productivity in natural grassland communities? *Ecology Letters* **10**, 680–689.

Grace, J.B., van Gardingen, P.R. & Luon, J. (1997) Tackling large scale problems by scaling-up. In: van Gardingen, P.R., Foody, G.M. & Curran P.J. (eds) *Scaling-up: from Cell to Landscape*, pp. 7–16. Society for Experimental Biology, Cambridge, UK.

Grant, P.R. (1972) Convergent and divergent character displacement. *Biological Journal of the Linnean Society* **4**, 39–68.

Grant, P.R. & Grant, B.R. (2008) *How and Why Species Multiply: the Radiation of Darwin's Finches*. Princeton University Press, Princeton, USA.

Green, J.L., Bohannan, B.J.M. & Whitaker, R.J. (2008) Microbial biogeography: from taxonomy to traits. *Science* **320**, 1039–1043.

Greenslade, P.J.M. (1972) Evolution in the staphylinid genus *Priochirus* (Coleoptera). *Evolution* **26**, 203–220.

Greenslade, P.J.M. (1983) Adversity selection and the habitat templet. *American Naturalist* **122**, 352–365.

Greiner, T., Frohns, F., Kang, M., Van Etten, J. L., Käsmann, A., Moroni, A., Hertel, B. & Theil, G. (2009) *Chlorella* viruses prevent multiple infections by depolarizing the host membrane. *Journal of General Virology* **90**, 2033–2039.

Griffiths, B.S., Ritz, K., Bardgett, R.D., Cook, R., Christensen, S., Ekelund, F., Sørensen, S.J., Bååth, E., Bloem, J., de Ruiter, P.C., Dolfing, J. & Nicolardot, B. (2000) Ecosystem response of pasture soil communities to fumigation-induced microbial diversity reductions: an examination of the biodiversity-ecosystem function relationship. *Oikos* **90**, 279–294.

Griffiths, H., Helliker, B., Roberts, A., Haslam, R.P., Girnus, J., Robe, W.E., Borland, A.M. & Maxwell, K. (2002) Regulation of Rubisco activity in crassulacean acid metabolism plants: better late than never. *Functional Plant Biology* **29**, 689–696.

Grime, J.P. (1965a) Comparative experiments as a key to the ecology of flowering plants. *Ecology* **45**, 513–515.

Grime, J.P. (1965b) Shade tolerance in flowering plants. *Nature* **208**, 161–163.

Grime, J.P. (1973a) Competitive exclusion in herbaceous vegetation. *Nature* **242**, 344–347.

Grime, J.P. (1973b) Control of species density in herbaceous vegetation. *Journal of Environmental Management* **1**, 151–167.

Grime, J.P. (1974) Vegetation classification by reference to strategies. *Nature* **250**, 26–31.

Grime, J.P. (1977) Evidence for the existence of three primary strategies in plants and its relevance to ecological and evolutionary theory. *American Naturalist* **111**, 1169–1194.

Grime, J.P. (1979) *Plant Strategies and Vegetation Processes*. Wiley, Chichester, UK.

Grime, J.P. (1984) The ecology of species, families and communities of the contemporary British flora. *New Phytologist* **98**, 15–33.

Grime, J.P. (1985) Towards a functional classification of vegetation. In: White, J. (ed.) The Population Structure of Vegetation. pp. 503–514. Dr. W. Junk, Dordrecht.

Grime, J.P. (1988a) The C-S-R model of primary strategies – origins, implications and tests. In: Gottlieb, L.D. & Jain, S.K. (eds). *Plant Evolutionary Biology*, pp. 371–393. Chapman & Hall, London, UK.

Grime, J.P. (1988b) Fungal strategies in ecological perspective. *Proceedings of the Royal Society of Edinburgh B* **94**, 167–169.

Grime, J.P. (1988c) Appendix 39. Memorandum submitted by Professor J.P. Grime, NERC, Unit of Comparative Plant Ecology, University of Sheffield. *Agriculture Committee. Second Report. Chernobyl: The Government's Reaction. Volume II*. Minutes of Evidence and Appendices, pp. 399–403. HMSO, London, UK.

Grime, J.P. (1993) Ecology sans frontiers. *Oikos* **68**, 385–392.

Grime, J.P. (1998) Benefits of plant diversity to ecosystems: immediate, filter and founder effects. *Journal of Ecology* **86**, 901–910.

Grime, J.P. (2001, 2nd edn) *Plant Strategies, Vegetation Processes and Ecosystem Properties*. Wiley, Chichester, UK.

Grime, J.P. (2002) Declining plant diversity: empty niches or functional shifts? *Journal of Vegetation Science* **13**, 457–460.

Grime, J.P. (2003) Plants hold the key: ecosystems in a changing world. *Biologist* **50**, 1–5.

Grime, J.P. (2006) Trait convergence and trait divergence in the plant community: mechanisms and consequences. *Journal of Vegetation Science* **17**, 255–260.

Grime, J.P. & Blythe, G.M. (1968) An investigation of the relationships between snails and vegetation at the Winnats Pass. *Journal of Ecology* **57**, 45–66.

Grime, J.P., Blythe, G.M. & Thornton, J.D. (1970) Food selection by the snail *Cepaea nemoralis*. In: Watson, A. (ed.) *Animal Populations in Relation to their Food Resources*, pp. 73–99. Blackwell, Oxford, UK.

Grime, J.P., Brown, V.K., Thompson, K., Masters, G.J., Hillier, S.H., Clarke, I.P., Askew, A.P., Corker, D., Kielty, J.P. (2000) The response of two contrasting limestone grasslands to simulated climate change. *Science* **289**, 762–765.

Grime, J.P., Fridley, J.D., Askew, A.P., Thompson, K., Hodgson, J.G. & Bennett, C.R. (2008) Long-term resistance to simulated climate change in an infertile grassland. *Proceedings of the National Academy of Science* **105**, 10028–10032.

Grime, J.P. & Jeffrey, D.W. (1965) Seedling establishment in vertical gradients of sunlight. *Journal of Ecology* **53**, 621–642.

Grime, J.P., Hodgson, J.G. & Hunt, R. (1988) *Comparative Plant Ecology: A Functional Approach to Common British Species.* Allen & Unwin, London, UK.

Grime, J.P., Hodgson, J.G. & Hunt, R. (2007, 2nd edn) *Comparative Plant Ecology: A Functional Approach to Common British Species.* Castlepoint Press, Colvend, UK.

Grime, J.P. & Hunt, R. (1975) Relative growth rate: its range and adaptive significance in a local flora. *Journal of Ecology* **63**, 393–422.

Grime, J.P. & Mackey, J.M.L. (2002) The role of plasticity in resource capture by plants. *Evolutionary Ecology* **16**, 299–307.

Grime, J.P., Mackey, J.M.L., Hillier, S.H. & Read, D.J. (1987) Floristic diversity in a model system using experimental microcosms. *Nature* **328**, 420–422.

Grime, J.P., MacPherson-Stewart, S.F. & Dearman, R.S. (1968) An investigation of leaf palatability using the snail *Cepaea nemoralis*. *Journal of Ecology* **56**, 405–420.

Grime, J.P., Thompson. K., Hunt, R., Hodgson, J.G., Cornelissen, J.H.C., Rorison, I.H., Hendry, G.A.F., Ashenden, T.W., Askew, A.P., Band, S.R., Booth, R.E., Bossard, C.C., Campbell, B.D., Cooper, J.E.L., Davison, A.W., Gupta, P.L., Hall, W., Hand, D.W., Hannah, M.A., Hillier, S.H., Hodkinson, D.J., Jalili, A., Liu, Z., Mackey, J.M.L., Matthews, N., Mowforth, M.A., Neal, A.M., Reader, R.J., Reiling, K., Ross-Fraser, W., Spencer, R.E., Sutton, F., Tasker, D.E., Thorpe, P.C. & Whitehouse, J. (1997) Integrated screening validates primary axes of specialization in plants. *Oikos* **79**, 259–281.

Grisebach, A. (1872) *Die vegetation der Erde nach ihrer klimatischen Anordnung: Ein Abriss der vergleichende Geographie der Pflanzen.* Wilhelm Engelmann, Leipzig, Germany.

Gros, R., Monrozier, L.J. & Faivre, P. (2006) Does disturbance and restoration of alpine grassland soils affect the genetic structure and diversity of bacterial and N_2-fixing populations? *Environmental Microbiology* **8**, 1889–1901.

Grubb, P.J. (1977) The maintenance of species-richness in plant communities: the importance of the regeneration niche. *Biological Reviews* **52**, 107–145.

Guo, Q. & Berry, W.L. (1998) Species richness and biomass: dissection of the hump-shaped relationships. *Ecology* **79**, 2555–2559.

Gupta, S. & Agrawal, S.C. (2006) Motility in *Oscillatoria salina* as affected by different factors. *Folia Microbiologica* **51**, 565–571.

Haberl, W. & Krystufek, B. (2003) Spatial distribution and population density of the harvest mouse *Micromys minutus* in a habitat mosaic at Lake Neusiedl, Austria. *Mammalia* **67**, 355–365.

Hadas, H., Einav, M., Fishov, I. & Zaritsky, A. (1997) Bacteriophage T4 development depends on the physiology of its host *Escherichia coli*. *Microbiology* **143**, 179–185.

Hallé, F., Oldmann, R.A.A. & Tomlinson, P.B. (1978) *Tropical Trees and Forests.* Springer Verlag, Berlin, Germany.

Halstead, B.W., Auerbach, P.S. & Campbell, D.R. (1990) *A Colour Atlas of Dangerous Marine Animals.* Wolfe Medical Publications Ltd, Ipswich, UK.

Harborne, J.B. (1997) Biochemical plant ecology. In: Dey, P.M. & Harborne, J.B. (eds) *Plant Biochemistry.* Academic Press, London, UK, 503–516.

Harley, J.L. & Harley, E.L. (1987) A checklist of mycorrhizas in the British Flora. *New Phytologist* **105(Suppl.)**, 1–102.

Harmon, L.J., Matthews, B., Des Roches, S., Chase, J.M., Shurin, J.B. & Schluter, D. (2009) Evolutionary diversification in stickleback affects ecosystem functioning. *Nature* **458**, 1167–1170.

Harper, J.L. (1977) *Population Biology of Plants.* Academic Press, London, UK.

Harper, J.L. (1982) After description. In: Newman, E.I. (ed.) *The Plant Community as a Working Mechanism*, pp. 11–25. Blackwell, Oxford, UK.

Harpole, T. (2005) Falling with the falcon. *Smithsonian Air & Space Magazine.* http://www.airspacemag.com/flight-today/falcon.html (last accessed 30 November 2011).

Harrison, K.A. & Bardgett, R.D. (2010) Influence of plant species and soil conditions on plant-soil feedback in mixed grassland communities. *Journal of Ecology* 98, 384–395.

Harvey, P.H., Read, A.F. & Nee, S. (1995) Why ecologists need to be phylogenetically challenged. *Journal of Ecology* 83, 535–536.

Heal, O.W. & Grime, J.P. (1991) Comparative analysis of ecosystems: past lessons and future directions. In: Cole, J., Lovett, G. & Findlay, S. (eds) *Comparative Analysis of Ecosystems: Patterns, Mechanisms and Theories,* pp. 7–23. Springer-Verlag, New York, USA.

Hector, A., Schmid, B., Beierkuhnlein, C., Caldeira, M.C., Diemer, M., Dimitrakopoulos, P.G., Finn, J.A., Freitas, H., Giller, P.S., Good, J., Harris, R., Högberg, P., Huss-Danell, K., Joshi, J., Jumpponen, A., Körner, C., Leadley, P.W., Loreau, M., Minns, A., Mulder, C.P.H., O'Donovan, G., Otway, S.J., Pereira, J.S., Prinz, A., Read, D.J., Scherer-Lorenzen, M., Schulze, E-D., Siamantziouras, A-SD., Spehn, E.M., Terry, A.C., Troumbis, A.Y., Woodward, F.I., Yachi, S. & Lawton, J.H. (1999) Plant diversity and productivity experiments in European grasslands. *Science* 286, 1123–1127.

Heino, M. & Kaitala, V. (1999) Evolution of resource allocation between growth and reproduction in animals with indeterminate growth. *Journal of Evolutionary Biology* 12, 423–429.

Helfman, G.S., Collette, B.B., Facey, D.E. & Bowen, B.W. (2009, 2nd edn) *The Diversity of Fishes.* Wiley-Blackwell, Oxford, UK.

Hendry, A.P. & Kinnison, M.T. (1999) The pace of modern life: measuring rates of contemporary microevolution. *Evolution* 53, 1637–1653.

Henschel, J.R. (1997) Psammophily in Namib Desert spiders. *Journal of Arid Environments* 37, 695–707.

Herrera, E.A. & Macdonald, D.W. (1989) Resource utilization and territoriality in group-living capybaras (*Hydrochoerus hydrochaeris*). *Journal of Animal Ecology* 58, 667–679.

Herrera, E.A. & Macdonald, D.W. (1993) Aggression, dominance, and mating success among capybara males (*Hydrochaeris hydrochaeris*). *Behavioral Ecology* 20, 624–632.

Hiatt, R.W. & Strasburg, D.W. (1960) Ecological relationships of the fish fauna on coral reefs of the Marshall Islands. *Ecological Monographs* 30, 65–127.

Hill, A.M. & Lodge, D.M. (1999) Replacement of resident crayfishes by an exotic crayfish: the roles of competition and predation. *Ecological Applications* 9, 678–690.

Hill, K. & Kaplan, H. (1999) Life-history traits in humans: theory and empirical studies. *Annual Review of Anthropology* 28, 397–430.

Hill, R.V., Witmer, L.M. & Norell, M.A. (2003) A new specimen of *Pinacosaurus grangeri* (Dinosauria: Ornithischia) from the Late Cretaceous of Mongolia: ontogeny and phylogeny of Ankylosaurs. *American Museum Novitates* 3395, 1–9.

Hillebrand, H., Bennet, D.M. & Cadotte, M.W. (2008) Consequences of dominance: a review of evenness effects on local and regional ecosystem processes. *Ecology* 89, 1510–1520.

Hirose, M., Katano, T. & Nakano, S.I. (2008) Growth and grazing mortality rates of *Prochlorococcus, Synechococcus* and eukaryotic picophytoplankton in a bay of the Uwa Sea, Japan. *Journal of Plankton Research* 30, 241–250.

Hodgson, J.G. (1986a) Commonness and rarity in plants with special reference to the Sheffield Flora. Part I. The identity, distribution and habitat characteristics of the common and rare species. *Biological Conservation* 36, 199–252.

Hodgson, J.G. (1986b) Commonness and rarity in plants with special reference to the Sheffield Flora. Part II. The relative importance of climate, soils and land use. *Biological Conservation* 36, 253–274.

Hodgson, J.G. (1991) The use of ecological theory and autecological datasets in studies of endangered plant and animal species and communities. *Pirineos* 138, 3–28.

Hodgson, J.G. (1993) Commonness and rarity in British butterflies. *The Journal of Applied Ecology* 30, 407–427.

Hodgson, J.G., Wilson, P.J., Hunt, R., Grime, J.P. & Thompson, K. (1999) Allocating CSR plant functional types: a soft approach to a hard problem. *Oikos* 85, 282–294.

Hoekstra, F.A., Golovina, E.A. & Buitink, J. (2001) Mechanisms of desiccation tolerance. *TRENDS in Plant Science* 6, 431–438.

Holdgate, M.W. (ed.) (1986) Quantitative aspects of the ecology of biological invasions. *Philosophical Transactions of the Royal Society of London B* 314, 653–654.

Holtum, J.A.M. & Winter, K. (1999) Degrees of crassulacean acid metabolism in tropical epiphytic and lithophytic ferns. *Australian Journal of Plant Physiology* 26, 749–757.

Horner, J.R. (2000) Dinosaur reproduction and parenting. *Annual Review of Earth and Planetary Sciences* 28, 19–45.

Horner, J.R., De Ricqlès, A. & Padian, K. (1999) Variation in dinosaur skeletochronology indicators: implications for age assessment and physiology. *Paleobiology* 25, 295–304.

Horner, J.R., De Ricqlès, A. & Padian, K. (2000) Long bone histology of the hadrosaurid dinosaur *Maiasaura peeblesorum*: growth dynamics and physiology based on an ontogenetic series of skeletal elements. *Journal of Vertebrate Paleontology* 20, 115–129.

Horner-Devine, M.C., Leibold, M.A., Smith, V.H. & Bohannan, B.J.M. (2003) Bacterial diversity patterns along a gradient of primary productivity. *Ecology Letters* 6, 613–622.

Hourdez, S., Lallier, F.H., De Cian, M-C., Green, B.N., Weber, R.E. & Toulmond, A. (2000) Gas transfer system in *Alvinella pompejana* (Annelida Polychaeta, Terebellida): functional properties if intracellular and extracellular hemoglobins. *Physiological and Biochemical Zoology* 73, 365–373.

Houseman, G.R. & Gross, K.L. (2006) Does ecological filtering across a productivity gradient explain variation in species pool-richness relationships? *Oikos* 115, 148–154.

Hubbell, S.P. (2001) *The Unified Neutral Theory of Biodiversity and Biogeography*. Monographs in Population Biology 32. Princeton University Press, Princeton, USA.

Hubbell, S.P. (2005) Neutral theory in community ecology and the hypothesis of functional equivalence. *Functional Ecology* 19, 166–172.

Hugenholtz, P. (2007) Riding giants. *Environmental Microbiology* 9, 5.

Hughes, N.F. (1992) Growth and reproduction of the Nile perch, *Lates niloticus*, an introduced predator, in the Nyanza gulf, Lake Victoria, east Africa. *Environmental Biology of Fishes* 33, 299–305.

Hummel, J., Gee, C.T., Südekum, K-H., Sander, P.M., Nogge, G. & Clauss, M. (2008) *In vitro* digestibility of fern and gymnosperm foliage: implications for sauropod feeding ecology and diet selection. *Proceedings of the Royal Society B.* 275, 1015–1021.

Hunt, R. & Colasanti, R.L. (2007) Self-assembling plants and integration across ecological scales. *Annals of Botany* 99, 1023–1034.

Huston, M.A. (1997) Hidden treatments in ecological experiments: re-evaluating the ecosystem function of biodiversity. *Oecologia* 110, 449–460.

Huston, M.A., Aarsen, L.W., Austin, M.P., Cade, B.S., Fridley, J.D., Garnier, E., Grime, J.P., Hodgson, J.G., Lauenroth, W.K., Thompson, K., Vandermeer, J.H. & Wardle, D.A. (2000) No consistent effect of plant diversity on productivity. *Science* 289, 1255–1257.

Hutchinson, T.C. (1967) Comparative studies of the ability to withstand prolonged periods of darkness. *Journal of Ecology* 55, 291–299.

Huxley, T.H. (1850) The Darwinian Hypothesis. *The Times*, 26 December.

IGFA (2001) *Database of IGFA angling records until 2001*. International Game Fish Association, Fort Lauderdale, USA.

Innes, L., Hobbs, P.J. & Bardgett, R.D. (2004) The impacts of individual plant species on rhizosphere microbial communities in soils of different fertility. *Biology & Fertility of Soils* 40, 7–13.

Itô, Y., Iwahashi, O., Yamane, So. & Yamane, Sk. (1985) Overwintering and nest reutilization in *Ropalidia fasciata* (Hymenoptera, Vespidae). *Kontyu* 53, 486–490.

Ivaldi, F. (1999) *Micromys minutus*: Eurasian harvest mouse. http://animaldiversity.ummz.umich.edu/site/accounts/information/Micromys_minutus.html (last accessed 30 November 2011).

Jackson, J.B.C. (1991) Adaptation and diversity of reef corals. *BioScience* 41, 475–482.

Jeffries, H.P. (1972) A stress syndrome in the hard clam, *Mercenaria mercenaria. Journal of Invertebrate Pathology* 20, 242–251.

Johnson, G.D. & Gill, A.C. (1994) Perciformi: ordine vasto e diversificato. In: Paxton, J.R. & Eschmeyer, W.N. (eds) *Pesci: caratteristiche, ambiente, comportamento*, pp. 181–196. Editoriale Giorgio Mondadori, Milan, Italy. (Italian language edition, original title: *Encyclopedia of Animals: Fishes*).

Jonasson, S. & Chapin, F.S. (1991) Seasonal uptake and allocation of the phosphorus in *Eriophorum vaginatum* L. measured by labelling with ^{32}P. *New Phytologist* 118, 349–357.

Jones, E.W. (1959) Biological Flora of the British Isles: *Quercus robur. Journal of Ecology* 47, 169–216.

Jones, K.E., Bielby, J., Cardillo, M., Fritz, S.A., O'Dell, J., Orme, D.L., Safi, K., Sechrest, W., Boakes, E.H., Carbone, C., Connolly, C., Cutts, M.J., Foster, J.K., Grenyer, R., Habib, M., Plaster, C.A., Price, S.A., Rigby, E.A., Rist, J., Teacher, A., Bininda-Emonds, O.R.P., Gittleman, J.L., Mace, G.M. & Pruvis, A. (2009) PanTHERIA: a species-level database of life-history, ecology and geography of extant and recently extinct mammals. *Ecology* 90, 2648.

Kajiura, S.M. & Holland, K.N. (2002) Electroreception in juvenile scalloped hammerhead and sandbar sharks. *Journal of Experimental Biology* 205, 3609–3621.

Karieva, P. (1994) Diversity begets productivity. *Nature* 368, 686–687.

Karieva, P. (1996) Diversity and sustainability on the prairie. *Nature* 379, 673–674.

Karraker, N.E., Richards, C.L. & Ross, H.L. (2006) Reproductive ecology of *Atelopus zeteki* and comparisons to other members of the genus. *Herpetological Review* 37, 284–288.

Katunzi, E.F.B., Van Densen, W.L.T., Wanink, J.H. & Witte, F. (2006) Spatial and seasonal patterns in the feeding habits of juvenile *Lates niloticus* (L.), in the Mwanza Gulf of Lake Victoria. *Hydrobiologia* 568, 121–133.

Kawasaki, T. (1980) Fundamental relations among the selections of life-history in the marine teleosts. *Bulletin of the Japanese Society of Scientific Fisheries* 46, 289–293.

Kearney, T.H. & Shantz, H.L. (1912) The water economy of dryland crops. In: *Yearbook of the United States Department of Agriculture 1911*, 351–62. USDA, Washington, DC, USA.

Keddy, P. (1990) Competitive hierarchies and centrifugal organisation in plant communities. In: Grace, J. & Tilman, D. (eds) *Perspectives on Plant Competition*, pp. 265–289. Academic Press, New York, USA.

Keddy, P. (1992) Assembly and response rules: two goals for predictive community ecology. *Journal of Vegetation Science* 3, 157–164.

Keddy, P. (2005) Putting the plants back into plant ecology: six pragmatic models for understanding and conserving plant diversity. *Annals of Botany* 96, 177–189.

Kellogg, D.E. & Hays, J.D. (1975) Microevolutionary patterns in Late Cenozoic Radiolaria. *Paleobiology* 1, 150–160.

Kerford, M. (2000) *The Ecology of the Bar Bellied Sea Snake* (Hydrophis elegans) *in Shark Bay, Western Australia*. Unpublished MSc Thesis, Simon Fraser University, Canada.

Kettler, G.C., Martiny, A.C., Huang, K., Zucker, J., Coleman, M.L., Rodrigue, S., Chen, F., Lapidus, A., Ferriera, S., Johnson, J., Steglich, C., Church, G.M., Richardson, P. & Chisholm, S.W. (2007) Patterns and implications of gene gain and loss in the evolution of *Prochlorococcus*. *PLoS Genetics* 3, e231.

Kettlewell, H.D.B. (1961) The phenomenon of industrial melanism in the Lepidoptera. *Annual Review of Entomology* 6, 245–262.

Kilinç, M., Karavin, N. & Kutbay, H.G. (2010) Classification of some plant species according to Grime's strategies in a *Quercus cerris* L. var. *cerris* woodland in Samsun, northern Turkey. *Turkish Journal of Botany* 34, 521–529.

King, T.J. (1977) The plant ecology of ant-hills in calcareous grasslands. 1: Patterns of species in relation to ant-hills in Southern England. *Journal of Ecology* 65, 235–256.

Kinnison, M.T. & Hairston, N.G. Jr (2007) Eco-evolutionary conservation biology: contemporary evolution and the dynamics of persistence. *Functional Ecology* 21, 444–454.

Kinnison, M.T. & Hendry A.P. (2001) The pace of modern life II: from rates of contemporary microevolution to pattern and process. *Genetica* 112–113, 145–164.

Kinnison, M.T., Unwin, M.J. & Quinn, T.P. (2008) Eco-evolutionary vs. habitat contributions to invasion in salmon: experimental evaluation in the wild. *Molecular Ecology* 17, 405–414.

Kinzig, A.P., Pacala, S.W. & Tilman, D. (2001) *The Functional Consequences of Biodiversity.* Princeton University Press, Princeton, USA.

Klütz, D. (2005) Molecular and evolutionary basis of the cellular stress response. *Annual Reviews in Physiology* 67, 225–257.

Kobayashi, Y, & Lü, J-C. (2003) A new ornithomimid dinosaur with gregarious habits from the Late Cretaceous of China. *Acta Palaeontologica Polonica* 48, 235–259.

König-Rinke M. (2008) *Bildung funktioneller Typgruppen des Phytoplanktons: Integration von Modell-, Frieland- und Laborarbeiten.* PhD Thesis, Universität Dresden, Germany.

Konopka, A. (1982) Buoyancy regulation and vertical migration by *Oscillatoria rubescens* in Crooked Lake, Indiana. *European Journal of Phycology* 17, 427–442.

Körner, K. & Jeltsch, F. (2008) Detecting general plant functional type responses in fragmented landscapes using spatially-explicit simulations. *Ecological Modelling* 210, 287–300.

Korpimäki, E. (1988) Effects of territory quality on occupancy, breeding performance and breeding dispersal in Tengmalm's owl. *Journal of Animal Ecology* 57, 97–108.

Krafft, B., Horel, A. & Julita, J.M. (1986) Influence of food supply on the duration of the gregarious phase of a maternal-social spider, *Coelotes terrestris* (Araneae, Agelenidae). *Journal of Arachnology* 14, 219–226.

Krebs, C.J. (2001) *Ecology.* Cummings, San Francisco, USA.

Kubodera, T. & Mori, K. (2005) First-ever observations of a live giant squid in the wild. *Proceedings of the Royal Society B* 272, 2583–2586.

Küchler, A.W. (1967) *Vegetation Mapping.* Ronald Press, New York, USA.

Kuypers, M.M.M. & Jørgensen, B.B. (2007) The future of single-celled environmental biology. *Environmental Microbiology* 9, 6.

Lack, D. (1947) *Darwin's Finches.* Cambridge University Press, Cambridge, UK.

Landeryou, J. (1999) *Marmota marmota*: alpine marmot. http://animaldiversity. ummz.umich.edu/site/accounts/information/Marmota_marmota.html (last accessed 30 November 2011).

Landman, N.H., Cochran, J.K., Cerrato, R., Mak, J., Roper, C.F.E. & Lu, C.C. (2004) Habitat and age of the giant squid (*Architeuthis sanctipauli*) inferred from isotopic analyses. *Marine Biology* 144, 685–691.

Lane, N. (2009) The cradle of life. *New Scientist* 204, 38–42.

Langworthy, T.A. (1977) Long-chain diglycerol tetraethers from *Thermoplasma acidophilum. Biochimica et Biophysica Acta* 487, 37–50.

Langworthy, T.A., Mayberry, W.R. & Smith, P.F. (1974) Long chain glycerol diether and polyol dialkyl glycerol triether lipids of *Sulfolobus acidocaldarius. Journal of Bacteriology* 119, 106–116.

Larcher, W. (2003, 4th edn) *Physiological Plant Ecology.* Springer-Verlag, Berlin, Germany.

Larson, G. (1992) *The Prehistory of the Far Side*. Andrews McMeel Publishing, Kansas City, USA.

Laughlin, D.C. & Moore, M.M. (2009) Climate-induced temporal variation in the productivity-diversity relationship. *Oikos* 118, 897–902.

Lavorel, S. & Garnier, E. (2002) Predicting changes in community composition and ecosystem functioning from plant traits: revisiting the Holy Grail. *Functional Ecology* 16, 545–556.

Lawrence, J.M. (1990) The effect of stress and disturbance on echinoderms. *Zoological Science* 7, 17–28.

Lawton, J.H. (2000) *Community Ecology in a Changing World*. Oldendorf/Luhe, Germany.

Lawton, J.L. (1995) Ecological experiments with model systems. *Science* 269, 328–331.

Lee, C.K., Cary, S.C., Murray, A.E. & Daniel, R.M. (2008) The eurythermalism of *Alvinella pompejana* and its episymbionts – an enzymic approach. *Applied and Environmental Microbiology* 74, 774–782.

Lee, P-F., Ding, T-S., Hsu, F-H. & Geng, S. (2004) Breeding bird species richness in Taiwan: distribution on gradients of elevation, primary productivity and urbanization. *Journal of Biogeography* 31, 307–314.

Lee, R.W. (2003) Thermal tolerances of deep-sea hydrothermal vent animals from the northeast Pacific. *Biological Bulletin* 205, 98–101.

Lehman, T.M. & Woodward, H.N. (2008) Modeling growth rates for sauropod dinosaurs. *Paleobiology* 34, 264–281.

Leibold, M.A. (1999) Biodiversity and nutrient enrichment in pond plankton communities. *Evolutionary Ecology Research* 1, 73–95.

Leigh, E.G. Jr (1999) *Tropical Forest Ecology: A View from Barro Colorado Island*. Oxford University Press, Oxford, UK.

Lenti Boero, D. (1994) Survivorship among young alpine marmots and their permanence in their natal territory in a high altitude colony. *Journal of Mountain Ecology* 2, 9–16.

Lenti Boero, D. (2001) Occupation of hibernacula, seasonal activity, and body size in a high altitude colony of Alpine marmots (*Marmota marmota*). *Ethology Ecology & Evolution* 13, 209–223.

Leps, J., Osbornova-Kosinova, J. & Rejmanek, K. (1982) Community stability, complexity and species life-history strategies. *Vegetatio* 511, 53–63.

Le Roux, X., Poly, F., Clays, A., Commeaux, C., Guillaumand, N., Lerondelle, C., Periot, C., Falcao-Salles, J., Schmid, B. & Weigelt, A. (2007) Effects of plant species richness versus plant functional diversity on soil nitrifiers and denitrifiers: results from a large plant-assemblage experiment. Poster presented at Rhizosphere 2 International Conference, Montpellier, France, 26–31 August 2007.

Leschine, S.B., Holwell, K. & Canale-Parola, E. (1988) Nitrogen fixation by anaerobic cellulolytic bacteria. *Science* 242, 1157–1159.

Lewontin, R.C. (1974) *The Genetic Basis of Evolutionary Change*. Columbia University Press, New York, USA.

Ley, R.E., Knight, R. & Gordon, J.I. (2007) The human microbiome: eliminating the biomedical/environmental dichotomy in microbial ecology. *Environmental Microbiology* 9, 3–4.

Li, J.S., Song, Y.L. & Zeng, Z.G. (2003) Elevational gradients of small mammal diversity on the northern slopes of Mt. Gilian, China. *Global Ecology & Biogeography* 12, 449–460.

Lima, J. de F, Abrunhosa, F. & Coelho, P.A. (2006) The larval development of *Pinnixa gracilipes* Coelho (Decapoda, Pinnotheridae) reared in the laboratory. *Revista Brasileira de Zoologia* 23, 480–489.

Lin, A.Y.M, Meyers, N.A. & Vecchio, K.S. (2006) Mechanical properties and structure of *Strombus gigas*, *Tridacna gigas*, and *Haliotis rufescens* sea shells: a comparative study. *Materials Science and Engineering* C26, 1380–1389.

a Linné, C. (1767, 13th edn) *Systema Naturæ*. http://books.google.com/books?id=Ix0A AAAAQAAJ&printsec=titlepage#PPA2,M1 (last accessed 30 November 2011).

a Linné, C. (1753) *Species Plantarum*.

Lirman, D. (2000) Fragmentation in the branching coral *Acropora palmata* (Lamarck): growth, survivorship, and reproduction of colonies and fragments. *Journal of Experimental Marine Biology and Ecology* 251, 41–57.

Locht, A., Yáñez, M. & Vázquez, I. (1999) Distribution and natural history of Mexican species of *Brachypelma* and *Brachypelmides* (Theraphosidae, Theraphosinae) with morphological evidence for their synonymy. *The Journal of Arachnology* 27, 196–200.

Lovett, D.L., Verzi, M.P., Burgents, C.A., Tanner, C.A., Glomski, K., Lee, J.J. & Towle, D.W. (2006) Expression profiles of Na^+,K^+-ATPase during acute and chronic hypoosmotic stress in the Blue Crab *Callinectes sapidus*. *Biological Bulletin* 211, 58–65.

Lubin, Y.D. (1991) Patterns of variation in female-biased colony sex ratios in a social spider. *Biological Journal of the Linnean Society* 43, 297–311.

Lüttge, U. (2004) Ecophysiology of Crassulacean Acid Metabolism (CAM). *Annals of Botany* 93, 629–652.

Lutz, H.J. (1928) Trends and silvicultural significance of upland forest successions in Southern New England. *Yale University School of Forestry Bulletin* 22, 1–68.

Luzzaro, A. (2005) *Plant strategies as a Tool for Describing and Interpreting Vegetation Dynamics on Alpine Glacier Forelands*. Unpublished PhD Thesis, University of Milan, Italy.

Luzzaro, A., Cerabolini, B., Caccianiga, M., Pierce, S. & Andreis, C. (2005) Strategie delle piante e tipi funzionali nello studio dinamica di vegetazione in ambienti periglaciale. *Informatore Botanico Italiano* 37, 224–225.

Ma, Y., Jiao, N.Z. & Zeng, Y.H. (2004) Natural community structure of cyanobacteria in the South China Sea as revealed by rpoC1 gene sequence analysis. *Letters in Applied Microbiology* 39, 353–358.

MacArthur, R.H. (1958) Population ecology of some warblers of northeastern coniferous forests. *Ecology* 39, 599–619.

MacArthur, R.H. (1964) Environmental patterns affecting bird species diversity. *American Naturalist* 98, 387–397.

MacArthur, R.H. (1968) The theory of the niche. In: Lewontin, R.C. (ed.) *Population Biology and Evolution*, pp. 159–176. Syracuse University Press, Syracuse, USA.

MacArthur, R.H. & MacArthur, J.W. (1961) On bird species diversity. *Ecology* 42, 594–598.

MacArthur, R.H. & Wilson, E.D. (1967) *The Theory of Island Biogeography*. Princeton University Press, Princeton, USA.

Machacek. H. (ed.) (2008) *World Records Freshwater Fishing*. http://www.fishingworldrecords.com (last accessed 5 October 2010).

MacGillivray, C.W., Grime, J.P. & The Integrated Screening Programme (ISP) Team. (1995) Testing predictions of resistance and resilience of vegetation subjected to extreme events. *Functional Ecology* 9, 640–649.

MacLeod, J. (1894) Over de bevruchting der bloemen in het Kempisch gedeelte van Vlaanderen. Deel II. *Botanische Jaarboek* 6, 119–511.

Maddison, D.R., Schulz, K-S. & Maddison, W.P. (2007) The tree of life web project. *Zootaxa* 1668, 19–40.

Madigan, M.T., Martinko, J.M. & Parker, J. (2003, 10th edn) *Brock Biology of Microorganisms*. Pearson Education, Inc., Upper Saddle River, New Jersey, USA.

Magurran, A.E. (2005) *Evolutionary Ecology: the Trinidadian guppy*. Oxford Series in Ecology and Evolution. Oxford University Press, Oxford, UK.

Mahmoud, A. (1973) *A laboratory approach to ecological studies of the grasses,* Arrhenatherum elatius, Agrostis capillaris and Festuca ovina. Unpublished PhD Thesis, University of Sheffield, UK.

Marc, P., Canard, A. & Ysnel, F. (1999) Spiders (Araneae) useful for pest limitation and bioindication. *Agriculture, Ecosystems and Environment* 74, 229–273.

Marcon, J.L. & Wilhelm, D. (1999) Antioxidant processes of the wild tambaqui, *Colossoma macropomum* (Osteichthyes, Serrasalmidae) from the Amazon. *Comparative Biochemistry and Physiology C – Toxicology & Pharmacology* 123, 257–263.

Marcov, M.V. (1985) Research on permanent quadrats in the USSR. In: White, J. (ed.) *The Population Structure of Vegetation*, pp.111–119. Dr. W. Junk Publishers, Dordrecht, The Netherlands.

Marks, C.O. & Muller-Landau, H.C. (2007) Comment on 'From plant traits to plant communities: a statistical mechanistic approach to biodiversity'. *Science* 316, 1425c.

Marschner, H. (1995, 2nd edn) *Mineral Nutrition of Higher Plants*. Academic Press, San Diego, USA.

Martin, A.J. (2009) Dinosaur burrows in the Otway Group (Albian) of Victoria, Australia, and their relation to Cretaceous polar environments. *Cretaceous Research* 30, 1223–1237.

Martin, E.M. (1996) Fitness costs of resource overlap among coexisting bird species. *Nature* 380, 338–340.

Martin, R.A. (2009) *Polar Seas: Life Under the Ice. Greenland Shark.* http://www.elasmo-research.org/education/ecology/polar-greenland.htm (last accessed 5 October 2010).

Martin, R.A., Rossmo, D.K. & Hammerschlag, N. (2009) Hunting patterns and geographic profiling of white shark predation. *Journal of Zoology* 279, 111–118.

Massant, W., Godefroid, S. & Koedam, N. (2009) Clustering of plant life strategies on meso-scale. *Plant Ecology* 205, 47–56.

Mateo, P., Douterelo, I., Berrendero, E. & Perona, E. (2006) Physiological differences between two species of cyanobacteria in relation to phosphorus limitation. *Journal of Phycology* 42, 61–66.

Matsumoto, T. (1998) Cooperative prey capture in the communal web spider, *Philoponella raffray* (Araneae, Uloboridae). *Journal of Arachnology* 26, 392–396.

Matsuo, A.Y., Wood, C.M. & Val, A.L. (2005) Effects of copper and cadmium on ion transport and gill metal binding in the Amazonian teleost tambaqui (*Colossoma macropomum*) in extremely soft water. *Aquatic Toxicology* 74, 351–364.

Matsuo, A.Y.O., Woodin, B.R., Reddy, C.M., Val, A.L. & Stegeman, J.J. (2006) Humic substances and crude oil induce cytochrome P450 1A expression in the Amazonian fish species *Colossoma macropomum* (Tambaqui). *Environmental Science & Technology* 40, 2851–2858.

Maxwell, C., Griffiths, H. & Young, A.J. (1994) Photosynthetic acclimation to light regime and water stress by the C3-CAM epiphyte *Guzmania monostachia*: gas-exchange characteristics, photochemical efficiency and the xanthophyll cycle. *Functional Ecology* 8, 746–754.

May, R.M. (1985) Evolutionary ecology and John Maynard Smith. In: Greenwood, P.J., Harvey, P.H. & Slatkin, M. (eds) *Essays in honour of John Maynard Smith*, pp. 107–116. Cambridge University Press, Cambridge, UK.

McComb, D.M., Tricas, T.C. & Kajiura, S.M. (2009) Enhanced visual fields in hammerhead sharks. *The Journal of Experimental Biology* 212, 4010–4018.

Miaud, C., Guyétant, R. & Elmberg, J. (1999) Variations in life-history traits in the common frog *Rana temporaria* (Amphibia: Anura): a literature review and new data from the French Alps. *Journal of Zoology, London* 249, 61–73.

Michalet, R., Brooker, R.W., Cavieres, L.A., Kikvidze, Z., Lortie, C.J., Pugnaire, F.I., Valiente-Banuet, A. & Callaway, R.M. (2006) Do biotic interactions shape both sides of the humped-back model of species richness in plant communities? *Ecology Letters* 9, 767–773.

Milborne, A. (2008) *Tadpoles and Frogs*. Usborne Publishing Ltd, London, UK.

Milne, A. (1961) Definition of competition among animals. In: Milnthorpe, F.L. (ed.) *Mechanisms in Biological Competition*, pp. 40–61. Cambridge University Press, Cambridge, UK.

Mitchell, M.G.E., Cahill, J.F. Jr & Hik, D.S. (2009) Plant interactions are unimportant in a subarctic-alpine plant community. *Ecology* 90, 2360–2367.

Mittelbach, G.G., Steiner, C.F., Scheiner, S.M., Gross, K.L., Reynolds, H.L., Waide, R.B., Willig, M.R., Dodson, S.I. & Gough, L. (2001) What is the observed relationship between species richness and productivity? *Ecology* 82, 2381–2396.

Mkumbo, O.C. & Ligtvoet, W. (1992) Changes in the diet of Nile perch, *Lates niloticus* (L.), in the Mwanza gulf, Lake Victoria. *Hydrobiologia* 232, 79–83.

Mokany, K., Ash, J. & Roxburgh, S. (2008) Functional identity is more important than diversity in influencing ecosystem processes in a temperate native grassland. *Journal of Ecology* 96, 884–893.

Molino, J.F. & Sabatier, D. (2001) Tree diversity in tropical rain forests: a validation of the intermediate disturbance hypothesis. *Science* 294, 1702–1704.

Mooney, H.A. (1974) Plant forms in relation to environment. In: Strain, B.R. & Billings,W.D. (eds) *Handbook of vegetation science. Part IV. Vegetation and Environment*, pp. 111–22. Junk, The Hague, The Netherlands.

Mooney, H.A. & Dunn, E.L. (1970) Convergent evolution of Mediterranean-climate evergreen sclerophyll shrubs. *Evolution* 24, 292–303.

Moore, D.R.J. & Keddy, P.A. (1989) The relationship between species richness and standing crop in wetlands: the importance of scale. *Vegetatio* 79, 99–106.

Moreno, A., Pereira, J. & Cunha, M. (2005) Environmental influences on age and size at maturity of *Loligo vulgaris*. *Aquatic Living Resources* 18, 377–384.

Morin, P.A., Archer, F.I., Foote, A.D., Vilstrup, J., Allen, E.E., Wade, P., Durban, J., Parsons, K., Pitman, R., Li, L., Bouffard, P., Abel Nielsen, S.C., Rasmussen, M., Willerslev, E., Gilbert, M.T.P. & Harkins, T. (2010) Complete mitochondrial genome phylogenetic analysis of killer whales (*Orcinus orca*) indicates multiple species. *Genome Research* 20, 908–916.

Morrison, C. & Hero, J-M. (2003) Geographic variation in life-history characteristics of amphibians: a review. *Journal of Animal Ecology* 72, 270–279.

Murdoch, T.J.T. (2007) *A functional group approach for predicting the composition of hard coral assemblages in Florida and Bermuda.* Unpublished PhD Thesis, University of South Alabama, USA.

Mustard, M., Standing, D., Aitkenhead, M., Robinson, D. & McDonald, A. (2003) The emergence of primary strategies in evolving plant populations. *Evolutionary Ecology Research* 5, 1067–1081.

Myers, N. (1996) The biodiversity crisis and the future of evolution. *The Environmentalist* 16, 1614–1174.

Naeem, S., Thompson, L.J., Lawlor, S.P., Lawton, J.H. & Woodfin, R.M. (1994) Declining biodiversity can alter the performance of ecosystems. *Nature* 368, 734–737.

Nakamura, Y., Kaneko, T., Sato, S., Ikeuchi, M., Katoh, H., Sasamoto, S., Watanabe, A., Iriguchi, M., Kawashima, K., Kimura, T., Kishida, Y., Kiyokawa, C., Kohara, M., Matsumoto, M., Matsuno, Ai., Nakazaki, N., Shimpo, S., Sugimoto, M., Takeuchi, C., Yamada, M. & Tabata, S. (2002) Complete genome structure of the thermophilic cyanobacterium *Thermosynechococcus elongatus* BP-1. *DNA Research* 9, 123–130.

Nantwig, W. (1985) Social spiders catch larger prey: a study of *Anelosimus eximius* (Araneae: Theridiidae). *Behavioural Ecology and Sociobiology* 17, 79–85.

Nasci, C., Da Ros, L., Campesan, G., van Vleet, E.S., Salizzato, M., Sperni, L. & Pavoni, B. (1999) Clam transplantation and stress-related biomarkers as useful tools for assessing water quality in coastal environments. *Marine Pollution Bulletin* 39, 255–260.

Navas, M-L., Roumet, C., Bellmann, A., Laurent, G. & Garnier, E. (2010) Suites of plant traits in species from different stages of a Mediterranean secondary succession. *Plant Biology* 12, 183–196.

Navas, M-L. & Violle, C. (2009) Plant traits related to competition: how do they shape the functional diversity of communities? *Community Ecology* 10, 131–137.

Newbold, A.J. & Goldsmith, F.B. (1981) The regeneration of oak and beech; a literature review. *Discussion papers on conservation No.33.* University College London, London, UK.

Newton, I. & Marquiss, M. (1976) Occupancy and success of nesting territories in the European sparrowhawk. *Journal of Raptor Research* 10, 65–71.

Nix-Stohr, S., Moshe, R. & Dighton, J. (2008) Effects of propagule density and survival strategies on establishment and growth: further investigations in the phylloplane fungal model system. *Microbial Ecology* 55, 38–44.

Noble, I.R. & Slatyer, R.O. (1980) The use of vital attributes to predict successional changes in plant communities subject to recurrent disturbances. *Vegetatio* 43, 5–21

Nogues-Bravo, D., Araujo, M.B., Romdal, T. & Rahbek, C. (2008) Scale effects and human impact on the elevational species richness gradients. *Nature* **453**, 216–218.

Norell, M.A., Clark, J.M., Chiappe, L.M. & Dashzeveg, D. (1995) A nesting dinosaur. *Nature* 378, 774–776.

Nuñez, M.A, Horton, T.R. & Simberloff, D. (2009) Lack of belowground mutualisms hinders Pinaceae invasions. *Ecology* 90, 2352–2359.

O'Connor, P.M. & Claessens, L.P.A.M. (2005) Basic avian pulmonary design and flow-through ventilation in non-avian theropod dinosaurs. *Nature* 436, 253–256.

O'Dor, R.K. & Webber, D.M. (1986) The constraints on cephalopods: why squid aren't fish. *Canadian Journal of Zoology* 64, 1591–1605.

Odum, E.P. (1953) *Fundamentals of Ecology*. Saunders, Philadelphia, USA.

Ogutuohwayo, R. (1993) The effects of predation by Nile perch, *Lates niloticus* L., on the fish of Lake Nabugabo, with suggestions for conservation of endangered endemic cichlids. *Conservation Biology* 7, 701–711.

Oksanen, L. (1981) Exploitation ecosystems in gradients of primary productivity. *American Naturalist* 118, 240–261.

Olde Venterink, H., Wassen, M.J., Belgers, J.D.M. & Verhoeven, J.T.A. (2001) Control of environmental variables on species density in fens and meadows: importance of direct effects and effects through community biomass. *Journal of Ecology* 89, 1033–1040.

Oliveira, A.C.B., Martinelli, L.A., Moreira, M.Z., Soares, M.G.M. & Cyrino, J.E.P. (2006) Seasonality of energy sources of *Colossoma macropomum* in a floodplain lake in the Amazon – Lake Camaleao, Amazonas, Brazil. *Fisheries Management and Ecology* 13, 135–142.

Olsen, C. (1958) Iron uptake in different plant species as a function of the pH value of the nutrient solution. *Physiologia Plantarum* 11, 889–905.

Olsen, G.J. (1994) Microbial ecology – Archaea, Archaea, everywhere. *Nature* 371, 657–658.

Orphan, V.J., Turk, K.A., Green, A.M. & House, C.H. (2009) Patterns of ^{15}N assimilation and growth of methanotrophic ANME-2 archaea and sulfate-reducing bacteria within structured syntrophic consortia revealed by FISH-SIMS. *Environmental Microbiology* 11, 1777–1791.

Orr, M.R. & Smith, T.B. (1998) Ecology and speciation. *Trends in Ecology and Evolution* 13, 502–506.

Orwin, K.H., Buckland, S.M., Johnson, D., Turner, B.L., Smart, S., Oakley, S. & Bardgett, R.D. (2010) Linkages of plant traits to soil properties and the functioning of temperate grassland. *Journal of Ecology* 98, 1074–1083.

Packard, G.C., Boardman, T.J. & Birchard, G.F. (2009) Allometric equations for predicting body mass of dinosaurs. *Journal of Zoology* 279, 102–110.

Padian, K., Horner, J.R. & De Ricqlès. A. (2004) Growth in small dinosaurs and pterosaurs: the evolution of Archosaurian growth strategies. *Journal of Vertebrate Paleontology* 24, 555–571.

Palkovacs, E.P. & Hendry, A.P. (2010) Eco-evolutionary dynamics: intertwining ecological and evolutionary processes in contemporary time. *F1000 Biology Reports* 2, 1.

Pandolfi, J.M. & Greenstein, B.J. (1997) Preservation of community structure in death assemblages of deep-water Caribbean reef corals. *Limnology & Oceanography* 42, 1505–1516.

Parsons, R.F. (1968) The significance of growth rate comparisons for ecology. *American Naturalist* 102, 595–595.

Partensky, F., Hess, W.R. & Vaulot, D. (1999) *Prochlorococcus*, a marine photosynthetic prokaryote of global significance. *Microbiol Mol Biol Rev* **63**, 106–127.

Patra, A.K., Abbadie, L., Calys-Josserand, A., Degrange, V., Grayston, S.J., Guillaumaud, N., Loiseau, P., Louault, F., Mahmood, S., Nazaret, S., Philippot, L., Poly, F., Prosser, J.I. & Le Roux, X. (2006) Effect of management regime and plant species on the enzyme activity and genetic structure of N-fixing, denitrifying and nitrifying bacterial communities in grassland soils. *Environmental Microbiology* **8**, 1005–1016.

Paul, J.H., Alfreider, A. & Wawrik, B. (2000) Micro- and macrodiversity in *rbcL* sequences in ambient phytoplankton populations from the southeastern Gulf of Mexico. *Marine Ecology Progress Series* **198**, 9–18.

Pearsall, W.H. (1950) *Mountains and Moorlands*. Bloomsbury Books, London, UK.

Pecl, G.T. & Moltschaniwskyj, N.A. (2006) Life-history of a short-lived squid (*Sepioteuthis australis*): resource allocation as a function of size, growth, maturation and hatching season. *ICES Journal of Marine Science* **63**, 995–1004.

Pellegrino, G., Caimi, D., Noce, M.E. & Musacchio, A. (2005) Effects of local flower density and flower colour polymorphism on pollination and reproduction in the rewardless orchid *Dactylorhiza sambucina* (L.) Soò. *Plant Systematics and Evolution* **251**, 119–129.

Pelletier, F., Garant, D. & Hendry, A.P. (2009) Eco-evolutionary dynamics. *Philosophical Transactions of the Royal Society B* **364**, 1483–1489.

Pianka, E.R. (1970) On r- and K-selection. *American Naturalist* **104**, 592–597.

Pianka, E.R. (1992) Evolution and the ecosystem. In: Baba, S., Akerele, O. & Kawaguchi, Y. (eds) *Natural Resources and Human Health – Plants of medicinal and nutritional value*, pp. 9–19. Proceedings of the First World Health Organization Symposium on Plants and Health for all: Scientific Advancement, Kobe, Japan, 26–28 August 1991. Elsevier Science Publishers, Amsterdam, The Netherlands.

Pianka, E.R. (1997) Australia's thorny devil. *Reptiles* **5**, 14–23.

Pickett, S.T.A. & White, P.S. (eds) (1987) *The Ecology of Natural Disturbance and Patch Dynamics*. Academic Press, New York, USA.

Pierce, S. (2007) The jeweled armor of *Tillandsia* – multifaceted or elongated trichomes photoprotect leaf blades. *Aliso* **23**, 44–52.

Pierce, S. (2009) http://www.elasmo-research.org/education/ecology/rocky-spiny_dog.htm (last accessed 30 November 2011).

Pierce, S. (2011) *The Conservation of Terrestrial Orchids*. Kindle eBook edition (English language edition translated from the Italian print edition, *La Conservazione delle Orchidee Terrestri: dalle Alpi alla Pianura Padana Lombarda*). The Native Flora Centre of the Lombardy Region (CFA), Galbiate, Italy.

Pierce, S., Ceriani, R.M., De Andreis, R. & Cerabolini, B. (2007a) The leaf economics spectrum of Poaceae reflects variation in survival strategies. *Plant Biosystems* **141**, 337–343.

Pierce, S., Ceriani, R.M., Villa, M. & Cerabolini, B. (2006) Quantifying relative extinction risks and targeting intervention for the orchid flora of a natural park in the European pre-alps. *Conservation Biology* **20**, 1804–1810.

Pierce, S., Luzzaro, A., Caccianiga, M., Ceriani, R.M. & Cerabolini, B. (2007b) Disturbance is the principal α-scale filter determining niche differentiation, coexistence and biodiversity in an alpine community. *Journal of Ecology* **95**, 698–706.

Pierce, S., Maxwell, K., Griffiths, H. & Winter, K. (2001) Hydrophobic trichome layers and epicuticular wax powders in Bromeliaceae. *American Journal of Botany* **88**, 1371–1389.

Pierce, S., Vianelli, A. & Cerabolini, B. (2005) From ancient genes to modern communities: the cellular stress response and the evolution of plant strategies. *Functional Ecology* **19**, 763–776.

Pierce, S., Winter, K. & Griffiths, H. (2002a) The role of CAM in high rainfall cloud forests: an *in situ* comparison of photosynthetic pathways in Bromeliaceae. *Plant, Cell & Environment* **25**, 1181–1189.

Pierce, S., Winter, K. & Griffiths, H. (2002b) Carbon isotope ratio and the extent of daily CAM use by Bromeliaceae. *New Phytologist* **156**, 75–83.

Piper, R. (2007) *Extraordinary Animals: An Encyclopedia of Curious and Unusual Animals*. Greenwood Press, San Francisco, USA.

Pobrabsky, J.E. & Hand, S.C. (1999) The bioenergetics of embryonic diapause in an annual killifish, *Austrofundulus limnaeus*. *Journal of Experimental Biology* **202**, 2567–2580.

Pobrabsky, J.E. & Hand, S.C. (2000) Depression of protein synthesis during diapause in embryos of the annual killifish *Austrofundulus limnaeus*. *Physiological and Biochemical Zoology* **73**, 799–808.

Pobrabsky, J.E. & Somero, G.N. (2004) Changes in gene expression associated with acclimation to constant temperatures and fluctuating daily temperatures in an annual killifish *Austrofundulus limnaeus*. *Journal of Experimental Biology* **207**, 2237–2254.

Post, D.M. & Palkovacs, E.P. (2009) Eco-evolutionary feedbacks in community and ecosystem ecology: interactions between the ecological theatre and the evolutionary play. *Philosophical Transactions of the Royal Society B* **364**, 1629–1640.

Prangishvili, D. & Garrett, R.A. (2004) Exceptionally diverse morphotypes and genomes of crenarchaeal hyperthermophilic viruses. *Biochemical Society Transactions* **32**, 204–208.

Prangishvili, D., Forterre, P. & Garrett, R.A. (2006) Viruses of the Archaea: a unifying view. *Nature Reviews Microbiology* **4**, 837–848.

Prasad, V., Strömberg, C.A.E., Alimohammadian, H. & Sahni, A. (2005) Dinosaur coprolites and the early evolution of grasses and grazers. *Nature* **310**, 1177–1180.

Preston, C.D., Telfer, M.G., Arnold, H.R., Carey, P.D., Cooper, J.M., Dines, T.D., Hill, M.O., Pearman, D.A., Roy, D.B. & Smart, S.M. (2002) *The Changing Flora of the UK*. Defra, London, UK.

Promislow, D., Clobert, J. & Barbault, R. (1991) Life-history allometry in mammals and squamate reptiles: taxon-level effects. *Oikos* **65**, 285–294.

Prosser, J.I., Bohannan, B.J.M., Curtis, T.P., Ellis, R.J., Firestone, M.K., Freckleton, R.P., Green, J.L., Green, L.E., Killham, K., Lennon, J.J., Osborn, A.M., Solan, M., van der Gast, C.J. & Young, J.P.W. (2007) The role of ecological theory in microbial ecology. *Nature* **5**, 384–392.

Pugh, G.J.F. (1980) Strategies in fungal ecology. *Transactions of the British Mycological Society* **75**, 1–14.

P'yankov, V.I., Ivanov, L.A. & Lambers, H. (2001) Plant construction cost in the boreal species differing in their ecological strategies. *Russian Journal of Plant Physiology* **48**, 67–73.

Quested, H.M., Press, M.C., Callaghan, T.V. & Cornelissen, J.H.C. (2002) The hemiparasitic angiosperm *Bartsia alpina* has the potential to accelerate decomposition in sub-arctic communities. *Oecologia* **130**, 88–95

Quested, H.M., Press, M.C., Callaghan, T.V. & Cornelissen, J.H.C. (2003) Litter of the hemiparasite *Bartsia alpina* enhances plant growth: evidence for a functional role in nutrient cycling. *Oecologia* **135**, 606–614.

Quintana, R.D. (2002) Influence of livestock grazing on the capybara's trophic niche and forage preferences. *Acta Theriologica* **47**, 175–183.

Rabotnov, T.A. (1985) Dynamics of plant coenotic populations. In: White, J. (ed.) *The Population Structure of Vegetation*, pp. 122–142. Dr. W. Junk Publishers, Dordrecht, The Netherlands.

Ramenskii, L.G. (1938) *Introduction to the Geobotanical study of Complex Vegetation*. Selkozgiz, Moscow, Russia.

Ranjard, L., Lignier, L. & Chaussod, R. (2006) Cumulative effects of short-term polymetal contamination on soil bacterial community structure. *Applied and Environmental Microbiology* **72**, 1684–1687.

Raoult, D. & Forterre, P. (2008) Redefining viruses: lessons from Mimivirus. *Nature Reviews Microbiology* **6**, 315–319.

Rapson, G.L., Thompson, K. & Hodgson, J.G. (1997) The humped relationship between species richness and biomass – testing its sensitivity to sample quadrat size. *Journal of Ecology* **85**, 99–100.

Ratcliffe, L.M. & Grant, P.R. (1983) Species recognition in Darwin's finches (*Geospiza* Gould). I. Discrimination by morphological cues. *Animal Behaviour* **31**, 1139–1153.

Raunkiaer, C. (1907) *Planterigets livsformer og deres betydning for geografien*. Munksgaard, Copenhagen, Denmark.

Recher, H.F. (1969) Bird species diversity and habitat diversity in Australia and North Africa. *American Naturalist* **103**, 75–80.

Regel, R.H. (2003) *Phytoplankton and turbulence at selected sites*. Unpublished PhD Thesis, The University of Adelaide, Australia.

Regoes, R.R., Crotty, S., Antia, R. & Tanaka, M.M. (2005) Optimal replication of poliovirus within cells. *The American Naturalist* **165**, 364–373.

Reich, P.B., Ellsworth, D.S., Walters, M.B., Vose, J.M., Gresham, C., Volin, J.C., Bowman, W.D. (1999) Generality of leaf trait relationships: a test across six biomes. *Ecology* **80**, 1955–1969.

Reynolds, C.S. (1984) Phytoplankton periodicity: the interactions of form, function and environmental variability. *Freshwater Biology* **14**, 111–142.

Reynolds, C.S. (1991) Functional morphology and adaptive strategies of freshwater plankton. In: Sandgren, C.D. (ed.) *Growth and Reproductive Strategies of Freshwater Phytoplankton. Volume 1*, pp. 388–434. Cambridge University Press, Cambridge, UK.

Reynolds, H.L., Packer, A., Bever, J.D. & Clay, K. (2003) Grassroots ecology: plant-microbe-soil interactions as drivers of plant community structure and dynamics. *Ecology* **84**, 2281–2291.

Reynolds, J.E. & Greboval, D.F. (1988) *Socio-economic effects of the evolution of Nile perch fisheries in Lake Victoria: a review*. CIFA Technical paper 17, FAO. http://www.fao.org/docrep/005/T0037E/T0037E00.htm (last accessed 30 November 2011).

Reznick, D.N. & Ghalambor, C.K. (2001) The population ecology of contemporary adaptations: what empirical studies reveal about the conditions that promote adaptive evolution. *Genetica* **112–113**, 183–198.

Rich, J.J., Heichen, R.S., Bottomley, P.J., Cromack, J. Jr & Myrold, D.D. (2003) Community composition and functioning of denitrifying bacteria from adjacent meadow and forest soils. *Applied and Environmental Microbiology* **69**, 5974–5982.

Ricqlès, A. de, Padian, K., Horner, J.R. & Francillon-Viellot, H. (2000) Paleohistology of the bones of pterosaurs (Reptilia: Archosauria): anatomy, ontogeny, and biomechanical implications. *Zoological Journal of the Linnean Society* **129**, 349–385.

Rinderknecht, A. & Blanco, R.E. (2008) The largest fossil rodent. *Proceedings of the Royal Society B* **275**, 923–928.

Risk, M.J. (1972) Fish diversity on a coral reef in the Virgin Islands. *Atoll Research Bulletin* **153**, 1–7.

Ritchie, K.B. (2006) Regulation of microbial populations by coral surface mucus and mucus-associated bacteria. *Marine Ecology Progress Series* **322**, 1–14.

Robertson, C.E., Harris, J.K., Spear, J.R. & Pace, N.R. (2005) Phylogenetic diversity and ecology of environmental Archaea. *Current Opinion in Microbiology* **8**, 638–642.

Robison, B.H., Reisenbichler, K.R., Hunt, J.C. & Haddock, S.H.D. (2003) Light production by the arm tips of the deep-sea cephalopod *Vampyroteuthis infernalis*. *Biological Bulletin* **205**, 102–109.

Rodwell, J.S. (1991) *British Plant Communities. Volume 1. Woodlands and Scrub*. Cambridge University Press, Cambridge, UK.

Rogers, R.W. (1988) Succession and survival strategies in lichen populations on a palm trunk. *Journal of Ecology* **76**, 759–776.

Rogers, S.W. (1999) *Allosaurus*, crocodiles, and birds: evolutionary clues from spiral computed tomography of an endocast. *The Anatomical Record (New Anat.)* **257**, 162–173.

Rohwer, F. (2007) Real-time microbial ecology. *Environmental Microbiology* **9**, 10.

Rosa, R., Pereira, J. & Nunes, M.L. (2005) Biochemical composition of cephalopods with different life strategies, with special reference to a giant squid, *Architeuthis* sp. *Marine Biology* **146**, 739–751.

Rouse, G.W., Goffredi, S.K. & Vrijenhoek, R.C. (2004) *Osedax*: bone-eating marine worms with dwarf males. *Science* 305, 668–671.

Rowe, R.J. (2009) Environmental and geometric drivers of small mammal diversity along elevational gradients in Utah. *Ecography* 32, 411–422.

Rowell, D.M. & Avilés, L. (1995) Sociality in a bark-dwelling huntsman spider from Australia, *Delena cancerides* Walckenaer (Araneae: Sparassidae). *Insectes Sociaux* 42, 287–302.

Roxburgh, S.H. & Mokany, K. (2007) Comment on 'From plant traits to plant communities: a statistical mechanistic approach to biodiversity'. *Science* 316, 1425b.

Royer, D.L., Miller, I.M., Peppe, D.J. & Hickey, L.J. (2010) Leaf economic traits from fossils support a weedy habit for early angiosperms. *American Journal of Botany* 97, 438–445.

Russell, J. & Baldwin, R.L. (1979) Comparison of maintenance energy expenditures and growth yields among several rumen bacteria grown in continuous culture. *Applied Environmental Microbiology* 37, 537–543.

Ruttan, L.M. (1990) Experimental manipulations of dispersal in the subsocial spider, *Theridion pictum*. *Behavioural Ecology and Sociobiology* 27, 169–173.

Rypstra, A.L. (1979) Foraging flocks of spiders: a study of aggregate behaviour in *Cyrtophora citricola* Forskal (Araneae: Araneidae) in West Africa. *Behavioural Ecology and Sociobiology* 5, 291–300.

Salisbury, E.J. (1942) *The Reproductive Capacity of Plants*. Bell, London, UK.

Sánchez-Villagra, M.R., Aquilera, O. & Horovitz, I. (2003) The anatomy of the world's largest extinct rodent. *Science* 301, 1708–1710.

Santiago, L.S. (2007) Extending the leaf economics spectrum to decomposition: evidence from a tropical forest. *Ecology* 88, 1126–1131.

Santiago, L.S. & Wright, S.J. (2007) Leaf functional traits of tropical forest plants in relation to growth form. *Functional Ecology* 21, 19–27.

Schaller, G.B. & Crawshaw, P.G. (1981) Social dynamics of a capybara population. *Saugetierkundliche Mitteilungen* 29, 3–16.

Schimper, A.F.W. (1903) *Plant Geographer upon a Physiological Basis*. Oxford University Press, Oxford, UK.

Schlesinger, A., Goldshmid, R., Hadfield, M.G., Kramarsky-Winter, E. & Loya, Y. (2009) Laboratory culture of the aeolid nudibranch *Spurilla Neapolitana* (Mollusca, Opistho-branchia): life-history aspects. *Marine Biology* 156, 753–761.

Schluter, D. (1994) Experimental evidence that competition promotes divergence in adaptive radiation. *Science* 266, 798–801.

Schluter, D. (2000) *The Ecology of Adaptive Radiation. Oxford Series in Ecology and Evolution*. Oxford University Press, Oxford, UK.

Schluter, D. (2001) Ecology and the evolution of species. *Trends in Ecology and Evolution* 16, 372–380.

Schluter, D. (2009) Evidence for ecological speciation and its alternative. *Science* 323, 737–741.

Schmidt-Neilsen, K. (1997, 5th edn) *Animal Physiology*. Cambridge University Press, Cambridge, UK.

Schmitz, O.J. (2009) Effects of predator functional diversity on grassland ecosystem function. *Ecology* 90, 2339–2345.

Schoener, T.W. (2011) The newest synthesis: understanding the interplay of evolutionary and ecological dynamics. *Science* 331, 426–429.

Schofield, P.J. & Chapman, L.J. (1999) Interactions between Nile perch, *Lates niloticus*, and other fishes in Lake Nabugabo, Uganda. *Environmental Biology of Fishes* 55, 343–358.

Schulte, P., Alegret, L., Arenillas, I., Arz, J.A., Barton, P.J., Bown, P.R., Bralower, T.J., Christeson, G.L., Claeys, P., Cockell, C.S., Collins, G.S., Deutsch, A., Goldin, T.J., Goto, K., Grajales-Nishimura, J.M., Grieve, R.A.F., Gulick, S.P.S., Johnson, K.R., Kiessling, W., Koeberl, C., Kring, D.A., MacLeod, K.G., Matsui, T., Melosh, J., Montanari, A., Morgan, J.V., Neal, C.R., Nichols, D.J., Norris, R.D., Pierazzo, E., Ravizza,

G., Rebolledo-Vieyra, M., Uwe Reimold, W., Robin, E, Salge, T., Speijer, R.P., Sweet, A.R., Urrutia-Fucugauchi, J., Vajda, V., Whalen, M.T. & Willumsen, P.S. (2010) The Chicxulub asteroid impact and mass extinction at the Cretaceous-Paleogene boundary. *Science* 327, 1214–1218.

Schulze, E-D., Beck, E. & Müller-Hohenstein, K. (2005) *Plant Ecology*. Springer-Verlag, Berlin, Germany.

Schulze, E-D. & Mooney, H.A. (eds) (1993) *Biodiversity and Ecosystem Function.* Springer-Verlag, Berlin, Germany.

Schweitzer, M.H., Johnson, C., Zocco, T.C., Horner, J.R. & Starkey. J.R. (1997) Preservation of biomolecules in cancellous bone of *Tyrannosaurus rex*. *Journal of Vertebrate Palaeontology* 17, 349–359.

Schweitzer, M.H., Watt, J.A., Avci, R., Knapp, L., Chiappe, L., Norell, M.A. & Marshall. M. (1999) Beta-keratin specific immunological reactivity in feather-like structures of the Cretaceous Alvarezsaurid, *Shuvuuia deserti*. *Journal of Experimental Biology* 255, 146–157.

Schweitzer, M.H., Zheng, W., Organ, C.L., Avci, R., Suo, Z., Freimark, L.M., Leblau, V.S., Duncan, M.B., Vander Heiden, M.G., Neveu, J.M., Lane, W.S., Cottrell, J.S., Horner, J.R., Cantley, L.C., Kalluri, R. & Asara, J.M. (2009) Biomolecular characterization and protein sequences of the Campian Hadrosaur B. *Canadensis*. *Science* 324, 626–631.

Seckbach, J. (ed.) (2007) *Algae and Cyanobacteria in Extreme Environments*. Springer, Berlin, Germany.

Seibel, B.A., Thuesen, E.V., Childress, J.J. & Gorodezky, L.A. (1997) Decline in pelagic cephalopod metabolism with habitat depth reflects differences in locomotory efficiency. *Biological Bulletin* 192, 262–278.

Seibel, B.A., Thuesen, E.V. & Childress, J.J. (1998) Flight of the vampire: ontogenetic gait-transition in *Vampyroteuthis infernalis* (Cephalopoda: Vampyromorpha). *The Journal of Experimental Biology* 201, 2413–2424.

Seibel, B.A., Chausson, F., Lallier, F.H., Zal, F., & Childress, J.J. (1999) Vampire blood: respiratory physiology of the vampire squid (Cephalopoda: Vampyromorpha) in relation to the oxygen minimum layer. *Experimental Biology Online* 4, 1–10.

Semlitsch, R.D. & Moran, G.B. (1984) Ecology of the redbelly snake (*Storeria occipitomaculata*) using mesic habitats in South Carolina. *The American Midland Naturalist* 111, 33–40.

Sereno, P.C., Martinez, R.N., Wilson, J.A., Varricchio, D.J., Alcober, O.A. & Larsson, H.C.E. (2008) Evidence for avian intrathoracic air sacs in a new predatory dinosaur from Argentina. *PLoS ONE* 3, e3303.

Shaw, M.W. (1974) The reproductive characteristics of oak. In: Morris, M.G. & Perring, F.H. (eds) *The British Oak: its History and Natural History*, pp. 162–181. E.W. Classey, Faringdon, UK.

Sheehan, C., Kirwan, L., Connolly, J. & Bolger, T. (2007) The effects of earthworm functional group diversity on earthworm community structure. *Pedobiologia* 50, 479–487.

Shetty, S. & Shine, R. (2002) Philopatry and homing behaviour of sea snakes (*Laticauda colubrina*) from two adjacent islands in Fiji. *Conservation Biology* 16, 1422–1426.

Shin, D.S., DiDonato, M., Barondeau, D.P., Hura, G.L., Hitomi, C., Berglund, J.A., Getzoff, E.D., Cary, S.C. & Tainer, J.A. (2009) Superoxide dismutase from the eukaryotic thermophile *Alvinella pompejana*: structures, stability, mechanism, and insights into amyotrophic lateral sclerosis. *Journal of Molecular Biology* 385, 1534–1555.

Shine, R. (2010) The ecological impact of invasive Cane Toads (*Bufo marinus*) in Australia. *The Quarterly Review of Biology* 85, 253–291.

Shine, R., Bonnet, X., Elphick, M.J. & Barrott, E.G. (2004) A novel foraging mode in snakes: browsing by the sea snake *Emydocephalus annulatus* (Serpentes, Hydrophiidae). *Functional Ecology* 18, 16–24.

Shipley, B. (2010) *From Plant Traits to Vegetation Structure: Chance and Selection in the Assembly of Ecological Communities.* Cambridge University Press, Cambridge, UK.

Shipley, B., Vile, D. & Garnier, É. (2006) From plant traits to plant communities: a statistical mechanistic approach to biodiversity. *Science* **314**, 812–814.

Shipley, B., Vile, D. & Garnier, É. (2007) Response to comments on 'From plant traits to plant communities: a statistical mechanistic approach to biodiversity'. *Science* **316**, 1425d.

Sieben, E.J.J., Morris, C.D., Kotze, D.C. & Muasya, A.M. (2010) Changes in plant form and function across altitudinal and wetness gradients in the wetlands of the Maloti-Drakensberg, South Africa. *Plant Ecology* **207**, 107–119.

Silander, J.A. & Antonovics, J. (1982) Analysis of inter-specific interactions in a coastal plant community – a perturbation approach. *Nature* **298**, 557–560.

Simonová, D. & Lososová, Z. (2008) Which factors determine plant invasions in man-made habitats in the Czech Republic? *Perspectives in Plant Ecology, Evolution and Systematics* **10**, 89–100.

Sims, D.W. (1999) Threshold foraging behaviour of basking sharks on zooplankton: life on an energetic knife-edge? *Proceedings of the Royal Society of London B Biological Science* **266**, 1437–1443.

Sims, D.W., Southall, E.J., Richardson, A.J., Reid, P.C. & Metcalfe, J.D. (2003) Seasonal movements and behaviour of basking sharks from archival tagging: no evidence of winter hibernation. *Marine Ecology Progress Series* **248**, 187–196.

Sinervo, B. & Clobert, J. (2008) Life-history strategies, multidimensional trade-offs, and behavioural syndromes. In: Danchin, É., Giraldeau, L.A. & Cézilly, F. (eds) *Behavioural Ecology*, pp. 135–184. Oxford University Press, Oxford, UK.

Sma, R.F. & Baggaley, A. (1976) Rate of excretion of ammonia by the Hard Clam *Mercenaria mercenaria* and the American Oyster *Crassostrea virginica*. *Marine Biology* **36**, 251–258.

Smayda, T.J. & Reynolds, C.S. (2001) Community assembly in marine phytoplankton: application of recent models to harmful dinoflagellate blooms. *Journal of Plankton Research* **23**, 447–461.

Smith, C.L. (1997) *National Audubon Society field guide to tropical marine fishes of the Caribbean, the Gulf of Mexico, Florida, the Bahamas, and Bermuda.* Alfred A. Knopf, Inc., New York, USA.

Smith, T.J. (1987) Seed predation in relation to tree dominance and distribution in mangrove forests. *Ecology* **68**, 266–273.

Smith, V.H. (2007) Microbial diversity-productivity relationships in aquatic ecosystems. *FEMS Microbial Ecology* **62**, 181–186.

Smith, M.D. & Knapp, A.K. (2003) Dominant species maintain ecosystem function with natural patterns of species loss. *Ecology Letters* **6**, 509–517.

Smith, S.A., Beaulieu, J.M. & Donoghue, M.J. (2010) An uncorrelated relaxed-clock analysis suggests an earlier origin for flowering plants. *Proceedings of the National Academy of Sciences of the USA* **107**, 5897–5902.

Smith, S.W., Au, D.W. & Show, C. (1998) Intrinsic rebound potential of 26 species of Pacific sharks. *Marine and Freshwater Research* **49**, 663–678.

Snow, D.W. & Perrins, C.M. (eds) (1999) *The Complete Birds of the Western Palaearctic on CD-ROM.* Oxford University Press, Oxford, UK.

Song, T.Y., Martensson, L., Eriksson, T., Zheng, W.W. & Rasmussen, U. (2005) Biodiversity and seasonal variation of the cyanobacterial assemblage in a rice paddy field in Fujian, China. *FEMS Microbiology Ecology* **54**, 131–140.

Southwood, T.R.E. (1977) Habitat, the templet for ecological strategies? *Journal of Animal Ecology* **46**, 337–365.

Southwood, T.R.E. (1988) Tactics, strategies and templets. *Oikos* **52**, 3–18.

Spielman, A. & D'Antonio, M. (2001) *Mosquito.* Faber and Faber, London, UK.

Stearns, S.C. (1976) Life-history tactics: a review of the ideas. *Quarterly Review of Biology* **51**, 3–47

Stearns, S.C. (1977) The evolution of life-history traits. A critique of the theory and a review of the data. *Annual Review of Ecology and Systematics* **8**, 145–171.

Stevens, D.J., Hansell, M.H. & Monaghan, P. (2000) Developmental trade-offs and life-histories: strategic allocation of resources in caddis flies. *Proceedings of the Royal Society of London B* **267**, 1511–1515.

Stone, E.C. & Vasey, R.B. (1968) Preservation of coastal redwoods on alluvial flats. *Science* **159**, 157–161.

Storey, K.B. & Storey, J.M. (1984) Biochemical adaptations for freezing-tolerance in the wood frog *Rana sylvatica*. *Journal of Comparative Physiology B* **155**, 29–36.

Storey, K.B. & Storey, J.M. (1992) Natural freeze tolerance in ectothermic vertebrates. *Annual Review of Physiology* **54**, 619–637.

Sugiyama, S., Zabed, H.M. & Okubo, A. (2008) Relationships between soil microbial diversity and plant community structure in seminatural grasslands. *Grassland Science* **54**, 117–124.

Surmacki, A., Goldyn, B. & Tryjanowski, P. (2005) Location and habitat characteristics of the breeding nests of the harvest mouse (*Micromys minutus*) in the reed-beds of an intensively used farmland. *Mammalia* **69**, 5–9.

Suzan-Monti, M., La Scola, B., Barrassi, L., Espinosa, L. & Raoult, D. (2007) Ultrastructural characterization of the giant volcano-like virus factory of *Acanthamoeba polyphaga Mimivirus*. *PLoS ONE* **3**, e328.

Suzuki, H. & Murai, M. (1980) Ecological studies of *Ropalidia (sic) fasciata* in Okinawa Island. I. Distribution of single- and multiple-foundress colonies. *Researches on Population Ecology* **22**, 184–195.

Swingland, I.R. (1977) Reproductive effort and life-history strategy of the Aldabran giant tortoise. *Nature* **269**, 402–404.

Tansley, A.G. (1935) The use and abuse of vegetational concepts and terms. *Ecology* **16**, 284–307.

Taylor, D.R., Aarssen, L.W. & Loehle, C. (1990) On the relationship between r/K selection and environmental carrying capacity: a new habitat templet for plant life-history strategies. *Oikos* **58**, 239–250.

Tennyson, Lord A. (1849) *In Memoriam A.H.H.*

Terres, J.K. (1991) *The Audubon Society Encyclopedia of North American Birds*. Wings Books, New York, USA.

Thompson, K. (2010) *Do We Need Pandas? The Uncomfortable Truth about Biodiversity*. Green Books, Foxhole, Totnes, UK.

Thompson, K., Askew, A.P., Grime, J.P., Dunnett, N.P. & Willis, A.J. (2005) Biodiversity, ecosystem function and plant traits in mature and immature plant communities. *Functional Ecology* **19**, 355–358.

Tilman, D. (1982) *Resource Competition and Community Structure*. Princeton University Press, Princeton, USA.

Tilman, D. (1988) *Plant Strategies and the Dynamics and Structure of Plant Communities*. Princeton University Press, Princeton, USA.

Tilman, D. (1999) Ecological consequences of biodiversity: a search for general principles. *Ecology* **80**, 1455–1474.

Tilman, D. (2001) Effects of diversity and composition on grassland stability and productivity. In: Press, M.C., Huntly, N.J. & Levin, S. (eds) *Ecology: Achievement and Challenge*, pp. 183–207. Blackwell Science, Oxford, UK.

Tilman, D. & Downing, J.A. (1994) Biodiversity and stability in grasslands. *Nature* **367**, 363–365.

Tilman, D., Knops, L.J., Wedin, D., Reich, P., Ritchie, M. & Sieman, E. (1997) The influence of functional diversity and composition on ecosystem processes. *Science* **277**, 1300–1302.

Tinkle, D.W, Wilbur, H.M. & Tilley, S.G. (1970) Evolutionary strategies in lizard reproduction. *Evolution* **24**, 55–74.

Tinkler, E., Montgomery, I. & Elwood, R.W. (2009) Foraging ecology, fluctuating food availability and energetics of wintering brent geese. *Journal of Zoology* **278**, 313–323.

Tscherko, D., Hammesfahr, U., Zeltner, G., Kandeler, E. & Böcker, R. (2005) Plant succession and rhizosphere microbial communities in a recently deglaciated alpine terrain. *Basic and Applied Ecology* **6**, 367–383.

Turner, R.D. (1973) Wood-boring bivalves, opportunistic species in the deep sea. *Science* **180**, 1377–1379.

Uetz, G.W. (1986) Web-building and prey capture in communal orb weavers. In: Shear, W.A. (ed) *Spiders: Webs, Behaviour and Evolution*, pp. 207–231. Stanford University Press, Palo Alto, USA.

Uetz, G.W. (1989) The 'ricochet effect' and prey capture in colonial spiders. *Oecologia* **81**, 154–149.

Urcelay, C., Díaz, S., Gurvich, D.E., Chapin, F.S. III, Cuevas, E. & Domínguez, L.S. (2009) Mycorrhizal community resilience in response to experimental plant functional type removals in a woody ecosystem. *Journal of Ecology* **97**, 1291–1301.

van der Heijden, M.G.A., Wiemken, A. & Sanders, I.R. (2003) Different arbuscular mycorrhizal fungi alter coexistence and resource distribution between co-occurring plants. *New Phytologist* **157**, 569–578.

van der Heijden, M.G.A., Bardgett, R.D. & van Straalen, N.M. (2008) The unseen majority: soil microorganisms as drivers of plant diversity and productivity in terrestrial ecosystems. *Ecology Letters* **11**, 296–310.

van der Putten, W.H., van Dijk, C. & Peters, B.A.M. (1993) Plant specific, soil-borne diseases contribute to succession in foredune vegetation. *Nature* **362**, 53–56.

Vandewalle, M., de Bello, F., Berg, M.P., Bolger, T., Dolédec, S., Dubs, F., Feld, C.K., Harrington, R., Harrison, P.A., Lavorel, S., da Silva, P.M., Moretti, M., Niemelä, J., Santos, P., Sattler, T., Sousa, J.P., Sykes, M.T., Vanbergen, A.J. & Woodcock, B.A. (2010) Functional traits as indicators of biodiversity response to land use changes across ecosystems and organisms. *Biodiversity Conservation* **19**, 2921–2947.

van Doorn GS, Edelaar P & Weissing FJ. (2009) On the origin of species by natural and sexual selection. *Science* **326**, 1704–1707.

Van Etten, J.L. (2003) Unusual life style of giant *Chlorella* viruses. *Annual Review of Genetics* **37**, 153–195.

van Wijk, M.T., Williams, M., Gough, L., Hobbie, S.E. & Shaver, G.R. (2003) Luxury consumption of soil nutrients: a possible competitive strategy in above-ground and below-ground biomass allocation and root morphology for slow-growing arctic vegetation. *Journal of Ecology* **91**, 664–676.

Varricchio, D.J., Jackson, F., Borkowski, J.J. & Horner, J.R. (1997) Nest and egg clutches of the dinosaur Troodon formosus and the evolution of avian reproductive traits. *Nature* **385**, 247–250.

Varricchio, D.J., Martin, A.J. & Katsura, Y. (2007) First trace and body fossil evidence of a burrowing, denning dinosaur. *Proceedings of the Royal Society B: Biological Sciences* **274**, 1361–1368.

Venditti, C., Meade, A. & Pagel, M. (2010) Phylogenies reveal new interpretation of speciation and the Red Queen. *Nature* **463**, 349–352.

Villalobos, A.R.A. & Renfro, J.L. (2007) Trimethylamine oxide suppresses stress-induced alteration of organic anion transport in choroid plexus. *The Journal of Experimental Biology* **210**, 541–552.

Villarreal, L.P., Defilippis, V.R. & Gottlieb, K.A. (2000) Acute and persistent viral life strategies and their relationship to emerging diseases. *Virology* **272**, 1–6.

Vitt, L.J., Pianka, E.R., Cooper, W.E. Jr & Schwenk, K. (2003) History and the global ecology of squamate reptiles. *The American Naturalist* **162**, 44–60.

Vollrath, F. (1982) Colony foundation in a social spider. *Zeitschrift für Tierpsychologie* **60**, 313–324.

Vollrath, F. (1986) Eusociality and extraordinary sex ratios in the spider *Anelosimus eximius* (Araneae: Theridiidae). *Behavioral Ecology and Sociobiology* **18**, 283–287.

von Humboldt, A. (1806) *Ideen zu einer Physiognomik der Gewachse*. Tübingen.

Vrijenhoek, R.C., Collins, P. & Van Dover, C.L. (2008) Bone-eating marine worms: habitat specialists or generalists? *Proceedings of the Royal Society B* **275**, 1963–1964.

Walter, A. (1990) The evolution of marmot sociality: I. Why disperse late? *Behavioural Ecology and Sociobiology* 27, 229–237.

Wanink, J.H. & Goudswaard, K. (1994) Effects of the Nile perch (*Lates niloticus*) introduction into Lake Victoria, east Africa, on the diet of Pied Kingfishers (*Ceryle rudis*). *Hydrobiologia* 280, 367–376.

Ward, D.M., Weller, R. & Bateson, M.M. (1990) 16S rRNA sequences reveal numerous uncultured microorganisms in a natural community. *Nature* 345, 63–65.

Wardle, D.A., Bonner, K.I. & Nicholson, K.S. (1997) Biodiversity and plant litter: experimental evidence which does not support the view that enhanced species-richness improves ecosystem function. *Oikos* 79, 247–258.

Wardle, D.A. & Giller, K.E. (1997) The quest for a contemporary ecological dimension to soil ecology. *Soil Biology & Biochemistry* 28, 1549–1554.

Wardle, D.A., Bardgett, R.D., Klironomos, J.N., Setälä, H., van der Putten, W.H. & Wall, D.H. 2004. Ecological linkages between aboveground and belowground biota. *Science* 304, 1629–1633.

Wardle, D.A., Huston, M.A., Grime, J.P., Berendse, F., Garnier, E., Lauenroth, W.K., Setala, H. & Wilson, S.D. (2000) Biodiversity and ecosystem function: an issue in ecology. *The Ecological Society of America Bulletin* 81, 232–235.

Warming, E. (1884) *Om skudbygning, overvintring og foryngelse*. Festskr. Naturh. Foren., Copenhagen, Denmark.

Warming, E. (1909) *Oecology of Plants: an Introduction to the Study of Plant Communities*. Oxford University Press, Oxford, UK.

Wedel, M.J. (2003) Vertebral pneumaticity, air sacs, and the physiology of sauropod dinosaurs. *Paleobiology* 29, 243–255.

Wedin, D.A. (1995) Species, nitrogen and grassland dynamics: the constraints of stuff. In: Jones, C.G. & Lawton, J.H. (eds) *Linking Species and Ecosystems*, pp. 253–262. Chapman and Hall, New York, USA.

Weese, D.J., Schwartz, A.K., Bentzen, P., Hendry, A.P. & Kinnison, M.T. (2011) Eco-evolutionary effects on population recovery following catastrophic disturbance. *Evolutionary Applications* 4, 354–366.

Weiher, E. & Keddy, P. (eds) (1999) *Ecological Assembly Rules: Perspectives, Advances, Retreats*. Cambridge University Press, Cambridge, UK.

Weithoff, G., Walz, N. & Gaedke, U. (2001) The intermediate disturbance hypothesis – species diversity or functional diversity? *Journal of Plankton Research* 23, 1147–1155.

Welden, C.W. & Slauson, W.L. (1986) The intensity of competition versus its importance: an overlooked distinction and some implications. *The Quarterly Review of Biology* 61, 23–44.

Wertz, S., Degrange, V., Prosser, J.I., Poly, F., Commeaux, C., Guillaumaud, N. & Le Roux. X. (2007) Decline of soil microbial diversity and resilience of key soil microbial functional groups following a model disturbance. *Environmental Microbiology* 9, 2211–2219.

West, R.C. (2005) The Brachypelma of Mexico. *Journal of the British Tarantula Society* 20, 108–119.

Westoby, M. (1998) A leaf-height-seed (LHS) plant ecological strategy scheme. *Plant and Soil* 199, 213–227.

Westoby, M., Leishman, M. & Lord, J. (1995) Issues of interpretation after relating comparative datasets to phylogeny. *Journal of Ecology* 83, 892–893.

Wherry, T. & Elwood, R.W. (2009) Relocation, reproduction and remaining alive in an orb-web spider. *Journal of Zoology* 279, 57–63.

Whitehouse, M.E.A. & Lubin, Y. (2005) The functions of societies and the evolution of group living: spider societies as a test case. *Biological Reviews* 80, 1–15.

Whitlock, R., Grime, J.P., Burke, T.E. & Booth, R.E. (2007) The role of genotypic diversity in determining grassland community structure under constant environmental conditions. *Journal of Ecology* 95, 895–907.

Whittaker, R.H. & Goodman, D. (1979) Classifying species according to their demographic strategy. I. Population fluctuations and environmental heterogeneity. *American Naturalist* 113, 185–200.

Willey, N.J., Tang, S. & Watt, N.R. (2005) Predicting inter-taxa differences in plant uptake of Cesium-134/137. *Journal of Environmental Quality* 34, 1478–1489.

Willey, N.J. & Wilkins, J. (2008) Phylogeny and growth strategy as predictors of differences in cobalt concentrations between plant species. *Environmental Science & Technology* 42, 2162–2167.

Willson, J.D. & Dorcas, M.E. (2004) Aspects of the ecology of small fossorial snakes in the western Piedmont of North Carolina. *Southeastern Naturalist* 3, 1–12.

Wilmer, P., Stone, G. & Johnson. I. (2005, 2nd edn) *Environmental Physiology of Animals*. Blackwell Science, Oxford, UK.

Wilson, E.O. (1969) The species equilibrium. In: *Diversity and Stability in Ecological Systems*, pp. 38–47. Brookhaven Symposia in Biology 22, Brookhaven National Laboratory. Upton, New York, USA.

Winemiller, K.O. (1989) Patterns of variation in life-history among South American fishes in seasonal environments. *Oecologia* 81, 225–241.

Winemiller, K.O. (1992) Life-history strategies and the effectiveness of sexual selection. *Oikos* 63(2), 318–327.

Winemiller, K.O. (1995) Fish ecology. *Encyclopedia of Environmental Biology* 2, 49–65.

Winemiller, K.O. & Rose, K.A. (1992) Patterns of life-history diversification in North American fishes: implications for population regulation. *Canadian Journal of Fisheries and Aquatic Sciences* 49, 2196–2218.

Wisheu, I.C. & Keddy, P.A. (1996) Three competing models for predicting the size of species pools: a test using eastern North American wetlands. *Oikos* 76, 253–258.

Wolff, J. & Sherman, P.W. (2007) *Rodent Societies*. University of Chicago Press, Chicago, USA.

Wood, C.M., Wilson, R.W., Gonzalez, R.J., Patrick, M.L., Bergman, H.L., Narahara, A. & Val, A.L. (1998) Responses of an Amazonian teleost, the tambaqui (*Colossoma macropomum*), to low pH in extremely soft water. *Physiological Zoology* 71, 658–670.

Woodward, F.I. & Diament, A.D. (1991) Functional approaches to predicting the ecological effects of global change. *Functional Ecology* 5, 202–212.

Wootton, A. (1993) *Insects of the World*. Blandford, London, UK.

Wright, I.J., Reich, P.B., Westoby, M., Ackerley, D.D., Baruch, Z., Bongers, F., Cavender-Bares, J., Chapin, T., Cornelissen, J.H.C., Diemer, M., Flexas, J., Garnier, E., Groom, P.K., Gulias, J., Hikosaka, K., Lamont, B.B., Lee, T., Lee, W., Lusk, C., Midgley, J.J., Navas, M-L., Niinemets, Ü., Oleksyn, J., Osada, N., Poorter, H., Poot, P., Prior, L., Pyankov, V.I., Roumet, C., Thomas, S.C., Tjoelker, M.G., Veneklaas, E.J. & Villar, R. (2004) The worldwide leaf economics spectrum. *Nature* 428, 821–827.

Xiao, S., Michalet, R., Wang, G. & Chen, S-Y. (2009) The interplay between species' positive and negative interactions shapes the community biomass-species richness relationship. *Oikos* 118, 1343–1348.

Xu, X., Zhao, Qi., Norell, M., Sullivan, C., Hone, D., Erickson, G., Wang, X., Han, F. & Guo, Y. (2009) A new feathered maniraptoran dinosaur fossil that fills a morphological gap in avian origin. *Chinese Science Bulletin* 54, 430–435.

Yakimov, M.M., Cappello, S., Crisafi, E., Tursi, A., Savini, A., Corselli, C., Scarfi, S. & Giuliano, L. (2006) Phylogenetic survey of metabolically active microbial communities with the deep-sea coral *Lophelia pertusa* from the Apulian plateau, Central Mediterranean Sea. *Deep-Sea Research I* 53, 62–75.

Yeager, C.M., Kornosky, J.L., Housman, D.C., Grote, E.E., Belnap, J. & Kuste, C.R. (2004) Diazotrophic community structure and function in two successional stages of biological soil crusts from the Colorado Plateau and Chihuahuan desert. *Applied and Environmental Microbiology* 70, 973–983.

Yergeau, E., Newsham, K.K., Pearce, D.A. & Kowalchuk, G.A. (2007) Patterns of bacterial diversity across a range of Antarctic terrestrial habitats. *Environmental Microbiology* 9, 2670–2682.

Ylonen, H. (1990) Spatial avoidance between the bank vole *Clethrionomys glareolus* and the harvest mouse *Micromys minutus* – an experimental study. *Annales Zoologici Fennici* 27, 313–320.

Ysnel, F. (1991) Data points for a study of population dynamics of an orb-weaving spider (*Larinioides cornutus*, Araneae, Araneidae). *Bulletin de la Société des Sciences Naturelles de Neuchatel* 116, 269–278.

Zampiga, E., Gaibani, G., Csermely, D., Frey, H. & Hoi, H. (2006) Innate and learned aspects of vole urine UV-reflectance use in the hunting behaviour of the common kestrel *Falco tinnunculus*. *Journal of Avian Biology* 37, 318–322.

Zbinden, M., Martinez, I., Guyot, F., Compère, P., Cambon-Bonavita, M.A. & Gaill, F. (2000) Characteristics of mineral particles associated with the Pompeii worm tubes. *Biology of the Cell* 92, 382.

Zhang, F., Kearns, S.L., Orr, P.J., Benton, M.J., Zhou, Z., Johnson, D., Xu, X. & Wang, X. (2010) Fossilized melanosomes and the colour of Cretaceous dinosaurs and birds. *Nature* 463, 1075–1078.

Zhu, Y,, Jiang, Y., Liu, Q., Kang, M., Spehn, E.M. & Körner, Ch. (2009) Elevational trends of biodiversity and plant traits do not converge – a test in the Helan Range, NW China. *Plant Ecology* 205, 273–283.

Organism Index

The Evolutionary Strategies that Shape Ecosystems, First Edition. J. Philip Grime, Simon Pierce.
© 2012 John Wiley & Sons, Ltd. Published 2012 by John Wiley & Sons, Ltd.

Subject Index

The Evolutionary Strategies that Shape Ecosystems, First Edition. J. Philip Grime, Simon Pierce.
© 2012 John Wiley & Sons, Ltd. Published 2012 by John Wiley & Sons, Ltd.

Printed and bound by CPI Group (UK) Ltd, Croydon, CR0 4YY

27/10/2024

14580146-0001